Lecture Notes on Data Engineering and Communications Technologies 287

Series Editors

Fatos Xhafa, *Technical University of Catalonia, Barcelona, Spain*
Rolando Herrero, *Northeastern University, Boston, MA, USA*

The aim of the book series is to present cutting edge engineering approaches to data technologies and communications. It will publish latest advances on the engineering task of building and deploying distributed, scalable and reliable data infrastructures and communication systems.

The series will have a prominent applied focus on data technologies and communications with aim to promote the bridging from fundamental research on data science and networking to data engineering and communications that lead to industry products, business knowledge and standardisation.

Indexed by SCOPUS, INSPEC, EI Compendex.

All books published in the series are submitted for consideration in Web of Science.

Aboul Ella Hassanien · Abdelkrim Haqiq ·
M. A. Jabbar · Essaid Sabir · K. C. Santosh ·
Luis Orozco Barbosa
Editors

The 4th International Conference on Artificial Intelligence and Smart Data Science (AISDS'2025)

Volume 2 AI and Data Science

Springer

Editors
Aboul Ella Hassanien
Faculty of Computers and Artificial Intelligence
Cairo University
Giza, Egypt

M. A. Jabbar
Department of Computer Science and Engineering
Vardhaman College of Engineering
Hyderabad, Telangana, India

K. C. Santosh
Department of Computer Science
University of South Dakota
Vermillion, SD, USA

Abdelkrim Haqiq
Faculty of Sciences and Techniques
Hassan 1st University
Settat, Morocco

Essaid Sabir
TÉLUQ, University of Québec
Montreal, QC, Canada

Luis Orozco Barbosa
Research Institute of Informatics
Castilla-La Mancha University
Albacete, Spain

ISSN 2367-4512 ISSN 2367-4520 (electronic)
Lecture Notes on Data Engineering and Communications Technologies
ISBN 978-3-032-18926-4 ISBN 978-3-032-18927-1 (eBook)
https://doi.org/10.1007/978-3-032-18927-1

This Springer imprint is published by the registered company Springer Nature Switzerland AG
The registered company address is: Gewerbestrasse 11, 6330 Cham, Switzerland

Preface

It is our great pleasure to welcome you to the 4th International Conference on Artificial Intelligence and Smart Data Science (AISDS 2025), held in the vibrant city of Marrakesh, Morocco, from 04 to 06 December, 2025. AISDS'2025 is organized by the Computer, Networks, Mobility and Modeling Laboratory (IR2M), Faculty of Sciences and Techniques, Hassan 1st University, Settat, Morocco, in collaboration with the Scientific Research Group in Egypt (SRGE), Cairo University, Egypt.

AISDS has established itself as a premier forum for researchers, practitioners, and industry experts to exchange ideas, present innovations, and explore emerging trends in Artificial Intelligence (AI) and Smart Data Science.

The rapid evolution of AI technologies and data-driven solutions continues to transform every aspect of modern life—from healthcare and education to transportation, security, and sustainable development. AISDS 2025 reflects this dynamic landscape by featuring cutting-edge research across diverse domains, including machine learning, deep learning, computer vision, natural language processing, optimization, IoT security, and big data analytics.

This year's conference received an impressive number of submissions from around the globe, demonstrating the growing interest and relevance of AI and data science in addressing real-world challenges. After a rigorous peer-review process, the selected papers represent high-quality contributions that advance theory, methodology, and applications. Topics span from predictive modeling and sentiment analysis to autonomous systems, healthcare AI, and smart resource management, ensuring a rich and multidisciplinary program.

We extend our sincere gratitude to the authors for their innovative work, the reviewers for their invaluable feedback, and the organizing committee for their dedication in making AISDS 2025 a success. We also thank our sponsors and partners for their support in fostering collaboration and knowledge exchange.

We invite you to explore the proceedings and engage with the ideas presented. We hope AISDS 2025 will inspire new research directions, strengthen academic and industrial partnerships, and contribute to shaping the future of intelligent systems and smart data solutions.

Welcome to AISDS 2025 in Marrakech—where innovation meets opportunity!
AISDS 2025 Organizing Committee

Abdelkrim Haqiq
Aboul Ella Hassanien
Conference Chairs

Organization

Honorary Chairs

Aawatif Hayar	GREENTIC, ENSEM – Casablanca, former Minister of Solidarity, Social Integration and Family and former President of Hassan II University, Casablanca, Morocco
Abdellatif Moukrim	President of Hassan 1st University, Settat, Morocco
Abdelmajid Farchi	Dean of the Faculty of Sciences and Techniques, Settat, Morocco
Jamila El Alami	Director of the National Center for Scientific and Technical Research, Morocco
Mohamed Knidiri	President of "Association le Grand Atlas", President of the Private University of Marrakesh and former Minister of National Education, Morocco
Said Rakrak	Dean of the Faculty of Sciences and Techniques - Guéliz, Marrakesh, Morocco

General Chairs

Abdelkrim Haqiq	GREENTIC, FST, Hassan 1st University, Settat, Morocco
Aboul Ella Hassanien	Faculty of Computer and AI, Cairo University, Egypt

Conference Co-chairs

KC Santosh	Department of Computer Science, University of South Dakota, USA
Luis Orozco Barbosa	Albacete Research Institute of Informatics, Castilla-La Mancha University, Abacete, Spain

Technical Program Chairs

Akhil Jabbar Meerja	Vardhaman College of Engineering, Hyderabad, Telangana, India
Ashraf Darwish	Faculty of Science, Helwan University, Cairo, Egypt
Essaid Sabir	University of Quebec (TÉLUQ), Montréal, Canada
Parthasarathy Subashini	Avinashilingam Institute for Home Science and Higher Education for Women - Deemed University, Coimbatore, India

International Advisory Board

Adel Mohamed Alimi	National, Engineering School of Sfax, Tunisia
Ahmad Taher Azar	College of Computer & Information Sciences, Prince Sultan University, Riyadh, Saudi Arabia
Dabia Ahmed Abouâinainen	Abdulrahman Bin Faisal University, Saudi Arabia
Peter J. Tonelato	School of Medicine, University of Missouri, Columbia MO, USA
Rawya Rizk	Port Said University, Egypt
Sundaramurthy Gayathri Devi	Avinashilingam Institute for Home Science and Higher Education for Women, Deemed University, Coimbatore, India
Vaclav Snasel	VSB-Technical University of Ostrava, Czech Republic
Youssef F. Rashed	Cairo University, Egypt

Track Chairs

Track 1: Foundations of Artificial Intelligence and Machine Learning

Abdellah Adib	FST- Mohammedia, Hassan II University, Casablanca, Morocco
Said El Kafhali	FST, Hassan 1st University, Settat, Morocco
Sam Goundar	RMIT University, School of Computing, Hanoi, Vietnam

Track 2: Smart Data Engineering and Analytics

Adam Slowik	Koszalin University of Technology, Koszalin, Poland
Mohamed Nemiche	Polydisciplinary Faculty of Taza, Sidi Mohamed Ben Abdellah University, Morocco
Said Raghay	Faculty of Sciences and Techniques - Guéliz, Marrakesh, Morocco

Track 3: Natural Language Processing and Human-Centered AI

Brahim Ouhbi	National School of Arts and Crafts (ENSAM), Meknès, Morocco
Krishnaveni Marimuthu	Avinashilingam Institute for Home Science and Higher Education for Women, Deemed University, Coimbatore, India
Mohamed Hanini	FST, Hassan 1st University, Settat, Morocco

Track 4: AI Applications in Industry, Society and Sustainability

Aurelio Bermúdez Marín	Albacete Research Institute of Informatics, Castilla-La Mancha University, Abacete, Spain
Driss Bouzidi	ENSIAS, Mohammed V University, Rabat, Morocco
Nizar Rokbani	High Institute of Applied Sciences and Technology, University of Sousse, Tunisia

Track 5: Cybersecurity, Privacy and Trust in AI

Ashraf Matrawy	School of Information Technology, Carleton University, Ottawa, Canada
Ghizlane Orhanou	Faculty of Sciences, Rabat, Morocco
Yassine Maleh	National School of Applied Sciences, Khouribga, Morocco

Track 6: Computer Vision and Applications

Driss El Ouadghiri	Faculty of Sciences, Meknès, Morocco
Nabil Laachfoubi	FST, Hassan 1st University, Settat, Morocco
Rafael Casado González	Albacete Research Institute of Informatics, Castilla-La Mancha University, Abacete, Spain

Publicity Chairs

Abdellah Ouammou	FST, Hassan 1st University, Settat, Morocco
Adnane El Hanjri	EMSI, Tangier, Morocco
Azza Ahmed Abdo Ali	Abdulrahman Bin Faisal University, Saudi Arabia
Chakib Ben Njima	National Engineering School of Sousse, University of Sousse, Tunisia
Houssam Halmaoui	ISMAC, Rabat, Morocco
Kamal Bouhassoune	FST, Hassan 1st University, Settat, Morocco
Lavika Goel, Malaviya	National Institute of Technology, Jaipur, India
Mehrez Abdellaoui	National Engineering School of Sousse, University of Sousse, Tunisia
Mohamed Hossam Ahmed	School of Electrical Engineering and Computer Science, Ottawa, Canada

Local Arrangement Committee Chairs

Abdelghani Ben Tahar	FST, Hassan 1st University, Settat, Morocco
Abderrahim Marzouk	FST, Hassan 1st University, Settat, Morocco
Iman Elmir	I2S, Hassan 1st University, Settat, Morocco
Mohamed Hanini	FST, Hassan 1st University, Settat, Morocco
Said El Kafhali	FST, Hassan 1st University, Settat, Morocco

Organizing Committee Members

Abdellah Jamali	National School of Applied Sciences, Berrechid, Morocco
Abdellah Zaaloul	Faculty of Economics and Management, Settat, Morocco
Adil El Idriss	FST, Hassan 1st University, Settat, Morocco
Adnane Founoun	FST, Hassan 1st University, Settat, Morocco

Ahmed Boujnoui	FST, Moulay Ismaïl University, Errachidia, Morocco
Amine Benmakhlouf	FST, Moulay Ismaïl University, Errachidia, Morocco
Asma Boulkaid	FST, Hassan 1st University, Settat, Morocco
Bouchaib Nassereddine	Faculty of Sciences and Techniques, Settat, Morocco
Haitam Ettazi	FST, Hassan 1st University, Settat, Morocco
Hamid Taramit	Center for Computational Simulation, Universidad Politécnica de Madrid, Spain
Hibateallah Mohtadi	FST, Hassan 1st University, Settat, Morocco
Hind Mikram	Faculty of Sciences and Techniques, Settat, Morocco
Loubna Hamami	FST, Hassan 1st University, Settat, Morocco
Soukaina Mihi	FST, Hassan 1st University, Settat, Morocco
Wafae Azoulay	FST, Hassan 1st University, Settat, Morocco
Youssef Hilali	FST, Hassan 1st University, Settat, Morocco
Zakaria Soufiane Hafdi	FST, Hassan 1st University, Settat, Morocco

TPC Members of AISDS2025

Abdelghani Bentahar	FST, Hassan 1st University, Settat, Morocco
Abdelhay Haqiq	School of Information Sciences, Mohammed V University, Rabat, Morocco
Abdellah Adib	FST - Mohammedia, Hassan II University, Casablanca, Morocco
Abdellah Ezzati	FST, Hassan 1st University, Settat, Morocco
Abdellah Jamali	ENSA - Berrechid, Hassan 1st University, Settat, Morocco
Abdellah Najid	INPT, Rabat, Morocco
Abdellah Ouammou	FST, Hassan 1st University, Settat, Morocco
Abdellah Zaaloul	Faculty of Economics and Management, Settat, Morocco
Abdellatif Kobbane	ENSIAS, Mohammed V University, Rabat, Morocco
Abderrahim Marzouk	FST, Hassan 1st University, Settat, Morocco
Abdessamad Bellouchi	FST, Hassan 1st University, Settat, Morocco
Adam Slowik Koszalin	University of Technology, Koszalin, Poland
Adnane El Hanjri	EMSI, Tangier, Morocco
Adnane Founoun	FST, Hassan 1st University, Settat, Morocco
Adnane Latif	ENSA, Cadi Ayyad University, Marrakesh, Morocco

Hajar Mousannif	Faculty of Sciences Semlalia, Cadi Ayyad University, Marrakesh, Morocco
Hamadou Saliah-Hassane	TELUQ, University of Quebec, Montreal, Canada
Hamid Taramit	Center for Computational Simulation, Universidad Politécnica de Madrid, Spain
Hanan El Bakkali	ENSIAS, Mohammed V University, Rabat, Morocco
Hatem Ben Sta	University Tunis El Manar, Tunisia
Heba Mohamed Moustafa Askr	Faculty of Computers and Artificial Intelligence, University of Sadat City, Egypt
Heba M. Abdel-Atty	Port Said University, Egypt
Heba Mahmoud Afify	Higher Institute of Engineering in Shorouk Academy, Egypt
Heba Nashaat	Port Said University, Egypt
Heba Y. M. Soliman	Port Said University, Egypt
Hicham Ben Alla	FST, Hassan 1st University, Settat, Morocco
Hicham Tribak	Polydisciplinary Faculty, Ouarzazate, Morocco
Hind Mikram	FST, Hassan 1st University, Settat, Morocco
Houssam Halmaoui	ISMAC, Rabat, Morocco
Iman Elmir	I2S, Hassan 1st University, Settat, Morocco
Imane Hilal	ESI, Mohammed V University, Rabat, Morocco
Iyad Lahcen Cherif	INPT, Rabat, Morocco
Jamal El Hachimi	ENSIAS, Mohammed V University, Rabat, Morocco
Jaouad Dabounou	FST, Hassan 1st University, Settat, Morocco
Kamal Bouhassoune	FST, Hassan 1st University, Settat, Morocco
Karmela Aleksis Maslac	Zagreb School of Economics and Management, Croatia
KC Santosh	University of South Dakota, USA
Khadija Douzi	FST - Mohammedia, Hassan II University, Casablanca, Morocco
Khalid El Makkaoui	Polydisciplinary Faculty, Nador, Morocco
Khalid Zine-Dine	Faculty of Sciences, Mohammed V University, Rabat, Morocco
Krishnaveni Marimuthu	Deemed University, Coimbatore, India
Laila Damri	ENSA - Berrechid, Hassan 1st University, Settat, Morocco
Laila Fetjah	Facutly of Sciences Aïn Chock, Hassan II University, Casablanca, Morocco
Lamia Karim	ENSA - Berrechid, Hassan 1st University, Settat, Morocco
Lavika Goel	Malaviya, National Institute of Technology (NIT), Jaipur, India

Lobna Mohamed Abouelmagd	Misr Higher Institute for Engineering and Technology, Mansoura, Egypt
Loubna Hamami	FST, Hassan 1st University, Settat, Morocco
Maria Zemzami	National Graduate School of Arts and Crafts (ENSAM), Rabat, Morocco
Mehrez Abdellaoui	High Institute of Applied Sciences and Technologies, Kairouan, Tunisia
Mohamed Baslam	FST, Moulay Slimane University, Beni Mellal, Morocco
Mohamed Ben Halima	ISIMS, University of Sfax, Tunisia
Mohamed Chakraoui	Polydisciplinary Faculty of Khouribga, Morocco
Mohamed Chaouki Abounaima	FST, My Abdellah University, Fès, Morocco
Mohamed El Kamili	EST, Hassan II University, Casablanca, Morocco
Mohamed Et-Tolba	INPT, Rabat, Morocco
Mohamed Es-Sadek Zeriab	National Graduate School of Arts and Crafts (ENSAM), Rabat, Morocco
Mohamed Hanini	FST, Hassan 1st University, Settat, Morocco
Mohamed Hossam Ahmed	School of Electrical Engineering and Computer Science, Ottawa, Canada
Mohamed Kissi	FST - Mohammedia, Hassan II University, Casablanca, Morocco
Mohamed Moughit	ENSAM, Hassan II University, Casablanca, Morocco
Mohamed Nabil	Faculty of Sciences, Chouaïb Doukali University, El Jadida, Morocco
Mohamed Naimi	ENSA - Berrechid, Hassan 1st University, Settat, Morocco
Mohamed Nemiche	Polydisciplinary Faculty of Taza, Sidi Mohamed Ben Abdellah University, Morocco
Mohamed Sabbane	Faculty of Sciences, My Ismail University, Meknès, Morocco
Mostafa Bellafkih	INPT, Rabat, Morocco
Mouad Ben Mamoun	Faculty of Sciences, Mohammed V University, Rabat, Morocco
Moulay Lahcen Hasnaoui	EST, Moulay Ismail University, Meknès, Morocco
Mounir Rifi	EST, Hassan II University, Casablanca, Morocco
Nadif Sami Abdelaaziz	FST, Hassan 1st University, Settat, Morocco
Najib Naja	INPT, Rabat, Morocco
Nidhi Sharma	Cox Automotive Inc., New York, USA
Nizar Rokbani	High Institute of Applied Sciences and Technology, University of Sousse, Tunisia
Noha El-Shoafy	Faculty of Computers and Artificial Intelligence, Matrouh University, Matrouh, Egypt

Noreddine Gherabi	ENSA - Khouribga, Morocco
Omar Ait Oualhaj	INPT, Rabat, Morocco
Omar El Beggar	FST - Mohammedia, Hassan II University, Casablanca, Morocco
Omar El Beqqali	Faculty of Sciences, Fès, Morocco
Ouail Ouchetto	FSJES, Université Hassan II Casablanca, Morocco
Parthasarathy Subashini	Deemed University, Coimbatore, India
Rachid Dakir	Polydisciplinary Faculty, Ouarzazate, Morocco
Rachid Latif	ENSA, Ibn Zohr University, Agadir, Morocco
Rajendran Sobha Ajin Singh	Painary P.O mIdukki, Kerala, India
Rajesh Vayyala	PRA Group Inc, CN, USA
Rajgopal Devabhaktuni	Macy's, Géorgie, USA
Rania Ahmed Mohamed	Modern University for the Technology & Information, Egypt
Rasha Saleh Abouelyazid	Egyptian Chinese University, Cairo, Egypt
Safaa Mahrach	Higher School of Technology, Beni Mellal, Morocco
Said Ben Alla	ENSA - Berrechid, Hassan 1st University, Settat, Morocco
Said Broumi	Faculty of Science Ben M'Sik, Hassan II University, Casablanca, Morocco
Said El Kafhali	FST, Hassan 1st University, Settat, Morocco
Said Jai Andaloussi	FS Aïn Chock, Hassan II University, Casablanca, Morocco
Said Raghay	FST, Cadi Ayyad University, Marrakech, Morocco
Said Rakrak	FST, Cadi Ayyad University, Marrakech, Morocco
Salah El Hadaj	ENCG, Cadi Ayyad University, Marrakesh, Morocco
Salima Chantit	FST - Mohammedia, Hassan II University, Casablanca, Morocco
Samira Douzi	Faculty of Medicine and Pharmacy, Mohammed V University, Rabat, Morocco
Sathish Kumar Velayudam	Cox Automotive, Atlanta, USA
Shakir Ullah	Stratford University, USA
Sharan Babu Paramasivam Murugesan	Microsoft, USA
Sitaram Srivatsavai	CRM Engineering, Gilbert, Arizona, USA
Sofia Douda	FST, Hassan 1st University, Settat, Morocco
Soukaina Mihi	FST, Hassan 1st University, Settat, Morocco
Srikanth Baride	University of South Dakota, USA
Sundaramurthy Gayathri Devi	Deemed University, Coimbatore, India

Supraja Chinthala	Product Management in Data Products, Atlanta, USA
Vijaya Sai Munduru	Salesforce, USA
Tolga Ensari	Department of Computer and Information Science, Arkansas Tech University, USA
Wiame Ech-Chelfi	National School of Applied Sciences, Berrechid, Morocco
Youssef Saadi	FST, Sultan Moulay Slimane University, Beni Mellal, Morocco
Zahi Jarir	Faculty of Sciences Semlalia, Cadi Ayyad University, Marrakesh, Morocco
Zakaria Boulghasoul	ENSA, Cadi Ayyad University, Marrakesh, Morocco

Contents

Big Data and Analytics

Clustering Learner Trajectories to Identify Cognitive Fatigue: A Big Data Analysis of EdNet

Houda Amazal(✉)

Polydisciplinary Faculty of Sidi Bennour, Chouaib Doukkali University, El-Jadida, Morocco
houda.kamouss@gmail.com

Abstract. This study proposes a scalable approach for modeling cognitive fatigue in large-scale online learning environments using behavioral log data from the EdNet dataset. By analyzing over 750,000 learners and their temporal interaction patterns, we extract two key features: the average time between actions and the number of questions attempted. These metrics are processed with PySpark and clustered using K-Means to reveal latent behavioral profiles. Results identify three distinct learner groups: engaged learners, disengaged learners, and intermittent learners. Statistical and visual analyses confirm the discriminative power of time-based features for inferring potential signs of cognitive fatigue. Unlike traditional performance-based models, this approach leverages implicit behavioral signals at scale to detect early disengagement patterns. The study contributes to the development of fatigue-aware learning analytics, offering a foundation for proactive interventions and adaptive feedback mechanisms in digital learning.

Keywords: Big data · Clustering · Learning analytics · Cognitive fatigue

1 Introduction

In recent years, the proliferation of digital learning platforms has generated unprecedented volumes of data, transforming the landscape of educational research and practice. This datafication of learning has led to the emergence of Educational Data Mining. At the intersection of data science and pedagogy, these fields have opened new avenues for the diagnosis of cognitive states, prediction of learner success, and design of adaptive interventions in both formal and informal educational contexts (Romero and Ventura 2020).

While prior research has primarily focused on predicting academic outcomes or modeling knowledge acquisition processes, less attention has been devoted to latent constructs such as cognitive fatigue. Unlike observable indicators like test scores or performance errors, cognitive fatigue tends to manifest implicitly,

A. E. Hassanien et al. (Eds.): AISDS 2025, LNDECT 287, pp. 3–12, 2027.
https://doi.org/10.1007/978-3-032-18927-1_1

often through temporal markers such as prolonged inactivity, irregular pacing of tasks, or abrupt declines in interaction frequency (Baig et al. 2020).

The growing availability of big educational datasets, has created new opportunities to explore these latent cognitive patterns at scale. These datasets provide rich temporal data that can be leveraged to analyze learning behaviors over time. However, most existing analyses on such datasets have prioritized knowledge tracing and performance classification, with limited exploration of engagement dynamics or fatigue detection (Chen et al. 2018). Furthermore, the increasing integration of big data infrastructures in educational research, such as distributed processing frameworks, has made it feasible to operationalize complex behavioral models across millions of user records. These technical advancements now make it possible to detect subtle patterns of disengagement, such as intermittent usage or extended interaction gaps, using scalable unsupervised learning techniques (Molenaar and Wise 2022).

In this study, we address the underexplored challenge of modeling cognitive fatigue in online learning by leveraging large-scale behavioral logs from the EdNet platform (Choi et al. 2020). Specifically, we aim to: (1) identify time-based interaction patterns indicative of fatigue or disengagement, (2) segment learners into behavioral profiles using scalable clustering techniques, and (3) interpret these profiles in light of cognitive theories and self-regulated learning frameworks. By combining temporal analytics with big data processing and unsupervised learning, this work contributes to the growing body of research seeking to develop fatigue-aware, adaptive learning systems capable of responding proactively to early signs of learner disengagement (Deeva et al. 2022).

2 Related Work

The rapid expansion of digital learning platforms has led to the accumulation of massive educational datasets, giving rise to two closely related fields: Educational Data Mining (EDM) and Learning Analytics (LA). These disciplines aim to uncover latent patterns in learner behavior, often with the goal of improving student support and enhancing learning outcomes. Foundational works such as Papamitsiou and Economides (2014) have systematically reviewed the empirical evidence supporting EDM and LA, emphasizing the potential of fine-grained behavioral traces for actionable insights. Similarly, Gaševic et al. (2014) advocated for learner-centric approaches in analytics, arguing that technological sophistication must serve pedagogical relevance.

Several studies have sought to infer cognitive or motivational states from behavioral logs. Conati and Lallé (2023) demonstrated that user interaction sequences in open-ended environments can be predictive of cognitive engagement. Tempelaar et al. (2015) investigated the predictive power of feedback-sensitive features derived from interaction logs. Matcha et al. (2019) further explored the link between learning strategies and academic outcomes, identifying engagement trajectories associated with success or dropout.

At a theoretical level, Cognitive Load Theory (Plass et al. 2010) provides a framework for understanding learner fatigue as a function of excessive task complexity or insufficient self-regulation. This perspective supports the interpretation of temporal disengagement, such as extended pauses or decreasing task volume, as potential indicators of cognitive fatigue, especially in unassisted self-paced environments.

With the advent of big data technologies, researchers have been able to move beyond small-scale log analysis to extract insights at scale. Daniel (2017) provided a conceptual overview of big data's implications for higher education, while Chen et al. (2020) reviewed how artificial intelligence techniques are transforming educational systems, including through the use of big data analytics. Recent studies such as Liu et al. (2022) have used temporal clickstream features combined with deep learning models to predict student performance, reinforcing the relevance of time-based behavioral features in learner modeling.

The EdNet dataset used in the present study, represents a major advancement in large-scale educational data collection. Its hierarchical structure and fine-grained logging of timestamped learner actions allow for sophisticated temporal analysis. Prior work leveraging EdNet has focused on knowledge tracing and performance prediction, but few studies have explored its use in modeling latent cognitive states, such as fatigue or disengagement.

Maldonado-Mahauad et al. (2018) exemplified how process mining techniques can reveal self-regulated learning strategies from big data, particularly in MOOC environments. These approaches align with the broader shift toward theory-driven behavioral analytics (Wise and Shaffer 2015), emphasizing that data-intensive methods must remain grounded in psychological and pedagogical constructs.

Recent studies have continued to advance the use of temporal behavioral data in educational contexts. Winter et al. (2024) demonstrated how behavioral trace data from online learning environments can serve as reliable indicators of learning engagement, employing a triangulation approach that combines self-reports with observational data to capture both cognitive and behavioral dimensions of engagement. Their work reinforces the value of trace-based analytics for understanding learner patterns at scale. Similarly, Nakamura et al. (2024) applied change point detection algorithms to educational big data, revealing how temporal shifts in student participation rates correlate with teacher behavior patterns. Their methodology demonstrates the feasibility of automatically identifying significant engagement changes from large-scale interaction logs, aligning with the temporal analytics approach adopted in the present study. Looking forward, Giannakos et al. (2025) discussed the transformative potential of AI-driven analytics in education, emphasizing both opportunities and ethical considerations as the field evolves toward more sophisticated data-intensive interventions.

Despite these advancements, few studies have combined large-scale behavioral data, time-sensitive indicators, and unsupervised machine learning to derive fatigue-related learner profiles. This gap underscores the contribution of the present work, which leverages PySpark and clustering algorithms to model cognitive disengagement using over 750,000 student trajectories from EdNet.

3 Methodology

This study utilizes the EdNet dataset, one of the largest publicly available educational datasets, which comprises over 131 million interaction records collected from students using the Santa learning platform. To ensure both scalability and precision in the modeling process, we focused on the KT1 subset, which provides a detailed chronological log of student activities including timestamp, *problem_id*, *time_taken*, *student_id*, *skill_id*, *correct*, and *log_delta*. Given the high granularity and volume of data, preprocessing was performed using PySpark, allowing efficient transformation of over 750,000 student records. The raw data required multiple layers of preprocessing to convert timestamp logs into analyzable learning sequences. First, Unix timestamps were transformed into standard temporal formats, then sorted chronologically by learner ID to reconstruct sessional interaction timelines.

For each student, we computed the time between successive actions using Spark window functions. This metric, referred to as *avg_time_diff*, serves as a key proxy for identifying disengagement patterns or cognitive fatigue. The average time between interactions varied significantly, ranging from a few seconds to over 100 days. In addition, the total number of questions attempted by each student (*total_questions*) was calculated as a second behavioral feature. This dual-dimensional feature space (time gaps and activity volume) enables behavioral profiling with scalable clustering methods.

To ensure robustness, we excluded entries with missing or invalid timestamps and filtered out sequences with less than two events. Students with only one interaction were deemed insufficient for temporal analysis and removed. As well, we extracted two primary behavioral indicators per learner: (1) average time between actions (*avg_time_diff*): interpreted as a proxy for temporal engagement. Longer gaps may signal disengagement or cognitive fatigue, while shorter intervals suggest sustained attention or flow; (2) total questions attempted (*total_questions*): reflects engagement depth. Higher values are associated with persistence and motivation. These features were selected for their robustness, universality across students, and independence from content-specific attributes. Then, we applied K-Means clustering (via PySpark MLlib) to partition the learner population into behavioral profiles. Before clustering: All features were standardized, Null values were removed to preserve model integrity, and a VectorAssembler was used to structure the input features into a Spark-compatible format. The optimal number of clusters ($k = 3$) was determined empirically, based on interpretability, cluster separability, and silhouette scores.

Each cluster was then mapped to a descriptive behavioral label based on post-hoc statistical summaries and visual inspection of the distributions: Cluster-0 (Engaged learners) include students with short inter-action times and high question volumes, Cluster-1 (Disengaged learners) include students characterized by extremely long delays between actions and minimal activity, and Cluster-2 (Intermittent learners) include students with moderate activity and long but less extreme time gaps.

In the absence of physiological markers, we adopt a behavioral approximation of cognitive fatigue: patterns of increasing latency, low task repetition, or irregular engagement rhythms may reflect mental exhaustion, distraction, or disengagement. This approach aligns with prior literature in learning analytics, where temporal interaction logs are used as proxies for cognitive states. This methodology lays the foundation for scalable, non-intrusive detection of latent mental states across massive learner populations.

4 Results

This section presents the main outcomes of the clustering process and explores the behavioral patterns that emerge from the EdNet dataset. The analysis is structured into three parts: (i) descriptive statistics, (ii) behavioral mapping in a two-dimensional space, and (iii) temporal dispersion analysis. Together, these dimensions provide a multidimensional interpretation of learner engagement at scale.

Table 1 summarizes the descriptive statistics for each of the three identified clusters. Cluster-0 is by far the largest, comprising 754,241 learners, or 99.6% of the dataset. Learners in this cluster exhibit an average time between actions of approximately 23,796.49 s (median: 40.00 s), suggesting frequent and regular interactions with the platform. The average number of questions attempted per student in this cluster is 126.27, with a maximum value reaching 40,157 questions. This points to a highly active and consistent learning behavior.

Cluster-1, in contrast, consists of only 127 learners (approximately 0.02%) and exhibits extreme values for temporal inactivity. The average time between interactions is over 13 million seconds (13,302,580.23), which corresponds to approximately 154 days. The median time is similarly high (11,718,044.50 s). Moreover, the number of questions answered is minimal, averaging only 3.39 tasks. These statistics suggest that Cluster-1 includes disengaged users, possibly those who abandoned the platform after limited interaction.

Cluster-2 includes 2,576 users and represents an intermediate profile. The average time between actions is approximately 3,003,661.42 s (or $\sim$ 34.7 days), with a median of 2,554,605.83 s. Learners in this group completed an average of 10.06 tasks, suggesting irregular but non-trivial engagement. This cluster likely includes learners who interact periodically with the platform, possibly in bursts of activity followed by periods of inactivity.

These summary statistics provide an initial typology of learners in terms of activity volume and temporal pacing. However, numeric summaries alone cannot capture the structural relationships between learner behaviors. For this, visual exploration becomes essential.

Figure 1 presents a scatterplot of all clustered learners in a two-dimensional feature space, where the x-axis corresponds to the logarithm of the average time between actions and the y-axis represents the total number of questions completed. The logarithmic scale was applied to account for the extreme skew in inter-action times and to better separate the clusters visually.

Table 1. Descriptive statistics per cluster.

Cluster	Students (n)	Avg. Time Diff (s)	Total Questions	Median Time Diff (s)	Max Time Diff (s)
0	754,241	23,796.49	126.27	40.00	1,513,464
1	127	13,302,580.23	3.39	11,718,044.50	32,632,353
2	2,576	3,003,661.42	10.06	2,554,605.83	8,114,025.80

The plot reveals a clear spatial segmentation among the clusters. Cluster-0 dominates the central and left region of the plot, forming a dense and compact distribution. This region reflects the behavioral profile of consistently engaged learners, characterized by low latency and high task completion rates. The tight spatial concentration suggests both behavioral homogeneity and high platform fidelity.

Cluster-1 is situated at the far right of the scatterplot, close to the x-axis, separated from Cluster-0 by several orders of magnitude in both features. The sparse distribution and isolation of Cluster-1 indicate profound disengagement: learners in this group logged into the system, engaged in a small number of tasks, and then remained inactive for extremely long periods. Their spatial location visually reinforces their statistical outlier status.

Cluster-2 is distributed along a narrow band between Cluster-0 and Cluster-1, in the right portion of the graph. It reflects learners with moderate latency and low-to-moderate task completion volume. The wider spread in this cluster suggests greater behavioral variability. While some learners may follow cyclical or seasonal learning patterns, others may engage sporadically without discernible regularity. Importantly, the separation between Cluster-1 and Cluster-2 is maintained, underscoring the validity of the clustering procedure.

This spatial representation complements the descriptive statistics and highlights how behavioral traits, when jointly considered, reveal distinct learner profiles that would be difficult to isolate using univariate analyses alone.

To further investigate the temporal dynamics of learner engagement, Fig. 2 presents a boxplot of the average time between actions for each cluster on a logarithmic scale. This visualization provides insights into the dispersion, central tendency, and outliers of each group's temporal behavior.

Cluster-0 again displays a consistent pattern. The boxplot shows a narrow interquartile range and a median around 40 s, confirming that the majority of learners in this cluster maintain a tight and regular rhythm of interaction. Few outliers extend beyond this range, suggesting a homogenous temporal engagement pattern within this dominant group.

In contrast, Cluster-1 reveals a drastically different structure. The interquartile range spans several orders of magnitude, with the median exceeding 11 million seconds. The presence of numerous extreme outliers and a long upper tail signifies an erratic engagement pattern. Learners in this group show prolonged periods of inactivity, and the high variance reinforces their classification as outliers in the dataset.

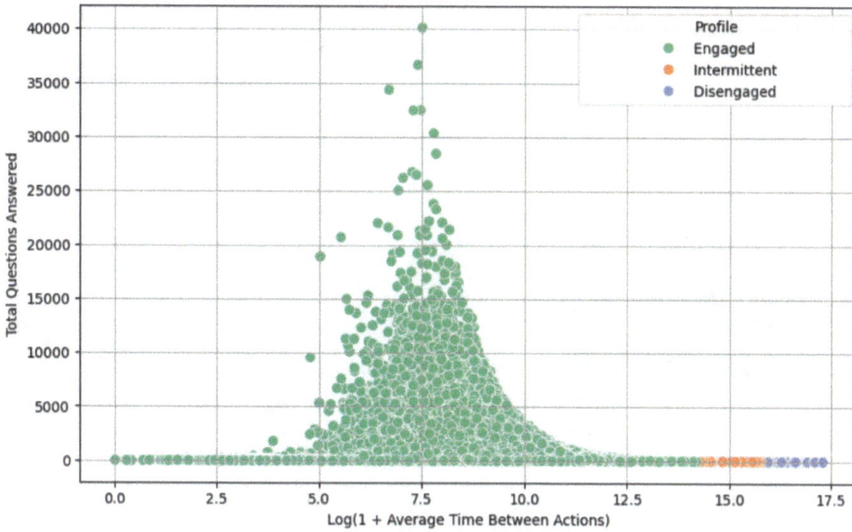

Fig. 1. Learner clusters by engagement latency and activity volume.

Cluster-2, though less extreme, presents a wider interquartile range than Cluster-0. The median time between actions is approximately 2.5 million seconds, and the overall spread indicates substantial heterogeneity. This is indicative of learners who might engage in concentrated bursts followed by long interruptions. The temporal patterns here are consistent with intermittent or on-demand learning behaviors, which are increasingly common in self-paced online platforms.

The boxplot thus provides a complementary view to the scatterplot by focusing exclusively on the temporal aspect of engagement. It confirms that interaction time is a robust and discriminative feature for identifying disengagement and behavioral fragmentation in large-scale educational datasets.

The combined insights from descriptive statistics (Table 1), spatial projection (Fig. 1), and temporal distribution (Fig. 2) lead to the identification of three robust behavioral profiles:

Cluster-0: Engaged learners. Learners in this group demonstrate regular, low-latency engagement with high task completion. Their behavior is stable, predictable, and indicative of sustained cognitive effort. This cluster likely includes the most successful and persistent learners.

Cluster-1: Disengaged learners. Characterized by extreme temporal delays and minimal task interaction, this group appears to include users who abandoned the platform shortly after enrolling. Their behavioral footprint suggests shallow engagement or exploratory use rather than meaningful participation.

Cluster-2: Intermittent learners. This group represents learners with irregular patterns of engagement, moderate in volume but highly variable in temporal

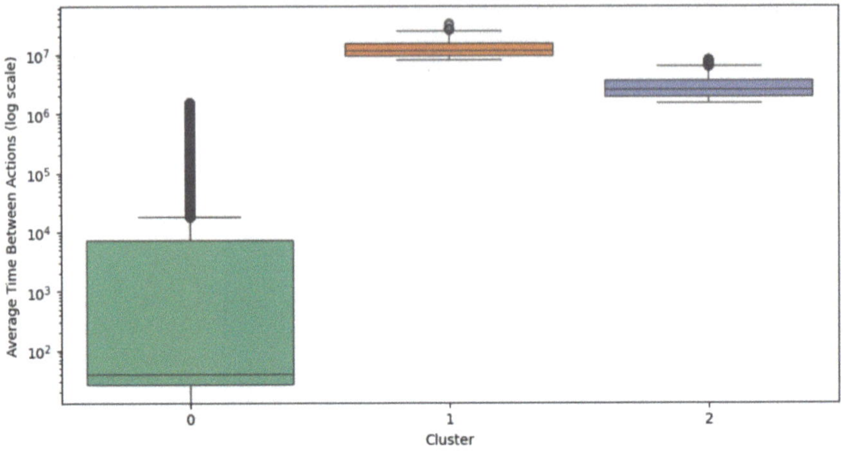

Fig. 2. Distribution of average time by cluster (log scale).

spacing. Their interaction style may be shaped by contextual constraints (e.g., limited availability, external obligations) or motivational factors.

These profiles are not only statistically grounded but also pedagogically meaningful. They can serve as the basis for adaptive interventions and targeted support mechanisms. For example, intermittent learners may benefit from nudging strategies or progress reminders, while disengaged users might require onboarding reinforcement or early outreach.

Overall, the clustering model demonstrates the value of temporal and behavioral data for understanding learner engagement in online environments. It also underscores the importance of scalable analytics pipelines capable of processing millions of interactions to extract meaningful pedagogical insights.

5 Conclusion

This study introduced a scalable, data-driven methodology to detect behavioral markers of cognitive fatigue in large-scale online learning environments. Using the EdNet dataset, one of the most comprehensive educational datasets to date, we analyzed interaction sequences from over 750,000 learners to extract temporal features indicative of engagement rhythms. Through unsupervised clustering using PySpark, we identified three distinct learner profiles: engaged learners, intermittent users, and disengaged learners. These profiles were validated through descriptive statistics and visualization techniques, revealing strong correlations between temporal spacing and task participation levels.

Unlike conventional models focused on performance prediction or knowledge tracing, this work emphasizes the value of temporal behavior as a non-intrusive proxy for underlying cognitive states, such as fatigue or disengagement. The proposed framework demonstrates how minimal but interpretable features, such as average time between actions, can be leveraged to segment learners and inform adaptive pedagogical strategies.

Future research should explore the integration of physiological or affective data to triangulate behavioral inferences, as well as longitudinal validation across diverse platforms and populations. The methodology presented here opens promising avenues for the development of fatigue-aware learning analytics systems, capable of offering timely interventions to at-risk students in real-world educational settings.

References

Baig, M.I., Shuib, L., Yadegaridehkordi, E.: Big data in education: a state of the art, limitations, and future research directions. Int. J. Educ. Technol. High. Educ. **17**(1), 1–23 (2020). https://doi.org/10.1186/s41239-020-00223-0

Chen, L., Chen, P., Lin, Z.: Artificial intelligence in education: a review. IEEE Access **8**, 75264–75278 (2020). https://doi.org/10.1109/ACCESS.2020.2988510

Chen, B., Knight, S., Wise, A.: Critical issues in designing and implementing temporal analytics. J. Learn. Anal. **5**(1), 1–9 (2018). https://doi.org/10.18608/jla.2018.51.1

Conati, C., Lallé, S.: Student modeling in open-ended learning environments. In: Handbook of Artificial Intelligence in Education, pp. 170–183. (2023). https://doi.org/10.4337/9781800375413.00018

Choi, Y., et al.: EdNet: a large-scale hierarchical dataset in education. In: Bittencourt, I.I., Cukurova, M., Muldner, K., Luckin, R., Millán, E. (eds.) AIED 2020. LNCS (LNAI), vol. 12164, pp. 69–73. Springer, Cham (2020). https://doi.org/10.1007/978-3-030-52240-7_13. https://doi.org/10.48550/arXiv.1912.03072

Daniel, B.K. (ed.): Big Data and Learning Analytics in Higher Education. Springer, Cham (2017). https://doi.org/10.1007/978-3-319-06520-5

Deeva, G., De Smedt, J., Saint-Pierre, C., Weber, R., DeWeerdt, J.: Predicting student performance using sequence classification with time-based windows. Expert Syst. Appl. **209**, 118182 (2022). https://doi.org/10.1016/j.eswa.2022.118182

Giannakos, M., et al.: The promise and challenges of generative AI in education. Behav. Inf. Technol. **44**(11), 2518–2544 (2025). https://doi.org/10.1080/0144929X.2024.2394886

Gašević, D., Dawson, S., Siemens, G.: Let's not forget: Learning analytics are about learning. TechTrends **59**(1), 64–71 (2014). https://doi.org/10.1007/s11528-014-0822-x

Liu, Y., Fan, S., Xu, S., Sajjanhar, A., Yeom, S., Wei, Y.: Predicting student performance using clickstream data and machine learning. Educ. Sci. **13**(1), 17 (2022). https://doi.org/10.3390/educsci13010017

Matcha, W., Gašević, D., Uzir, N.A., Jovanović, J., Pardo, A.: Analytics of learning strategies: associations with academic performance and feedback. In: Proceedings of the 9th International Conference on Learning Analytics & Knowledge, pp. 461–470 (2019). https://doi.org/10.1145/3303772.3303787

Maldonado-Mahauad, J., Pérez-Sanagustín, M., Kizilcec, R.F., Morales, N., Muñoz-Gama, J.: Mining theory-based patterns from big data: identifying self-regulated learning strategies in massive open online courses. Comput. Hum. Behav. **80**, 179–196 (2018). https://doi.org/10.1016/j.chb.2017.11.011

Molenaar, I., Wise, A.F.: Temporal aspects of learning analytics - grounding analyses in concepts of time. In: The Handbook of Learning Analytics, 2nd edn., pp. 66–76 (2022). https://doi.org/10.18608/hla22.007

Nakamura, K., Ishihara, M., Horikoshi, I., Ogata, H.: Uncovering insights from big data: change point detection of classroom engagement. Smart Learn. Environ. **11**(1), 31 (2024). https://doi.org/10.1186/s40561-024-00317-6

Papamitsiou, Z., Economides, A.A.: Learning analytics and educational data mining in practice: a systematic literature review of empirical evidence. J. Educ. Technol. Soc. **17**(4), 49–64 (2014)

Plass, J.L., Moreno, R., Brünken, R. (eds.): Cognitive Load Theory. Cambridge University Press, Cambridge (2010). https://doi.org/10.1017/CBO9780511844744

Romero, C., Ventura, S.: Educational data mining and learning analytics: an updated survey. Wiley Interdiscip. Rev. Data Min. Knowl. Discov **10**(3), e1355 (2020). https://doi.org/10.1002/widm.1355

Tempelaar, D.T., Rienties, B., Giesbers, B.: In search for the most informative data for feedback generation: learning analytics in a data-rich context. Comput. Hum. Behav. **47**, 157–167 (2015). https://doi.org/10.1016/j.chb.2014.05.038

Winter, M., et al.: Behavioral trace data in an online learning environment as indicators of learning engagement in university students. Front. Psychol. **15**, 1396881 (2024). https://doi.org/10.3389/fpsyg.2024.1396881

Wise, A.F., Shaffer, D.W.: Why theory matters more than ever in the age of big data. J. Learn. Anal. **2**(2), 5–13 (2015). https://doi.org/10.18608/jla.2015.22.2

Evaluating the Relationship Between Online Gaming and Mental Health Using Data Mining Techniques

Ghada Asim[1,2](✉), Mai K. Galab[1], and Eman K. Elsayed[2,3]

[1] Faculty of Computers & Artificial Intelligence - BFCAI, Benha University, Benha, Egypt
Ghadaasim25@gmail.com
[2] School of Computer Science, Canadian International College (CIC), New Cairo, Egypt
[3] Faculty of Science, Al-Azhar University (Girls), New Cairo, Egypt

Abstract. As online gaming becomes more and more popular worldwide, researchers are paying more attention to how it affects mental health. However, the connection between gaming behavior and mental health conditions like anxiety and depression is still unclear since there is a dearth of trustworthy, labeled data connecting digital behavior to psychological effects. This study investigates the relationship between online gaming behaviors and mental health using cutting-edge machine learning techniques. The collection contains demographic data, gaming activity indicators, and psychological tests like the Social Phobia Inventory (SPIN), the Generalized Anxiety Disorder Scale (GAD), and the Satisfaction with Life Scale (SWLS). Behavioral clusters were found using unsupervised models (K-Means and Agglomerative Clustering) and subsequently categorized using supervised models (Logistic Regression, Random Forest, and Deep Neural Network) following preprocessing (scaling, addressing missing values, and encoding). With an accuracy of 97%, an F1-score of 0.98, and a ROC-AUC of 0.976, the deep Neural Network outperformed the others and showed a great predictive capacity to identify mental health features from gaming activities.

Keywords: Machine Learning Classification · Deep Neural Networks Satisfaction with Life Scale · Generalised Anxiety Disorder scale · The Social Phobia Inventory

1 Introduction

One of the most important facets of total human health is mental health, which has a big impact on a person's everyday functioning and quality of life. It has to do with mental and emotional health, the capacity to manage stress, perform successfully at work, and make a positive contribution to society. Since gaming habits and behaviours can have a positive or negative impact on mental states, there has been a surge in interest in studying mental health in recent years, particularly as life's stresses have increased and online gaming has become more and more popular among young people. In order to support early diagnosis and encourage improved mental health among online gamers, it is now

A. E. Hassanien et al. (Eds.): AISDS 2025, LNDECT 287, pp. 13–21, 2027.
https://doi.org/10.1007/978-3-032-18927-1_2

crucial to analyse and comprehend the relationship between psychological behaviours and gaming habits using cutting-edge approaches like machine learning. The purpose of this project is to use data gathered from the Kaggle platform to analyse and categorise the mental health state of online gamers using machine learning techniques. In addition to psychological indicators determined by standardised and validated tools, the dataset comprises features like age, gender, and gaming hours. These tools include GAD_T [1], which measures the intensity of anxiety symptoms; SWL_T [2], which measures overall life satisfaction; and SPIN_T, which assesses social anxiety symptoms. The most accurate model was identified by comparing performance using a variety of classification algorithms.Random Forest [3], Logistic Regression [5], and DNN [5] were among these methods. The DNN model outperformed all other models in terms of classification performance, according to the results. This demonstrates its potential as a trustworthy instrument for evaluating mental health issues and identifying those who might require early psychological assistance. The study is divided into multiple primary sections:

1. Dataset Description: explains the dataset that was used in the research.
2. Literature Survey: Reviews previous research and relevant studies related to the topic.
3. Proposed Methodology: describes the steps involved in the research, such as model training, evaluation, data splitting, clustering, and preprocessing.
4. Result and Discussion: identifies the best categorisation model by discussing and contrasting its performance.
5. Conclusion and Future Work: summarises the main conclusions and makes recommendations for future study and improvement directions.

2 Literature Survey

Numerous studies have looked into the predictive capabilities and psychometric qualities of psychological measures linked to anxiety and life satisfaction. Using a German sample of 9,711 adults, Diener et al. [6] carried out a comprehensive study on the Satisfaction with Life Scale (SWLS). Descriptive statistics, confirmatory factor analysis (CFA), and reliability analysis (Cronbach's alpha) validated the scale's unidimensional structure and internal consistency in measuring subjective quality of life. The results showed a strong positive correlation with psychological well-being, optimism, and general life satisfaction, a negative correlation with worry and weariness, and great dependability. The study's cross-sectional design, however, hampered the capacity to draw conclusions about causality, and its concentration on Germans may have reduced the study's cultural generalizability. Likewise,Lietal [7]. used a large sample of 67,281 Chinese adolescents to investigate the psychometric properties of the Generalized Anxiety Disorder Scale (GAD7).The results showed strong reliability and substantial differences, with anxiety levels being higher among females and rising with age, according to exploratory and confirmatory factor analyses, item correlation, and group difference tests by gender and age.The focus solely on teenagers and cultural norms that can limit external alidity were among the limitations. Connor and associates [8] examined the Social Phobia Inventory's (SPIN) psychometric validity in both clinical and non-clinical samples. The results demonstrated good discriminant validity and outstanding reliability ($\alpha = 0.94$), effectively differentiating between people with and without social anxiety, using discriminant validity analysis, internal consistency testing, and sensitivity to symptom changes over time. However, the study ignored cultural differences that might have affected the results and failed to thoroughly assess long term predictive validity. Zhang and associates [9] used machine learning algorithms like Random Forest and Support Vector Machines, feature selection, and cross-validation to forecast life satisfaction levels based on demographic and psychological factors. Their models outperformed traditional regression techniques in identifying important predictive characteristics, such as stress levels and social support. However, the models' interpretability was constrained, and bias might have been introduced due to their reliance on self-reported data. Finally, Kim et al. [10] created a deep learning model that uses physiological information from wearable sensors, such as skin conductance and heart rate variability, to objectively identify generalized anxiety symptoms. The AI model outperformed conventional questionnaire-based techniques in terms of sensitivity and specificity under real-time monitoring conditions. However, restrictions such the small sample size, reliance on wearable technology, and narrow age range limited how broadly the results could be applied.

3 Dataset Description

Table 1. Summarizes the size, properties, and important features of the dataset used to analyses online gamers' mental health.

Attribute	Description
GAD1 to GAD7	Questions from the GAD-7 anxiety assessment scale
SWL1 to SWL5	Questions from the SWLS life satisfaction scale
SPIN1 to SPIN17	Questions from the SPIN social phobia scale
GAD_T	Total score for GAD
SWL_T	Total score for SWL
SPIN_T	Total score for SPIN
Game	Name of the participant's preferred game
Platform	Gaming platform (e.g., PC, Console)
Hours	Weekly hours spent playing games
earnings	Income generated from gaming
why play	Reason for playing games
Narcissism	Narcissism score
Gender	Participant's gender
Age	Participant's age
Work	Employment status
Degree	Educational background

This dataset, which has 14,464 rows and 55 columns, was acquired from Kaggle [11] and offers a solid basis for machine learning and statistical analysis applications. The information was gathered from players in different nations as part of an international survey. The survey comprised basic psychological questions that are frequently used to gauge life satisfaction, social anxiety, and anxiety. Typically, it comprises of multiple sets of questions used in psychological research. The collection includes answers to standardised psychological tests such the SPIN, the GAD, and the SWLS. Apart from psychological evaluations, the data comprises demographic details like age, gender, education, and work status, together with indicators of gaming behaviour like the number of hours played, the platform of choice, and gaming revenue. These various characteristics enable a thorough examination to comprehend the connection between gaming behaviours and mental health.

4 Proposed Methodology

The suggested methodology's four primary stages—the data preprocessing phase, the clustering phase, the training phase, and the model evaluation phase—are depicted in Fig. 1. The study's workflow, as shown in Fig. 1, Several consecutive phases make up the suggested analytical pipeline, which is intended to guarantee precise and dependable data processing. In the data preparation step, missing values were managed by imputing numerical characteristics with the median and categorical features with the mode. While the mode guarantees consistency by reflecting the most prevalent category for each variable, the median was selected because it is less impacted by outliers and helps maintain a steady data distribution. To maintain data quality and minimize bias, columns with a high percentage of missing values were eliminated. Following preprocessing, the GAD_T questionnaire was utilized to aggregate responses from seven items (GAD1 to GAD7) to determine the overall anxiety score (GAD7_Score). After that, participants were divided into two categories based on their level of anxiety: anxious (scores $\geq$10) and non-anxious (scores $<$10). To preserve class balance, the dataset was divided into 70% for training and 30% for testing using stratified sampling, and categorical variables were encoded using one-hot encoding. Prior to model training, all features were normalized using StandardScaler, and SMOTE (Synthetic Minority Over-sampling Technique) was applied to the training data to resolve class imbalance. A number of supervised classification models, such as a Deep Neural Network (DNN), Random Forest, Decision Tree, and Logistic Regression, were used and compared. The DNN is made up of several densely linked layers with ReLU activation, followed by a sigmoid activation for binary classification and batch normalization and dropout layers to prevent overfitting. To guarantee consistency, all models underwent the same preprocessing, training, and assessment processes. Accuracy, ROC-AUC, and the confusion matrix—which displays true positives, true negatives, false positives, and false negatives—were used to assess the model's performance.

Every visualization of a confusion matrix was kept for later examination and display.

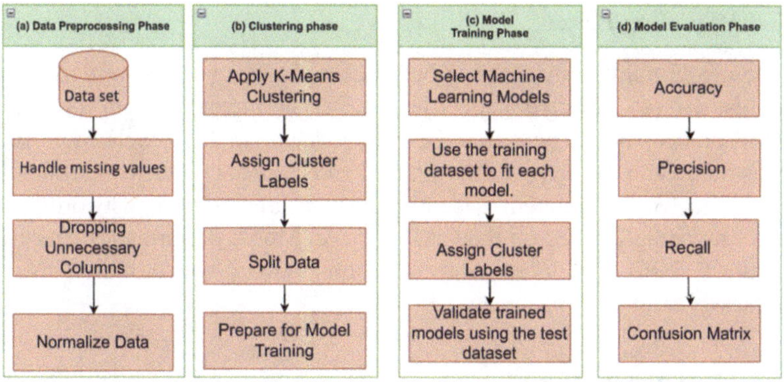

Fig. 1. The suggested approach, which includes data splitting, K-Means clustering, data preprocessing, and model evaluation

4.1 Data Preprocessing Phase

The accuracy and dependability of the model were built on the data preprocessing phase. To deal with missing values, an organized statistical method was used: categorical features were filled with the most frequent mode to preserve category consistency, while numerical features were imputed using the median to lessen the impact of outliers. In order to avoid bias and noise, columns like "highest league" that had more than 50% missing data were eliminated. In order to normalize the value ranges and boost the effectiveness of models that rely on distance measures, data standardization using StandardScaler was then carried out. By taking these actions, the dataset was best prepared for the modelling stage, guaranteeing a more solid and trustworthy basis for further analysis and forecasting.

4.2 Clustering Phase

Rather than employing unsupervised clustering approaches, the target variable used for classification during the data preparation stage was generated from the Generalized Anxiety Disorder scale (GAD_T). Each participant's responses to the seven GAD items (GAD1–GAD7) were added up to determine their GAD_T score. Participants with a total score of 10 or above were given a value of 1 (High Anxiety) based on known clinical thresholds, while those with a score of less than 10 were given a value of 0 (Low Anxiety). The dataset was transformed into a labeled format that was appropriate for supervised learning models. The dataset was subjected to a number of preprocessing procedures before model training, such as imputing missing values, encoding categorical variables using one-hot encoding, and standardizing numerical features using StandardScaler to guarantee that each feature contributed equally. SMOTE was utilized to address class imbalance in the training data, and a stratified train-test split was utilized to preserve class balance between datasets. These procedures made the dataset ready for models like Decision Tree, Random Forest, Logistic Regression, and Deep Neural Network (DNN) to accurately and consistently classify anxiety levels.

4.3 Model Training Phase

During the model training stage, a number of classification models were used to assess how well the dataset predicted anxiety levels. Decision trees, random forests, logistic regression, and deep neural networks (DNNs) were among the models employed. Each model was trained on the prepared training set and assessed on the test set that had not yet been viewed. This step sought to identify the most efficient and trustworthy classifier for identifying anxiety based on the GAD-T scale. Model performance was evaluated using accuracy, ROC-AUC scores, and confusion matrices, which offered insights into true and false predictions.

4.4 Model Evaluation Phase

In order to predict anxiety levels resulting from the GAD-T scale, four classification models were applied to the processed information during the Model Evaluation Phase. Among the models were a Deep Neural Network (DNN), Random Forest, Decision Tree, and Logistic Regression. After applying SMOTE, each classical model was trained on scaled and balanced training data, and its predictive reliability was assessed using accuracy, ROC-AUC, and confusion matrices on the test set. Logistic regression was used as a linear baseline for comparison, and the decision tree and random forest models were trained with regulated depth and ensemble settings to prevent overfitting. The Adam optimizer and binary cross-entropy loss were used to train the DNN model, which was constructed with several dense layers utilizing ReLU activations, batch normalization, and dropout for regularization. Its output layer predicted the presence or absence of anxiety by handling binary classification using a sigmoid activation. To compare the prediction power of different models, confusion matrices and performance reports were produced for each model and stored for further examination.

5 Result and Discussion

The results of Part 4 of the suggested method, which concentrated on the application and assessment of various categorization models for predicting anxiety levels based on the GAD-T scale, are presented in this part. A Deep Neural Network (DNN), Random Forest, Decision Tree, and Logistic Regression were the four models created and assessed. To guarantee equitable learning across classes, each classical model was trained using the balanced and scaled dataset (after using SMOTE and (StandardScaler). The prediction power and classification reliability of each model were then assessed using accuracy, ROC-AUC, and confusion matrices. Nonlinear correlations and hierarchical feature interactions were captured by the decision tree and random forest models, with logistic regression acting as a baseline linear model for comparison. To effectively handle binary anxiety classification, the DNN model was trained using the Adam optimizer and binary cross-entropy loss. It was composed of several dense layers with ReLU activations, batch normalization, and dropout regularization. To visualize model performance, confusion matrices were created for each model and stored in the results analysis subfolder. The outcomes were compared in order to ascertain which model offered the best accurate forecasts and to pinpoint methods for enhancing performance in the future.

5.1 Model Evaluation and Results Phase

Several assessment measures and confusion matrices were used in this phase to evaluate the performance of three classification models. Table 1 summarises the findings and shows each model's accuracy, precision, recall, and F1-score for both classes. The following are the formulas [10] for these metrics across all models:

$$\text{Accuracy} = \frac{(\text{TP} + \text{TN})}{(\text{TP} + \text{TN} + \text{FP} + \text{FN})} \text{x}100\% \tag{1}$$

$$\text{Precision} = \frac{\text{TP}}{(\text{TP} + \text{FP})} \text{x}100\% \tag{2}$$

$$\text{Recall} = \frac{\text{TP}}{(\text{TP} + \text{FN})} \text{x}100\% \tag{3}$$

$$\text{F1} - \text{Score} = 2 * \frac{(\text{Precision} * \text{Recall})}{(\text{Precision} + \text{Recall})} \text{x}100\% \tag{4}$$

With 97% accuracy and a ROC-AUC score of 0.9765, the DNN model performed the best overall, proving its superior capacity to differentiate between anxiety levels. Following closely after, the Random Forest classifier achieved 92% accuracy with an ROC-AUC of 0.9658, demonstrating balanced precision (0.95) and recall (0.78) for the positive class. Even with a lower accuracy of 83% and an ROC-AUC of 0.8070, the logistic regression model showed respectable discriminative potential, especially with a precision of 0.99 for the negative class. Table 2 summarizes the confusion matrix metrics, such as True Positives (TP), True Negatives (TN), False Positives (FP), and False Negatives (FN), to provide a better understanding of each model's classification capability. Additionally, a direct comparison of the classification performance and misclassification patterns of the three algorithms—logistic regression, random forest, and DNN—is made possible by Fig. 2's visual depiction of the confusion matrices (Table 3).

Table 2. The classification performance parameters for each model assessed in this study, such as accuracy, F1-score, recall, and precision.

Model	Precision		Recall		f1-score		Accuracy %
	Class 0	Class 1	Class 0	Class 1	Class 0	Class 1	
LogisticRegression	0.83%	0.71%	0.99%	0.07%	0.91%	0.12%	0.83%
Random Forest Classifier	0.95%	0.76%	0.95%	0.78%	0.95%	0.97%	0.92%
DNN	0.98%	0.93%	0.99%	0.90%	0.98%	0.91%	0.97%

Table 3. Confusion matrix results for each model, showing the counts of False Negatives (FN), False Positives (FP), True Negatives (TN), and True Positives (TP).

Model	True Positives (TP)	True Negatives (TN)	False Positives (FP)	False Negatives (FN)
LogisticRegression	47	3315	19	659
Random Forest Classifier	548	3165	169	158
DNN	633	3285	49	73

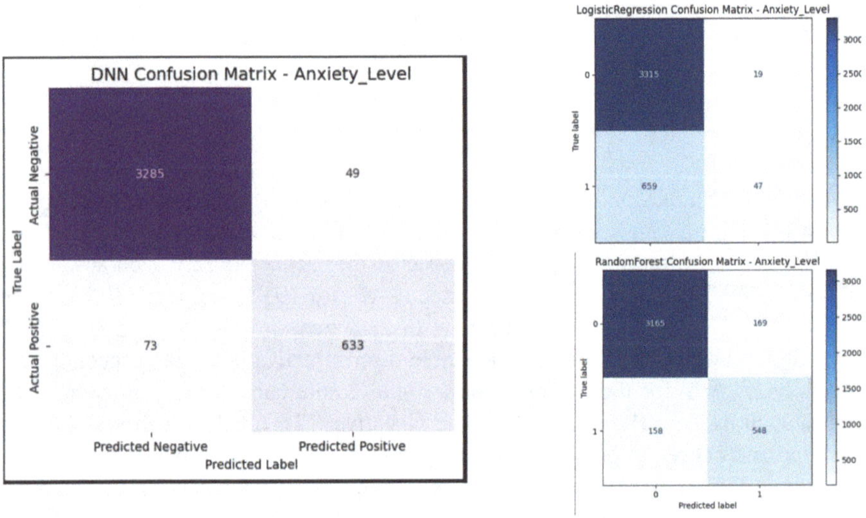

Fig. 2. Confusion matrices of different machine learning models (KNN, Decision Tree, Random Forest, Naive Bayes, and DNN) used in the study

6 Conclusion and Future Work

This study highlights the potential of machine learning techniques in understanding the complex relationship between online gaming behaviors and mental health. By leveraging psychological assessment scales (GAD, SWLS, SPIN) alongside demographic and behavioral indicators, we successfully developed a predictive model capable of classifying mental health conditions with high accuracy. Among the tested models, the Deep Neural Network achieved the best performance, reaching 97% accuracy, confirming its effectiveness in detecting psychological patterns based on behavioral data. The combination of unsupervised clustering techniques and supervised classification models enabled a comprehensive exploration of the dataset and revealed underlying patterns that may not be visible through traditional statistical methods. These findings demonstrate that AI can play a crucial role in mental health monitoring and open new doors for preventive intervention strategies in digital environments. Future Work: Future research could expand on this work by incorporating time-series data to observe mental health trends over longer periods, or by including physiological data (such as heart rate or sleep patterns) for a more holistic analysis. Additionally, integrating real-time gaming data and exploring cross-cultural or gender-based differences in mental health impacts could provide deeper insights. Developing an interactive dashboard for visualizing predictions and explanations could also enhance practical use for mental health professionals and game developers alike.

References

1. Wei, Q., et al.: Epidemiology of gaming disorder and its effect on anxiety and insomnia in Chinese ethnic minority adolescents. BMC Psychiatry. **22**(1), 260 (2022)
2. López-Ortega, M., Torres-Castro, S., Rosas-Carrasco, O.: Psychometric properties of the Satisfaction with Life Scale (SWLS): Secondary analysis of the Mexican Health.
3. Belgiu, M., Drăguţ, L.: Random forest in remote sensing: a review of applications and future directions. ISPRS J. Photogramm. Remote Sens. **114**, 24–31 (2016)
4. LaValley, M.P.: Logistic regression. Circulation. **117**(18), 2395–2399 (2008)
5. Hussain, H., Tamizharasan, P.S., Rahul, C.S.: Design possibilities and challenges of DNN models: a review on the perspective of end devices. Artif. Intell. Rev., 1–59 (2022)
6. Hinz, A., et al.: Psychometric properties of the satisfaction with life scale (SWLS), derived from a large German community sample. In: Quality of Life Research, vol. 27. Springer, Heidelberg (1661–1670)
7. Sun, J., Liang, K., Chi, X., Chen, S.: Psychometric properties of the generalized anxiety disorder scale-7 item (GAD-7) in a large sample of Chinese adolescents. In: Healthcare, vol. 9, p. 1709. MDPI (2021)
8. Connor, K.M., Davidson, J.R., Churchill, L.E., Sherwood, A., Weisler, R.H., Foa, E.: Psychometric properties of the social phobia inventory (SPIN): new self-rating scale. Br. J. Psychiatry. **176**(4), 379–386 (2000)
9. Alves, W.A., Gobber, C.F., Silva, D.J., Morimitsu, A., Hashimoto, R.F., Marcotegui, B.: Image segmentation based on ultimate levelings: from attribute filters to machine learning strategies. Pattern Recogn. Lett. **133**, 264–271 (2020)
10. Maxwell, A.E., Warner, T.A., Guillén, L.A.: Accuracy assessment in convolutional neural network-based deep learning remote sensing studies—part 1: literature review. Remote Sens. **13**(13), 2450 (2021)
11. Dataset on online gaming and anxiety data. Accessed Nov 01 2024 [Online]. Available: https://www.kaggle.com/datasets/divyansh22/online-gaming-anxiety-data/data

A Survey on Personal Health Knowledge Graphs: Construction, Applications, and Challenges

Hafsa Elkoumikhi[1(✉)], Abdellah Madani[1], and Moulay Driss El Ouadghiri[2]

[1] LAROSERI Laboratory, Faculty of Sciences, Chouaib Doukkali University, El Jadida, Morocco
elkoumikhi.h114@ucd.ac.ma

[2] IA Laboratory, Faculty of Sciences, Moulay Ismail University, Meknes, Morocco

Abstract. Personal Health Knowledge Graphs (PHKGs) are transforming the landscape of personalized healthcare by integrating a variety of data sources, including electronic health records (EHRs), clinical databases, wearable device data, and environmental information, into comprehensive, semantically rich frameworks. This article provides an in-depth survey of the methodologies behind PHKG construction, highlighting the integration of diverse data, the use of standardized ontologies, and the application of advanced knowledge extraction techniques. It examines the significant applications of PHKGs in clinical decision support, health monitoring, personalized recommendations, and disease prediction. By identifying key challenges, synthesizing current advancements and mapping future research opportunities, this study aims to establish PHKGs as a cornerstone for achieving precision medicine and proactive care.

Keywords: PHKG · Knowledge Graphs · Healthcare · Artificial intelligence · Clinical Decision Support

1 Introduction

The digital transformation of healthcare has ushered in a new era of data-driven medicine, with Electronic Health Records (EHRs) at its core. EHRs facilitate the systematic storage, retrieval, and sharing of patient information, revolutionizing healthcare delivery by improving data accessibility and supporting clinical decision-making. However, as the volume, complexity, and diversity of health data grow, challenges related to integration, analysis, and utilization become increasingly apparent, underscoring the need for innovative solutions to achieve personalized and precise care.

Globally, the adoption of EHRs has seen a remarkable upward trajectory. According to the Organisation for Economic Co-operation and Development (OECD), among 27 surveyed countries, a significant number have implemented

A. E. Hassanien et al. (Eds.): AISDS 2025, LNDECT 287, pp. 22–31, 2027.
https://doi.org/10.1007/978-3-032-18927-1_3

national EHR systems, although challenges like system fragmentation persist [15]. In China, for example, the average annual EHR adoption rate in hospitals was 6.1% between 2007 and 2018, reflecting steady progress [16]. Similarly, Türkiye reported an increase in basic EHR usage from 25.5% in 2014 to 39.1% in 2017 [17,18].

Morocco also is headed in this direction. Although EHR adoption is only starting, initiatives like the Moroccan Electronic Health Record System (MEHRS) [19] and implementation of advanced Hospital Information Systems (HIS) demonstrate an increasing focus on computerization of health services. Nevertheless, issues of poor interoperability, fragmented systems, and weak legal frameworks continue to be formidable obstacles to wholesale adoption [20].

To address these constraints, Personal Health Knowledge Graphs (PHKGs) have been proposed as a groundbreaking alternative. PHKGs offer a robust and logical framework for the harmonization of heterogenous data sources like EHRs, wearables, genomic and lifestyle data. PHKGs allow medical experts to provide comprehensive actionable knowledge by integrating different data sets into a patient-focused system, enabling more precise diagnosis, better treatment plans, and preventive therapy. For example, combining genomic data, clinical data, and data from wearable monitors provides an integrative picture of a patient's health to support precision care and prevention. Although promising, PHKGs face issues such as data heterogeneity, privacy concerns, and scalability. Additionally, their integration into existing healthcare infrastructures requires usability, accessibility, and ethics.

The following sections have a precise explanation of the elements involved in the development of PHKG, including the integration of heterogeneous data sources and the application of standardized ontologies. Then, certain knowledge extraction techniques, statistical and natural language processing (NLP), are performed to build such graphs. The survey also touches on the various uses that PHKGs can have, ranging from clinical decision-making to health monitoring and providing patient-specific advice. Aside from their potential, we also discuss the challenges, limitations and difficulties that they have, including data privacy concerns, interoperability, scalability, and the need for validation methods. Finally, we await the future of PHKGs, discussing prospective developments in AI-based inference, real-time data fusion and additional standardization with a view to best affecting healthcare.

2 Construction of PHKG

The construction of personal health knowledge graphs (PHKG) involves the integration of heterogeneous data sources, the use of standardized ontologies, and the use of advanced knowledge extraction methods. Each of these factors plays a major role in constructing solid and semantically enriched knowledge graphs with the capability to serve a broad range of health applications, as shown in Fig. 1. The next section discusses in explicit detail the factors involved.

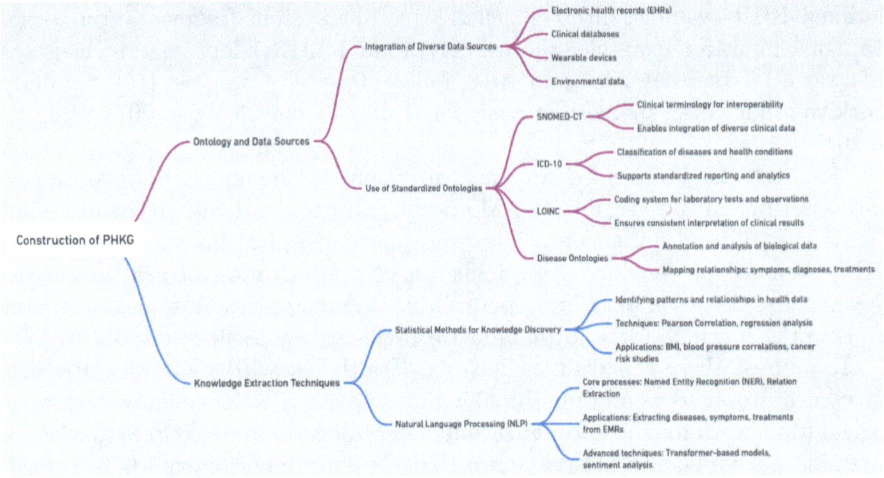

Fig. 1. Construction process of a Personal Health Knowledge Graph (PHKG).

2.1 Ontology and Data Sources

2.1.1 Integration of Diverse Data Sources

One of the needs for constructing PHKG includes the integration of heterogeneous data sources. These are those structured and unstructured forms of data such as clinical databases, wearable outputs of devices, electronic health records (EHRs), and even environment data. Their integration enables an omniscient view of patient health to be considered. For instance, the kHealth Framework describes how the integration of clinical records, wearable health signals, and environment IoT devices enable monitoring of patient health and creation of actionable intelligence at the level of real-time [11]. Similarly, others note the integration of EHRs diagnostic data with external knowledge of medicine with the objective of constructing diagnostic reasoning pathways and hence increasing accuracy and result interpretability [8].

Integrative approaches especially in wearable device data are crucial in personalized medicine. Wearables track real-time physiological indicators like heart rate, blood pressure, and activity levels that, when coupled with clinical information, offer a more comprehensive picture of patient well-being.

2.1.2 Use of Standardized Ontologies

Standard ontologies are essential for the harmonization and enrichment of data in PHKGs. Ontologies like SNOMED-CT, ICD-10, and LOINC ensure semantic interoperability, and data from different sources can therefore be seamlessly integrated. These ontologies also convey semantic richness, enabling further analysis and inference.

For example, disease ontologies are being heavily applied in data annotation, integration, and analysis of biological and clinical data. They enable the organized structuring of knowledge so that researchers and clinicians can browse complicated datasets easily [13]. The *RNA-KG* study is a case in point, illustrating how RNA-specific ontologies can be applied to organize complex biological data by compiling information across public databases [14].

In addition, ontologies offer a basis to map relations among entities, for example, symptoms, diagnoses, treatments, and outcomes, to improve interpretability and usability of PHKGs for clinical practice.

2.2 Knowledge Extraction Techniques

2.2.1 Statistical Methods for Knowledge Discovery

Statistical methods underlie knowledge discovery in Personal Health Knowledge Graphs (PHKGs) since they enable a person to identify meaningful relationships and patterns in health information. The most common methods include the application of Pearson Correlation and regression analysis to identify correlations between health indicators, i.e., between determinants of lifestyle and disease consequences. For example, one study comparing a set of physical examination measurements, including BMI, blood lipid concentration, and blood pressure, had significant correlations that describe the complex relationship between lifestyle activity and physiological indicators [21]. Likewise, a study of health-protective behaviors discovered a positive correlation between such behaviors and health hardness, indicating that individuals with higher hardness would tend to possess healthier lifestyles [22]. These results, combined with investigations regarding whether cancer morbidity is affected by aspects of lifestyle and death, have significant implications for behavior intervention because conformity to behaviors that foster health has been found to substantially lower the risk of cancer [23]. These findings demonstrate the worth of statistical analysis in enriching PHKGs with actionable knowledge of health dynamics.

2.2.2 Natural Language Processing (NLP)

Natural Language Processing (NLP) techniques are required to mine entities and relationships from unstructured sources, e.g., clinical notes, discharge summaries, and medical literature. NLP techniques transform raw text data into structured formats that can be integrated into Personal Health Knowledge Graphs (PHKGs). Core NLP processes, i.e., Named Entity Recognition (NER) and Relation Extraction, are extensively utilized to identify key entities such as symptoms, medications, and diagnoses, and extract their relations.

Advanced NLP techniques, including transformer-based deep learning models, have also enhanced the performance and scalability of such operations. A pipeline of NER and Relation Extraction was also shown to match entities in clinical text to biomedical knowledge graphs, again enhancing semantic interpretation of unstructured text [24].

Furthermore, sentiment analysis and topic modeling broaden NLP applications to infer health trends and patient-reported outcomes from health forums and social media. Such data provides additional insight to the knowledge regarding health concerns and increases the completeness of PHKGs.

3 Applications of Personal Health Knowledge Graphs (PHKGs)

3.1 Clinical Decision Support

Integration of personal health knowledge graphs (PHKG) into clinical decision support systems (CDSS) facilitates precision medicine through individualized treatment regimens, prediction of adverse reactions, and improved diagnosis precision. Several research studies have demonstrated its utility:

- **Explainable Decision Support**: Medical knowledge graphs provide comprehensible decision support, assisting clinicians in prompt diagnosis and treatment suggestions. Yet, there are challenges like disjointed patient data across multiple hospitals and data security issues can restrict their use. Creating avenues to incorporate knowledge graph systems into multicenter, information-sensitive healthcare environments is the key to successful decision support while not sacrificing data privacy and security [1].
- **Knowledge Graph-Based CDSS Framework**: A new framework has been put forward for clinical decision support systems (CDSS) that use knowledge graphs. The goal is to lay the groundwork for more research in this area. These frameworks can really help in creating CDSS that make better use of knowledge graphs for decision-making [2].
- **Patient-Centric Knowledge Graphs**: We're seeing an important change in healthcare with patient-centric knowledge graphs. These focus on providing care that's customized to individual patients by mapping out their specific health data. By integrating these knowledge graphs into CDSS, we can offer more personalized recommendations [3].

3.2 Health Monitoring

PHKGs facilitate ongoing health monitoring through the integration of wearable sensor and patient self-report data, among others. Integration supports real-time monitoring, detection of abnormalities at an early stage, and pre-emptive chronic disease management. Wearables, for example, offer long-term objective measures that are more precise compared to self-reported data [4]. Studies have also suggested architectures for building PHKGs from heterogeneous sources to facilitate monitoring and query efficiency [5]. Through real-time data processing, PHKGs are able to notify the user in case of threshold crossing, such as in COPD monitoring systems that provided alerts from abnormal values to urgent care need [6]. Through the integration of disparate data with real-time data, PHKGs facilitate patient management, empower both patients and clinicians, and improve overall quality of care.

3.3 Health Recommendations

PHKGs enable personalized health counseling by correlating medical history, behavior, genetics, and environment information. Personalized advice increases patient involvement and adherence. For example, library-based mobile health platforms have provided personalized reminders in the form of promotion of self-care in diabetic patients [10]. Similarly, a PHKG model incorporating medical data, social data, and day-to-day observation with the aid of a Personal Health Ontology has been used to provide personalized food suggestions and thus improve compliance and results for chronically ill patients [7].

The capability of PHKGs to incorporate context information greatly adds a relativity to health advice. It was part of the research that highlighted the importance of apps that incorporate a patient's health status as well as context-aware data from environmental sensors and other sources. By aggregating this information together in a PHKG, health professionals can offer advice both in terms of medical history as well as prevailing environmental conditions that affects health [5]. For example, a recommendation system can provide advice on lifestyle based on air quality levels for individuals who have asthma, thereby providing context-relevant advice.

3.4 Disease Prediction

PHKGs are used to create emerging models for classifying new models that evaluate health risk and forecast potential disease. Research suggested a heterogeneous knowledge graph that uses data from the National Health and Nutrition Examination Survey (NHANES) to give patients' health risk classifications. The model derived medical domain knowledge from big data and demonstrated strong performance in forecasting health hazard of different diseases [9]. With the execution of this classification model, healthcare professionals are able to detect patients who are at greater risk for certain diseases, making intervention in a timely manner possible. Another significant development in PHKGs is illustrated within the GRAPHCARE framework [12], which increases disease prediction by combining external biomedical knowledge and evidence from EHRs. GRAPHCARE constructs patient-specific knowledge graphs for every patient by combining concept-specific disease, procedure, and medication graphs with visit information based on time. Patient-specific graphs are passed through a Bi-Attention Augmented Graph Neural Network (BAT-GNN), which is able to process efficiently emphasize important patterns for prediction.

Through the utilization of patient-specific and external knowledge, systems such as GRAPHCARE enhance increasingly precise and personalized disease prediction, facilitating proactive and efficient healthcare interventions.

4 Challenges and Limitations of PHKG

4.1 Data Quality and Privacy

PHKGs face issues of inconsistent, incomplete, or outdated data due to varied sources like EHRs, wearables, and insurance records. Such problems can reduce credibility. Privacy is also critical, as personal health data is highly sensitive. Balancing data sharing with strict privacy and security requirements remains a major challenge, especially with decentralized sources like wearables.

4.2 Scalability

Another major challenge is scaling knowledge graphs to handle large datasets. The rapid expansion of personal health data resulting from the increasing adoption of EHRs and wearable health devices requires scalable approaches. Scalability challenges arise because large collections of data need to be handled efficiently and stored, all while maintaining good performance. The computational overhead associated with merging large-scale data from a wide variety of sources can also be significant. It is also likely that the integration of knowledge graphs with different technologies would be needed, such as cloud and edge computing, which would bring additional complexities in terms of data management, and processing.

4.3 Interoperability

One of the greatest challenges in bringing heterogeneous data sources and systems together is interoperability. Personal health information comes from numerous different sources, and each of these sources can use a different standard, format, and structure for the data. Overcoming significant interoperability obstacles is necessary for the smooth integration of diverse data sources such as these. For instance, variation in ontology and data schemas may prevent the data from being properly integrated and analyzed. Furthermore, the fact that integrating data from structured and unstructured sources makes it challenging also discourages it. The solution to such issues is to have good data integration architectures and standards which can facilitate ease of sharing as well as data integration.

5 Future Directions

These following areas require additional study and advancement in order to fully realize the potential of PHKGs:

- **Advancements in AI and Machine Learning:** The combination of generative AI and deep learning models techniques can help improve automatic knowledge extraction inference algorithms, and real-time updates, making PHKGs more intelligent and adaptive.
- **Real-Time Health Data Integration:** With the expansion of 5G and IoT enabled health devices, the incorporation of continuous multimodal data streams into PHKGs will enable more accurate predictive analytics and real-time patient monitoring.
- **Interoperability and Standardization:** Standardized ontologies and secure protocols will be necessary for seamless data exchange across health care systems for mass deployment.

- **Security and Privacy Enhancements:** In development techniques that protect privacy, like federated learning and blockchain technologies, will help address patient confidentiality concerns and data security issues.
- **Scalability and Computational Efficiency:** As PHKGs increase in complexity, distributed computing and edge computing technologies will be needed to maximize computational resources while keeping the performance and efficiency levels sustained.

By taking actions on these future opportunities and challenges, PHKG will not only evolve, but it will also be a continued mainstay of modern medical practice, and toward more personalized, data driven and proactive medical interventions.

6 Conclusion

Personal Health Knowledge Graphs (PHKGs) have emerged as a powerful way for integrating, analyzing, and applying health data for personalized health care. Thanks to standardized ontologies and modern data extraction methods, PHKGs have the ability to provide clinical decision support, health management, and personalized recommendations, ultimately improving the treatment of patients. The role of PHKGs in precision medicine and the management of chronic diseases shows their transformative power by enabling proactive, personalized, and tailored health intervention.

However, PHKGs must navigate the challenges of data privacy, scale, or interoperability. Addressing these challenges will be central to determining their scalability and ubiquity within clinical care studies. Such challenges notwithstanding, the capacity of PHKGs to improve treatment plans, identify health risks, and enhance overall healthcare outcomes across large and diverse patient populations should not be undermined as we move towards a health systems approach to precision medicine.

References

1. Shang, Y., Tian, Y., Lyu, K., et al.: Electronic health record-oriented knowledge graph system for collaborative clinical decision support using multicenter fragmented medical data: design and application study. J. Med. Internet Res. (2024). https://doi.org/10.2196/54263
2. Xiayu, X., Zhongru, W., Yan, J., Binxing, F.: Knowledge graph-based clinical decision support system reasoning: a survey. In: Proc. IEEE Int. Conf. Data Sci. Cyberspace (DSC), pp. 373–380 (2019). https://doi.org/10.1109/DSC.2019.00063
3. Al Khatib, H.S., Neupane, S., et al.: Patient-centric knowledge graphs: a survey of current methods, challenges, and applications. Front. Artif. Intell. **7** (2024). https://doi.org/10.3389/frai.2024.1388479
4. Huhn, S., Axt, M., Gunga, H.C., et al.: The impact of wearable technologies in health research: scoping review. JMIR Mhealth Uhealth **10** (2022). https://doi.org/10.2196/34384

5. Rastogi, N., Zaki, M.: Personal health knowledge graphs for patients. arXiv preprint (2020). https://doi.org/10.48550/arXiv.2004.00071
6. Bloor, D., Ugwuoke, N., Taylor, D., Lewis, K., Mur, L., Lu, C.: Towards a personal health knowledge graph framework for patient monitoring. arXiv preprint (2023). https://doi.org/10.48550/arXiv.2311.06524
7. Seneviratne, O., Harris, J., Chen, C.-H., McGuinness, D.L.: Personal health knowledge graph for clinically relevant diet recommendations. arXiv preprint (2021). https://doi.org/10.48550/arXiv.2110.10131
8. Xin, R., Chen, J., et al.: A personal healthcare knowledge graph framework for diagnosis of pelvic masses diseases (2024). https://doi.org/10.3233/SHTI231207
9. Tao, X., et al.: Mining health knowledge graph for health risk prediction. World Wide Web **23**(4), 2341–2362 (2020). https://doi.org/10.1007/s11280-020-00810-1
10. Ammar, N., Bailey, J.E., et al.: Using a personal health library–enabled mhealth recommender system for self-management of diabetes among underserved populations: use case for knowledge graphs and linked data. JMIR Mhealth Uhealth **9** (2021). https://doi.org/10.2196/24738
11. Gyrard, A., Gaur, M., et al.: Personalized health knowledge graph. In: CEUR Workshop Proc. (2018)
12. Jiang, P., Xiao, C., Cross, A., Sun, J.: GraphCare: enhancing healthcare predictions with personalized knowledge graphs (2024). https://doi.org/10.48550/arXiv.2305.12788
13. Kurbatova, N., Swiers, R.: Disease ontologies for knowledge graphs. BMC Bioinform. **22** (2021). https://doi.org/10.1186/s12859-021-04173-w
14. Cavalleri, E., Cabri, A., Soto-Gomez, M., et al.: An ontology-based knowledge graph for representing interactions involving RNA molecules. Sci. Data **11** (2024). https://doi.org/10.1038/s41597-024-03673-7
15. Slawomirski, L., et al.: Progress on implementing and using electronic health record systems: developments in OECD countries as of 2021. OECD Health Working Papers, no. 160 (2023). https://doi.org/10.1787/4f4ce846-en
16. Liang, J., Li, Y., Zhang, Z., et al.: Adoption of Electronic Health Records (EHRs) in China during the past 10 years: consecutive survey data analysis and comparison of Sino-American challenges and experiences. J. Med. Internet Res. **23**(2), e24813 (2021). https://doi.org/10.2196/24813
17. Köse, İ, Cece, S., Yener, S., et al.: Basic electronic health record (EHR) adoption in Türkiye is nearly complete but challenges persist. BMC Health Serv. Res. **23**, 987 (2023). https://doi.org/10.1186/s12913-023-09859-w
18. Köse, İ, Rayner, J., Birinci, S., et al.: Adoption rates of electronic health records in Turkish hospitals and the relation with hospital sizes. BMC Health Serv. Res. **20**(1), 967 (2020). https://doi.org/10.1186/s12913-020-05767-5
19. Benbrahim, H., Hachimi, H., Amine, A.: Moroccan electronic health record system. In: Proceedings of the International Conference on Industrial Engineering and Operations Management, Paris, France (2018)
20. Parks, R., Wigand, R.T., et al.: Electronic health records implementation in Morocco: challenges of silo efforts and recommendations for improvements. Int. J. Med. Inform. **129**, 430–437 (2019). https://doi.org/10.1016/j.ijmedinf.2019.05.026
21. Wang, H., Shuai, P., et al.: A correlation-based feature analysis of physical examination indicators can help predict the overall underlying health status using machine learning. Sci. Rep. **12**, 19626 (2022). https://doi.org/10.1038/s41598-022-20474-3

22. Hosseini, Z., Pourjalil, F., Homayuni, A.: Investigating the correlation between health-promoting lifestyle and health hardiness with quality of life. BMC Psychol. **12**, 511 (2024). https://doi.org/10.1186/s40359-024-02012-0
23. Zhang, Y.-B., Pan, X.-F., et al.: Combined lifestyle factors, incident cancer, and cancer mortality: a systematic review and meta-analysis of prospective cohort studies. Br. J. Cancer **122**, 1085–1093 (2020). https://doi.org/10.1038/s41416-020-0741-x
24. Haq, H.U., Kocaman, V., Talby, D.: Deeper clinical document understanding using relation extraction. arXiv (2021). https://arxiv.org/abs/2112.13259

Introducing and Analyzing the Fake Relationship Attack in Social IoT

Mohcine Baalla(✉) and Driss Bouzidi

Smart Systems Laboratory (SSL), ENSIAS, Mohammed V University in Rabat, Rabat, Morocco
{mohcine_baalla,driss.bouzidi}@um5.ac.ma

Abstract. The Social Internet of Things (SIoT) is an extension of traditional IoT by enabling devices to form social relationships, supporting resource sharing, service discovery, and trust-based interactions. While these social connections enhance efficiency and context-awareness, they also introduce new security challenges unique to the social layer. In this paper, we focus on the Fake Relationship Attack (FRA), a SIoT-specific threat where malicious devices forge or manipulate social ties to illegitimately gain trust, elevate privileges, or access sensitive resources. We analyze the capabilities and objectives of potential attackers, including the creation of Sybil devices, exploitation of friend-of-a-friend links, and strategic temporal manipulation to maximize influence within the network. Through realistic attack scenarios, we show how FRA can subvert trust propagation mechanisms and compromise system security in ways that traditional IoT defenses, focused on device integrity or communication security, cannot prevent. By formally characterizing FRA, this work highlights the vulnerabilities inherent in the SIoT social layer and establishes a foundation for future research on detection and mitigation strategies. Our analysis emphasizes the importance of considering relationship-based threats when designing SIoT systems and provides guidance for understanding the scope, impact, and potential consequences of attacks targeting social trust.

Keywords: SIoT · IoT · Attacks in SIoT · Fake Relationship Attack

1 Introduction

The Internet of Things (IoT) has transformed how smart devices communicate, collaborate, and provide services across domains such as healthcare, transportation, and smart homes.

Building on this, the SIoT has emerged as the convergence of social networks and IoT [1], allowing devices to form social relationships inspired by human interactions like ownership, co-location, co-work, and friendship. These relationships play a central role in service discovery, collaboration, and especially trust evaluation, introducing a social trust layer that supports device interactions and distinguishes SIoT from classical IoT and social networks.

While this social layer brings scalability, efficiency, and context awareness, it also expands the attack surface, introducing vulnerabilities absent in traditional IoT [2].

A. E. Hassanien et al. (Eds.): AISDS 2025, LNDECT 287, pp. 32–42, 2027.
https://doi.org/10.1007/978-3-032-18927-1_4

Unlike conventional IoT, where attacks target device or communication flaws, SIoT is exposed to threats exploiting trust and relationships. Malicious entities can manipulate social ties to inflate trust, spread false recommendations, or gain unauthorized access. Social semantics such as friend-of-a-friend links and co-work ties create opportunities for dishonest recommendation, bad-mouthing, ballot-stuffing, self-promotion, and collusion, compromising SIoT trust integrity [3].

These SIoT-specific threats show that trust is not only essential but a main attack target. If an attacker fabricates a social tie with a trusted device, it can inherit trust and access sensitive services like locks, cameras, or data. Sybil-based attacks are especially potent since fake relationships amplify trust propagation. Dishonest recommendations can cascade through the network once fraudulent ties exist, while traditional IoT security, focused on integrity and communication, cannot protect against such relationship-based threats.

In this context, we introduce the FRA, a novel SIoT-specific threat where attackers forge or fabricate social relationships with legitimate devices to exploit trust in the SIoT graph. Unlike Sybil or impersonation attacks targeting identities, FRA attacks the relationship-based trust fabric itself. It combines identity theft, Sybil attacks, impersonation, and social engineering, but is distinct for exploiting trust propagation, for instance, when a malicious device claims to be a "friend-of-a-friend" of a trusted device to gain unauthorized access.

The main contributions of this paper are summarized as follows:

- Formalization of FRA: We introduce and formally define this novel SIoT-specific attack that directly targets relationship-based trust.
- Threat Model: We describe the mechanisms by which adversaries forge social connections to manipulate trust propagation and gain unauthorized privileges.
- Analysis of Attack Scenarios: We explore realistic FRA scenarios and their implications for trust propagation and system security.

The remainder of this paper is structured as follows. Section 2 surveys related works on security challenges and trust management in SIoT. Section 3 introduces the Fake Relationship Attack and presents its threat model and scenarios. Section 4 provides a detailed analysis and discussion of FRA's implications. Section 5 describes potential directions for detection and mitigation. Finally, Sect. 6 concludes the paper with limitations and directions for future research.

2 Related Works

This section surveys the most relevant recent literature across four threads that directly inform and contrast with our work: (1) Trust Management and Social-Graph Attacks in SIoT, (2) identity/Sybil defenses and ledgered attestations tailored to SIoT, and (3) security and privacy techniques in SIoT. For each thread we summarize representative papers, highlight what they do, and explain how they relate to but leave open the specific problem of analyzing relationship authenticity and trust manipulation in SIoT (the gap our paper addresses).

2.1 Trust Management and Social-Graph Attacks in SIoT

Trust management remains central to SIoT research because the social relationships that give SIoT its utility are also the channels through which trust flows and decisions are made. Several surveys synthesize the state of the art, classify trust attributes and update rules, and highlight attack classes that exploit social semantics. The authors in [4] provide a taxonomy of trust computation, aggregation, and update strategies for SIoT, and catalogue attacks (self-promotion, bad-mouthing, ballot-stuffing, etc.) that threaten reputation systems in social IoT graphs; they emphasize the need for context-aware, decentralized trust models that adapt to dynamic social links and resource constraints. In [2], the authors similarly review trustworthiness management across SIoT, stressing that its reliance on socially derived inputs (owner links, co-location, friendship) makes classical IoT defenses insufficient and raises unique privacy concerns when trust information is shared. These surveys underline that trust in SIoT is both an enabler and a major vulnerability, motivating adaptive, context-aware models able to reason over evolving social graphs. Several studies also show how adversaries exploit social graphs to distort reputation. Works surveyed in [5] analyze malicious behaviors including collusion and dishonest recommendations, noting that detection usually depends on statistical or community-based anomaly analysis. Graph-embedding techniques such as Trust2Vec [6] detect large-scale manipulation through signed-network representations, yet assume authentic social ties as input, so even advanced detection remains limited when relationships themselves are forged, as in FRA.

2.2 Sybil, Identity-Binding and Ledgered Attestation Approaches in SIoT

Because many relationship attacks in SIoT begin with identity abuse (fake identities, mass device creation), a line of work focuses on Sybil detection and on stronger identity/ownership bindings in social IoT graphs. SybilPSIoT [7] is a recent SIoT-specific method that blends a web-of-trust model with smart-contract anchors to prevent Sybil things and to bind objects to owners; their hybrid prevention/detection approach shows that signed relationships and lightweight ledger anchoring reduce the feasibility of large Sybil campaigns in social IoT settings.

2.3 Zero-Knowledge Proofs, Selective Disclosure and Privacy Primitives for IoT

Cryptographic selective disclosure and zero-knowledge proof (ZKP) techniques are rapidly entering the IoT space as tools to reconcile verification with privacy. Surveys on zero-knowledge authentication for IoT review how ZKPs provide credential-style authentication without identity leakage and discuss performance trade-offs and proof composition (Schnorr-type proofs, Bulletproofs, zk-SNARKs) [8].

System proposals now show concrete IoT use cases: the zk-IoT framework [9] applies ZKPs to attest firmware integrity and computations on ledgers, demonstrating their feasibility in IoT ecosystems; similarly, zk-PoL [10] presents privacy-preserving proofs of co-location without revealing exact coordinates, a primitive directly relevant to SIoT relationship predicates.

Taken together, this literature shows that ZKPs and selective disclosure are mature enough for relationship-predicate proofs, yet few works explicitly design an SIoT relationship-validation protocol to prevent FRA while keeping trust propagation semantics.

2.4 Hybrid Defenses: When Ledgers, ML, Proximity and Crypto Meet

To better situate our contribution, we summarize recent works addressing trust and security in the SIoT, from blockchain-based trust management to hybrid blockchain/ML defenses and cryptographic privacy methods. Table 1 compares their contribution, methodology, strengths, and limitations, showing that despite progress, most defenses still focus on general IoT or emphasize blockchain/ML integration without addressing SIoT-specific threats such as FRA.

Table 1. Comparative summary of recent research on trust and security.

Approach	Paper	Technique	Strengths	Limitations	SIoT Relevance
ML-based trust	[11]	GNN + Knowledge Graph	Detects anomalous clusters	Needs training data	Social trust modeling
ML-based	[12]	GNN + RNN hybrid	Learns dynamic trust patterns	High computation	Detecting fake ties
Blockchain Survey	[13]	Ledger-anchored trust	Transparency, revocation	Energy, overhead	Broad SIoT trust
Hybrid ML + Ledger	[14]	Blockchain + ML	Combines audit + detection	Coordination overhead	Hybrid resilience

2.5 Gaps and Positioning of This Work

While prior literature addresses SIoT trust, Sybil detection, and social graph security, there is no comprehensive study that formalizes and analyzes the FRA. Existing works rarely explore the multi-stage dynamics of FRA, its propagation effects on trust, or realistic scenarios in which it can occur. This paper addresses this gap by thoroughly characterizing FRA, providing a foundation for future mitigation strategies.

3 FRA and Threat Model

In this section, we present a detailed analysis of the FRA, a threat that exploits the social layer of SIoT. We describe its conceptual foundations and characteristics, highlighting how it differs from traditional IoT or social network attacks. We then formalize the threat model, outlining the attacker's capabilities, goals, and potential scenarios. Understanding these aspects is essential for designing defenses that protect trust propagation without compromising privacy or network functionality.

3.1 Introduction to FRA

The FRA occurs when an attacker device creates in a network with legitimate nodes (see Fig. 1) illegitimate social ties. It is a composite attack combining:

- Identity Theft: impersonating a legitimate device.
- Sybil Attacks: creating multiple fake identities to amplify trust.
- Social Engineering: exploiting human-mediated interactions.
- Impersonation: masquerading as a trusted friend-of-a-friend.

Unlike traditional attacks, FRA targets relationship-based trust rather than device identities or communication channels. For example, in a smart home, a fake device claiming to be a friend-of-a-friend of a trusted lock can gain access to locks, or personal data, exploiting trust propagation without using conventional IoT vulnerabilities.

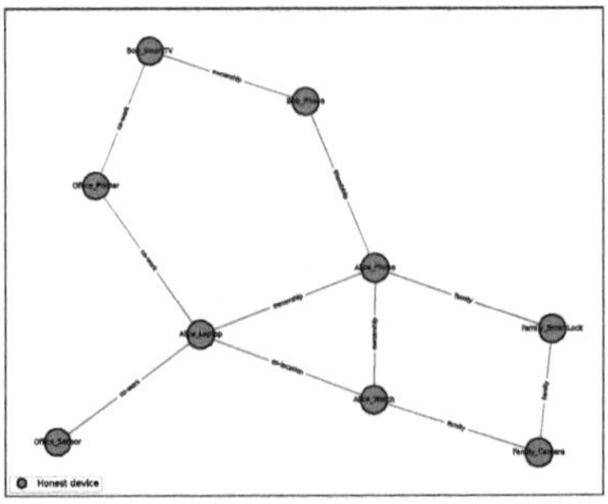

Fig. 1. Benign SIoT Network

3.2 FRA as a Composite

FRA should be understood as a composite, multi-stage campaign rather than a single exploit. At the lowest level the adversary must gain the ability to claim or establish relationships: this can be done by onboarding a fraudulent device, stealing or spoofing an identity token, exploiting lax onboarding policies, or creating low-cost virtual devices (Sybil nodes). Once footholds exist, the attacker performs edge-level manipulations, creating edges to high-trust nodes, forming chains (friend-of-a-friend) that bridge into trusted communities, or coordinating compromised honest nodes to act as endorsers. Finally, the adversary leverages the network's trust aggregation rules to inflate attacker trust until it crosses service thresholds. As each stage (onboarding, edge creation, endorsement manipulation) maps to concrete system primitives, defenders must consider protections at multiple layers, credential issuance, relationship verification, recommendation weighting, and revocation, to prevent FRA end-to-end.

3.3 Threat Model

In the FRA, the adversary targets the social and trust layers of SIoT networks. The attacker can onboard devices, legitimate or virtual, using identity spoofing to impersonate existing ones or create pseudo-identities, enabling fraudulent social edges. These devices can claim direct or indirect relationships such as "friend-of-a-friend" connections, exploiting trust transitivity. Trust can be further amplified through a Sybil strategy, where multiple controlled devices interconnect and endorse each other, inflating credibility. The adversary may also compromise honest devices to validate forged relationships. By leveraging trust propagation, interaction history, and reputation metrics, indirect trust paths can be manipulated to gain unauthorized access and elevated privileges. The attacker's primary goals are to illegitimately acquire trust, access services, manipulate recommendations, and maintain persistent presence through sustained indirect trust. This model shows how FRA exploits SIoT's structural and functional characteristics, revealing critical attack vectors and vulnerabilities.

3.4 Realistic Attack Scenarios and Mechanics

FRA manifests in several ways depending on the target SIoT relationships: Friend-of-a-friend exploit (single-edge FRA): An attacker introduces a device claiming a link to a neighbor of a trusted target. Indirect trust allows access to low-risk services, which can then be escalated by extending links or exploiting service logic. Sybil-amplified FRA: Multiple low-cost Sybil devices claim links to high-trust anchors, interconnect, and endorse each other, inflating trust metrics above acceptance thresholds. Temporal strategic FRA: Attackers initially behave well to build interaction history, then inject forged edges at high-leverage points. This strategy effectively combines behavioral trust manipulation—where nodes blend early compliance with later malicious actions—with identity-based fabrication to artificially inflate their standing, a severe composite threat similarly recognized in dynamic vehicular networks [15]. By the time anomalies are detected, sufficient indirect trust has been accrued to bypass audits. FRA is a staged, composite threat exploiting social ties and trust propagation, demonstrating why relationship authenticity and transitive trust are key SIoT vulnerabilities.

4 Simulations and Discussion

To better understand the practical implications of FRA on Social IoT networks, we conduct a series of simulations using SIoT graphs that capture realistic device relationships. These simulations allow us to observe how trust propagation behaves under FRA scenarios and to quantify the potential impact on the network.

4.1 Simulation Setup

We construct the SIoT relationship graph from the SmartSantander dataset [16], where nodes are physical IoT objects (sensors, actuators, smart appliances, personal devices) and edges represent social or functional links such as co-location, co-ownership, or dependency (e.g., a thermostat depends on a temperature sensor). To illustrate realistic radio behavior and node placement, we import a topology snapshot into the Cooja emulator [17], mapping each logical node to a Contiki node with IEEE 802.15.4 parameters; the Unit Disk Graph Medium (UDGM) models range and interference. Figure 2 shows the representative 50-node snapshot. Cooja is used for visualization to show that the logical relationships we attack can exist in a radio environment, while trust computations and FRA manipulations are performed in Python. The trust in the simulated SIoT network is calculated using [18], integrating multiple social and interaction-based features into a comprehensive trust score that assesses direct trust from interaction history and indirect trust from recommendations by neighboring objects deemed trustworthy.

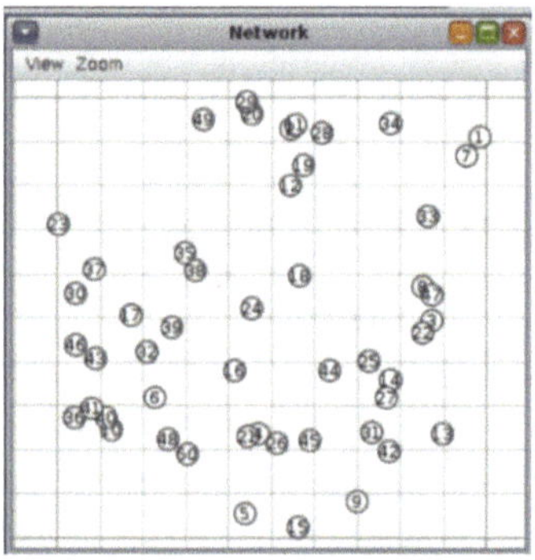

Fig. 2. SIoT simulation using Cooja

The model quantifies social similarities through shared community memberships, friendships, and co-work relationships. Community-of-interest similarity is based on the proportion of shared communities, while friendship similarity reflects ties with neighboring objects. These features are combined into a single trust score via a weighted sum, with weights adjusted to context, reputation, and environmental factors. Trust scores evolve dynamically as interactions and relationships change, enabling detection of misbehavior or reliability shifts. Experiments confirm that this multi-faceted approach effectively distinguishes trustworthy from untrustworthy objects.

4.2 Benign Simulation

In the first scenario we evaluate the SIoT network under benign conditions without any adversarial manipulations. Trust values are computed for each pair of objects according to the multi-feature trust model described in Sect. 4.1. To provide a global view of the system, we calculate two metrics over the entire network at each timestep: Mean trust to honest nodes ($_{hon}$), defined as the average trust assigned by all honest nodes toward other honest nodes, and Mean trust to attackers ($_{att}$), which in this scenario is evaluated only against a attacker set after the attack, serving as the baseline reference. The benign results show that $_{hon}$ steadily increases (see Fig. 3) from its initialization and stabilizes around a high value reflecting the reliability of legitimate social IoT devices. This establishes the expected trust dynamics in a cooperative SIoT environment and provides the benchmark against which the FRA can be measured.

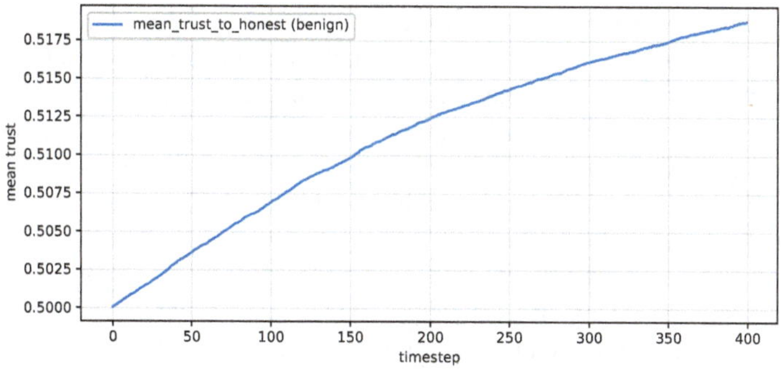

Fig. 3. Mean Trust in Benign Simulation

4.3 Network with FRA Attack

We next introduce the FRA by injecting forged social edges into the SmartSantander-based SIoT graph. Malicious nodes advertise counterfeit friendships, co-location links, or shared community memberships in order to appear more tightly integrated into the social fabric of the network.

Once these fabricated links are introduced, we observe an immediate deviation in the trust metrics (see Fig. 4). Specifically, $_{\text{att}}$ increases sharply as honest nodes incorporate forged relationships into their trust computations. This effect is amplified when sybil helpers or multiple forged edges are introduced, causing $_{\text{att}}$ to approach or even surpass T_{hon}. In parallel, T_{hon} may decline slightly due to recommendation pollution, as malicious nodes indirectly distort honest-to-honest trust propagation.

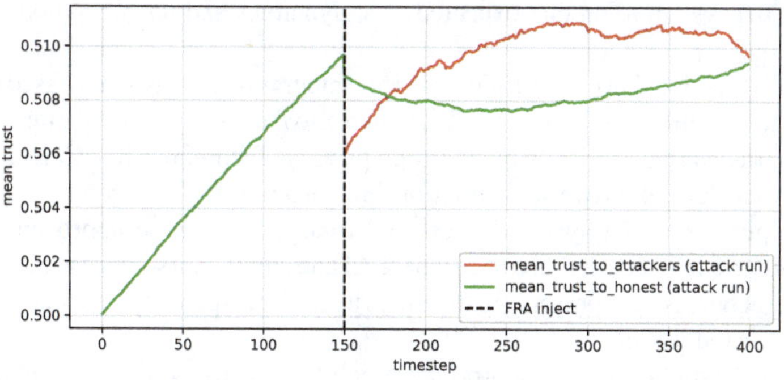

Fig. 4. Mean Trust in Simulation With FRA

4.4 Key Takeaways

The experiments reveal several key insights into the behavior of Social IoT networks under FRA. A clear adversarial signature emerges: the mean trust to attackers ($_{\text{att}}$) rises sharply as soon as forged relationships appear in the SIoT graph. This increase, distinct from the gradual and stable trust evolution under benign conditions, provides a direct indicator of compromise.

Beyond this, the experiments show the attack's severity. By examining the change in attacker trust values before and after FRA, and the ratio between attacker and honest trust levels, it becomes evident that adversaries can quickly approach or reach parity with legitimate nodes. In some cases, ${}_{att}$ nearly equals Thon, meaning malicious nodes become indistinguishable from trustworthy devices. This underscores FRA's ability to undermine the foundations of trust-based interaction in SIoT systems.

Another key observation is that the attack not only elevates adversarial trust but also indirectly degrades trust between honest nodes. Through recommendation pollution, forged relationships distort trust paths and reduce the accuracy of honest-to-honest assessments, weakening the reliability of the trust fabric and showing that FRA's harm extends beyond the immediate success of attackers.

5 Towards Detection and Mitigation

In this paper, we primarily tried to focus on defining and analyzing the FRA attack, but the results naturally lead to explore intelligent mechanisms for the detection and mitigation of this severe attack. Future research by us or other researchers should aim to bridge trust management and artificial intelligence, to help the SIoT systems autonomously recognize and adapt to deceptive social behaviors.

One promising direction to the solution involves trust analysis driven by the data, where we use machine-learning models trained to identify abnormal trust evolution patterns or inconsistencies in social relationship formation. Approaches based on graphs, such as Graph Neural Networks (GNNs) or representation learning on trust graphs, could help capture structural anomalies and hidden dependencies indicative of fake relationships. Similarly, temporal trust modeling using recurrent or attention-based models may allow the system to predict expected trust dynamics and flag deviations caused by manipulation.

Another direction that can be followed is the integration of reasoning based on context and federated intelligence, allowing distributed SIoT devices to collaborate and detect suspicious trust variations while maintaining privacy. Techniques like Zero-Knowledge Proofs can let devices prove their relationships are real without showing any private details, helping keep the system honest and transparent. These approaches reflect a broader shift in SIoT security research, where AI and smart data science are used not only for prediction but also for preserving the integrity and transparency of social interactions among connected devices.

Other future studies should incorporate benchmarking frameworks to compare detection accuracy, false-positive rates, and computational cost across diverse SIoT environments. This will ensure that proposed approaches are scalable and interpretable, reinforcing the practical value of trust analysis using AI in mitigating emerging attacks such as FRA.

6 Conclusion

This work has examined the impact of FRA on Social IoT networks. By constructing realistic SIoT graphs from the SmartSantander dataset and instantiating them in Cooja, we evaluated how adversarially forged social edges distort trust propagation. Our simulations showed that under benign conditions, trust dynamics evolve smoothly, with honest devices converging toward high mutual trust while attacker credibility remains negligible. However, once FRA links are introduced, attacker trust values rise sharply, in some cases reaching parity with honest nodes. This not only enables malicious devices to gain undeserved credibility but also indirectly degrades honest-to-honest trust through recommendation pollution.

The results highlight both the severity and the detectability of FRA. Simple global metrics, namely mean trust toward attackers and honest nodes, are sufficient to reveal adversarial signatures and track the progression of compromise. These findings underline the need for SIoT solutions to incorporate robust defenses against fabricated relationships, such as stricter validation of social edges, anomaly detection on trust dynamics, or resilience mechanisms that reduce the influence of polluted recommendations.

References

1. Elghomary, K., Bouzidi, D., Daoudi, N.: A comparative analysis of OSN and SIoT trust models for a trust model adapted to MOOCs platforms. In: Proceedings of the 2nd International Conference on Networking, Information Systems & Security, pp. 1–8 (2019)
2. Sagar, S., Mahmood, A., Sheng, Q.Z., Zhang, W.E., Zhang, Y., Pabani, J.K.: Understanding the trustworthiness management in the social internet of things: a survey. Comput. Netw. **251**, 110611 (2024). https://doi.org/10.1016/j.comnet.2024.110611
3. Kumari, S., Kumar, S.: Trust Management in Social Internet of things: challenges and future directions. Int. J. Comput. Digit Syst. **14**, 899–920 (2023). https://doi.org/10.12785/ijcds/140170
4. Alam, S., Zardari, S., Noor, S., Ahmed, S., Mouratidis, H.: Trust Management in Social Internet of things (SIoT): a survey. IEEE Access. **10**, 108924–108954 (2022). https://doi.org/10.1109/ACCESS.2022.3213699
5. Mulahuwaish, A., et al.: A survey of social cybersecurity: techniques for attack detection, evaluations, challenges, and future prospects. Comput. Hum. Behav. Rep. **18**, 100668 (2025)
6. Dhelim, S., Aung, N., Kechadi, T., Ning, H., Chen, L., Lakas, A.: Trust2Vec: large-scale IoT trust management system based on signed network Embeddings. IEEE Internet Things J. **10**, 553–562 (2023). https://doi.org/10.1109/JIOT.2022.3201772
7. Dayyani, A., Abbaspour, M.: SybilPSIoT: preventing Sybil attacks in signed social internet of things based on web of trust and smart contract. IET Commun. **18**, 258–269 (2024). https://doi.org/10.1049/cmu2.12734
8. Chen, Z., Jiang, Y., Song, X., Chen, L.: A survey on zero-knowledge authentication for internet of things. Electronics. **12**, 1145 (2023). https://doi.org/10.3390/electronics12051145
9. Ramezan, G., Meamari, E.: zk-IoT: securing the internet of things with zero-knowledge proofs on blockchain platforms. In: 2024 IEEE International Conference on Blockchain and Cryptocurrency (ICBC), pp. 1–7. IEEE (2024)
10. Wu, W., Liu, E., Gong, X., Wang, R.: Blockchain based zero-knowledge proof of location in IoT. In: ICC 2020–2020 IEEE International Conference on Communications (ICC), pp. 1–7 (2020)

11. Yu, Z., et al.: KGTrust: evaluating trustworthiness of SIoT via knowledge enhanced graph neural networks. In: Proceedings of the ACM Web Conference 2023, pp. 727–736 (2023)
12. Anciline Jenifer, J.: Deep learning models for trust management in social internet of things. J. Inf. Syst. Eng. Manag. **10**, 226–232 (2025). https://doi.org/10.52783/jisem.v10i4s.493
13. Jmal, R., Masmoudi, M., Amous Ben Amor, I., Sedes, F.: Blockchain-powered trust management methodology in SIoT: a survey. Peer—Peer Netw. Appl. **18** (2025). https://doi.org/10.1007/s12083-025-02015-7
14. Taherdoost, H.: Blockchain and machine learning: a critical review on security. Information. **14**, 295 (2023). https://doi.org/10.3390/info14050295
15. Baalla, M., Bouzidi, D.: Zero knowledge proof in vehicular networks. In: 2025 12th International Conference on Wireless Networks and Mobile Communications (WINCOM), pp. 1–6. IEEE (2025)
16. Marche, C., Atzori, L., Pilloni, V., Nitti, M.: How to exploit the social internet of things: query generation model and device profiles' dataset. Comput. Netw. **174**, 107248 (2020). https://doi.org/10.1016/j.comnet.2020.107248
17. Velinov, A., Mileva, A.: Running and testing applications for Contiki OS using Cooja simulator (2016)
18. Sagar, S., Mahmood, A., Sheng, Q.Z., Zaib, M., Sufyan, F.: Can we quantify trust? Towards a trust-based resilient SIoT network. Computing. **106**(2), 557 (2023)

Revealing Bots in Crypto-Focused Social Media Communities via a Data-Driven Framework

Achraf Benmoumen(✉), Abdelhay Haqiq, and Najima Daoudi

ITQAN Team, LyRICA Laboratory, School of Information Sciences ESI, Rabat, Morocco
achraf.benmoumen@esi.ac.ma

Abstract. The rapid expansion of social media, in the absence of strict regulatory oversight, has significantly transformed online communication dynamics. This evolution has also facilitated the proliferation of automated and malicious accounts, commonly referred to as bots. Cryptocurrency-centered communities, which constitute the focus of this study, are particularly exposed to such threats, often manifesting through scams, misleading investment schemes, and disinformation campaigns. This research presents a methodological framework for the proactive identification of suspicious accounts in multilingual cryptocurrency-related discussions. The proposed approach combines behavioral indicators with textual features, leveraging machine learning paradigms. A multilingual dataset comprising over 10,000 Facebook comments was curated and enriched with linguistic and behavioral metadata. Two complementary architectures were evaluated: fastText lexical embeddings coupled with a Light classifier and DistilBERT contextual embeddings combined with a Multi-Layer Perceptron (MLP). After optimization, the first model achieved an AUC of 96%, while the second reached 95%, offering enhanced semantic and contextual understanding, particularly valuable in multilingual environments. Feature analysis underscores the necessity of combining behavioral and textual signals for reliable author-level detection across social platforms. The proposed framework thus enables proactive monitoring and supports the creation of safer, more trustworthy digital ecosystems by mitigating the influence of fraudulent accounts. It represents a foundational step toward addressing the growing challenge of bot detection in thematic online communities.

Keywords: Bot detection · Social media bots · Malicious accounts · Spam detection · Fake profiles · Automation detection

1 Introduction

The exponential growth of social media platforms such as Facebook, Twitter, and Telegram has profoundly transformed the way individuals communicate, exchange information, and engage within online communities. These platforms have evolved into vital spaces for discourse, collaboration, and knowledge sharing [1]. However, this rapid expansion has also been accompanied by a parallel rise in malicious activities, including spam campaigns, coordinated misinformation, and the proliferation of automated or semi-automated accounts [2]. Such vectors of the social web can distort public opinion and undermine the reliability of online content, posing serious challenges to community integrity and user safety [3].

A. E. Hassanien et al. (Eds.): AISDS 2025, LNDECT 287, pp. 43–55, 2027.
https://doi.org/10.1007/978-3-032-18927-1_5

Cryptocurrency-focused social media communities are particularly vulnerable to these threats [3]. The financial incentives associated with such spaces, ranging from investment discussions to trading advice, make them prime targets for scams, phishing attacks, and deceptive promotional schemes [4]. Within these communities, bots can amplify fraudulent messages, disseminate misleading narratives, and cultivate an atmosphere of distrust among genuine users [5].

In this context, the automatic detection of malicious accounts has become a critical task to ensure secure and trustworthy online ecosystems. Traditional moderation mechanisms, which rely heavily on heuristic rules, are often insufficient to effectively address the scale and complexity of bot behavior [6]. Moreover, the scarcity of publicly available and domain-specific data further complicates research and model development in this field. As a result, data-driven and automated detection techniques have emerged as essential tools for enhancing threat intelligence and advancing social media analysis [7].

In the same vein, several research gaps remain. Few studies focus specifically on platforms like Facebook, and even fewer examine thematic niches such as cryptocurrency-related discussions. Moreover, the available multilingual datasets are often outdated, limiting the application of more advanced models across different languages. These gaps, which have been addressed in our research and form the foundation of our contribution, highlight the need for hybrid approaches that integrate feature-based methods with embedding-based techniques to robustly identify coordinated malicious accounts within specialized social media communities. Such frameworks are particularly crucial for detecting campaigns that disseminate spam for specific objectives, especially when characterized by massive interaction patterns and strong behavioral indicators that significantly increase the probability of successful bot detection.

This paper focuses on the identification of suspicious behavioral patterns in user-generated comments within cryptocurrency-oriented discussions on social media. Following a structured research breakdown, comprising an introduction, literature review, methodology, and experimentation (see Fig. 1), it proposes a detection framework that integrates feature engineering with supervised learning models, encompassing traditional machine learning approaches. The overarching goal is to capture both behavioral indicators and textual features associated with potential bot activity.

Specifically, the second section presents the state of the art, the third section covers related works, and the fourth section details the methodology along with the proposed framework, before concluding with results and discussion.

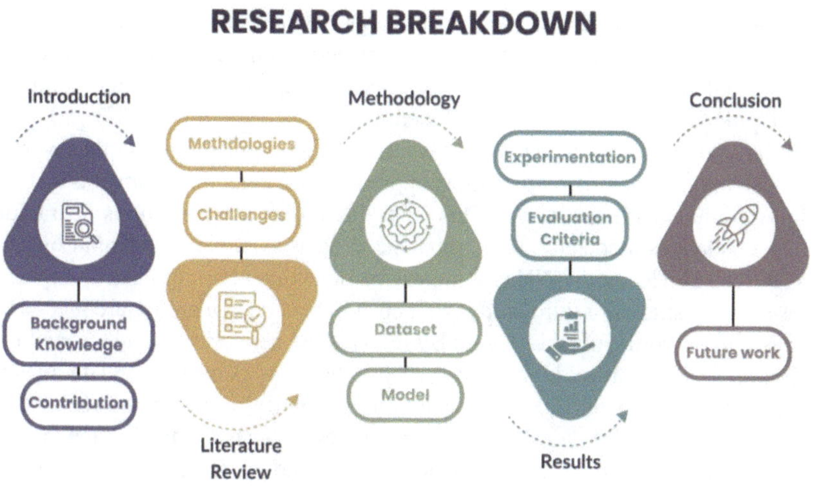

Fig. 1. Conceptual Workflow of the Research Process.

2 State of Art

2.1 Bot Detection on Social Media

Bot detection on social media platforms has attracted considerable research attention over the past decade. Traditional approaches largely rely on feature-based methods, which utilize profile characteristics, content, and embedded links to identify automated or malicious accounts [8]. These methods are often effective for detecting simple or rule-based bots but may struggle with sophisticated or context aware behaviors.

2.2 Novel Data Collection and Its Importance for Bot Detection

The advancement of bot detection models is critically dependent on the availability of relevant, high-quality datasets. A novel, domain-specific dataset is paramount, as it directly addresses the evolving tactics of malicious actors and fills existing gaps in the literature [9]. Such a dataset, particularly one focused on a specific niche discussion on a specific platform, provides several key advantages. It enables the training of more robust and generalizable models, allows for the validation of new detection techniques in a realistic setting, and opens new research perspectives. Specifically, it facilitates the exploration of cross-lingual model performance, the analysis of behavioral patterns in specialized communities, and the development of advanced models that effectively combine feature engineering with modern ML learning approaches, thereby pushing the boundaries of current capabilities.

3 Related Works

The following works are directly relevant to research on detecting automated bots, which frequently propagate agenda-driven content via spam. First, this study introduces TIMBRE, a time-aware methodology designed to detect and remove social bots in political discussions, particularly during election campaigns. By combining keyword-based classification with temporal analysis, the approach effectively distinguishes between human and automated activity, revealing how bot interference can distort public opinion and influence polarization patterns on social media [10].

Similarly, building on the need for advanced detection methods, another work provides a systematic review of deep learning approaches for detecting spam bots and cyberbullying in online social networks. This study highlights the limitations of traditional methods, challenges associated with imbalanced datasets, and the potential of models such as RNNs, LSTMs, and CNNs for improving detection accuracy [11]. Complementing these survey studies, a third study presents a hybrid deep learning architecture that combines GloVe word embedding with BiGRU-LSTM networks to detect social media bots. This approach achieves high performance on benchmark datasets and demonstrates the effectiveness of semantic text representations in identifying automated accounts [12].

Additionally, a comprehensive survey examines spam bot detection techniques in online social networks by analyzing features extracted from user profiles and interactions. It compares the effectiveness of various machine learning and deep learning approaches, evaluates commonly used datasets, identifies limitations in existing methods, and outlines future directions for enhancing the accuracy and robustness of automated bot detection systems [13].

Finally, this research aims to develop a fast, efficient, and deployable anti-spam bot capable of detecting and managing unwanted messages on social media and messaging platforms through a combination of machine learning algorithms [14].

These works illustrate both the progress made in bot detection research and the ongoing need for robust, data-driven, and hybrid approaches capable of addressing the diverse strategies employed by social bots.

4 Methodology

This fourth section presents the methodology adopted in this study (See Fig. 2), encompassing dataset collection, feature engineering, text embedding, modeling approaches, and evaluation procedures. The procedures were designed to extract and quantify textual and behavioral information from multilingual social media comments, providing a structured framework for the identification of potentially suspicious or spam activity.

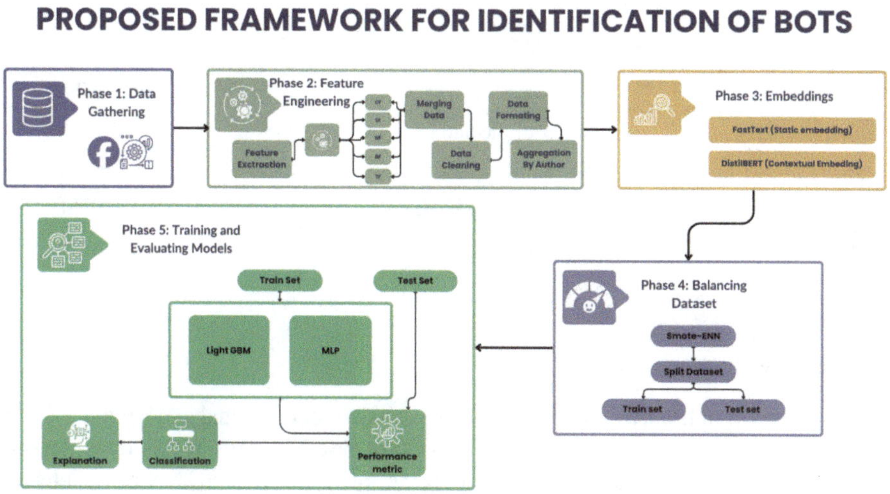

Fig. 2. Our Proposed framework for identification of bots.

4.1 Dataset Collection

The dataset used in this study was collected from public posts related to cryptocurrency, trading, and forex. A thematic filtering process was applied using relevant keywords associated with these domains to ensure that only comments directly related to the target topics were captured. This approach enabled the construction of a dataset containing more than 10,000 comments. The final dataset comprises over 10,700 multilingual comments contributed by multiple authors. Each entry includes structured information such as Comment ID, Author ID, text content, author profile details, creation date, and associated URLs, providing a comprehensive foundation for both textual and behavioral analysis (see Table 1). Facebook was specifically chosen as the platform for data collection due to its widespread use for cryptocurrency discussions and the richness of user interaction data it provides, which are critical for studying bot activity and user behavior. When collecting data, ethical considerations were taken into account to ensure user privacy. All data were anonymized and some informations was masked to protect the identities of profils while allowing for meaningful analysis.

Table 1. Extracted Features from Dataset for Bot Detection.

Feature	Description	Feature Type
Comment ID	Unique identifier for the comment	Content
Comment Text	Text of the comment	Content
Author Name	Name of the comment author	Profile
Author ID	Unique identifier of the comment author	Profile
Author Profile URL	URL of the author's profile	Profile

(continued)

Table 1. *(continued)*

Feature	Description	Feature Type
Author Gender	Gender of the author	Profile
Author Profile Image	URL to the author's profile picture	Profile
Comment Creation Date	Timestamp when the comment was posted	Behavioral
Comment URL	URL to the comment	Content

4.2 Feature Engineering

Leveraging Python libraries, we extracted an extensive set of textual and behavioral features to characterize user activity and content. Text-based features included text length, word count, emoji count, number of links, presence of crypto-related keywords, detected language, and Flesch Reading Ease score. Temporal features, capturing activity patterns such as posting hour and day of the week, were also computed. Notably, the dataset we collected accounts for the multilingual dimension (see Fig. 3).

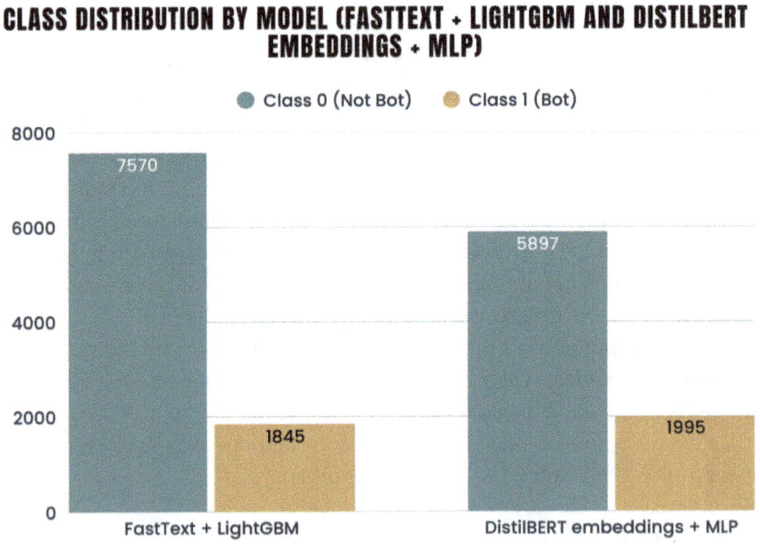

Fig. 3. Multilingual Dimension: Dataset Distribution by Language.

In addition, several derived indicators, including activity score, suspicion score, text density, and unique domain count, were introduced to quantify various aspects of user behavior. These indicators served as Key Performance Indicators (KPIs) during the feature extension process, enabling a more refined understanding of content conciseness, posting frequency, and potential spam-like behavior.

- Text Density: ratio of words per character, measuring text conciseness.

- Activity Score: combination of comment volume and uppercase usage, highlighting aggressive or spam-like posting behavior.
- Unique Domain Count: number of distinct domains shared by an author, reflecting content diversity.
- Suspicion Score: composite metric integrating link sharing, keyword presence, and domain diversity, designed to flag suspicious posting patterns.

Lastly, all extracted and derived features were aggregated at the author level, producing the final dataset used for word embedding generation, model training, and evaluation (see Table 2).

Table 2. Features of the Dataset Aggregated by Author ID after Feature Extension.

Feature	Description	Feature Type
Author ID	Unique identifier for the author	Profile
Clean text	Concatenated or representative text of author's comments	Content
Num comments	URL to the author's profile	Profile
Text length	Average length of comments (in characters)	Profile
Word count	Average number of words per comment	Content
Emoji count	Average number of emojis per comment	Content
Num links	Average number of links per comment	Content
Contains keywords	Average proportion of comments containing target keywords	Content
Uppercase ratio	Average ratio of uppercase letters in comments	Content
Flesch reading ease	Average Flesch Reading Ease score of the comments	Content
Has profile pic	Indicates if the author has a profile picture	Profile
Domain count	Total number of domains linked in comment	Behavioral
Hour	Average posting hour	Behavioral
Weekday	Average posting weekday	Behavioral
Author Name	Name of the author	Profile
Author Profile URL	URL to the author's profile	Profile
Author Gender	Gender of the author (if available)	Profile
Name length	Number of characters in author name	Profile
Profile url length	Number of characters in profile URL	Profile
Text density	Ratio of meaningful text to total content	Content
Activity score	Score combining comment volume and uppercase usage	Behavioral
domains	Number of unique domains shared by the author	Profile
Unique domain count	Number of unique domains linked	Profile

(continued)

Table 2. (*continued*)

Feature	Description	Feature Type
Suspicion score	Composite suspicion score based on: num links, Link sharing - contains_keywords: Presence of target keywords- unique_domain_count >3: Penalty if more than 3 unique domains	Behavioral

4.3 Text Embedding and Modeling Approaches

Following the feature engineering phase, a third stage was conducted to design and evaluate the modeling framework. In this study, two complementary paradigms based on machine learning algorithms were selected to leverage their respective advantages while addressing the multilingual and heterogeneous nature of the dataset.

As for data annotation, it was performed automatically, based on the suspicion score field, a composite indicator reflecting the level of suspicious behavior for each profile. This score was computed from multiple behavioral features, such as the number of shared links, the presence of targeted keywords, and the count of unique domains used. In the first version, a threshold of 1.0 was applied to generate the binary labels, where scores above 1 indicated abnormal behavior. In the optimized version, the threshold was lowered to 0.5 to make the detection more sensitive, allowing the model to capture earlier signs of automation or suspicious activity, while maintaining a fully automated, rule-based annotation process (see Fig. 4 for the class distribution).

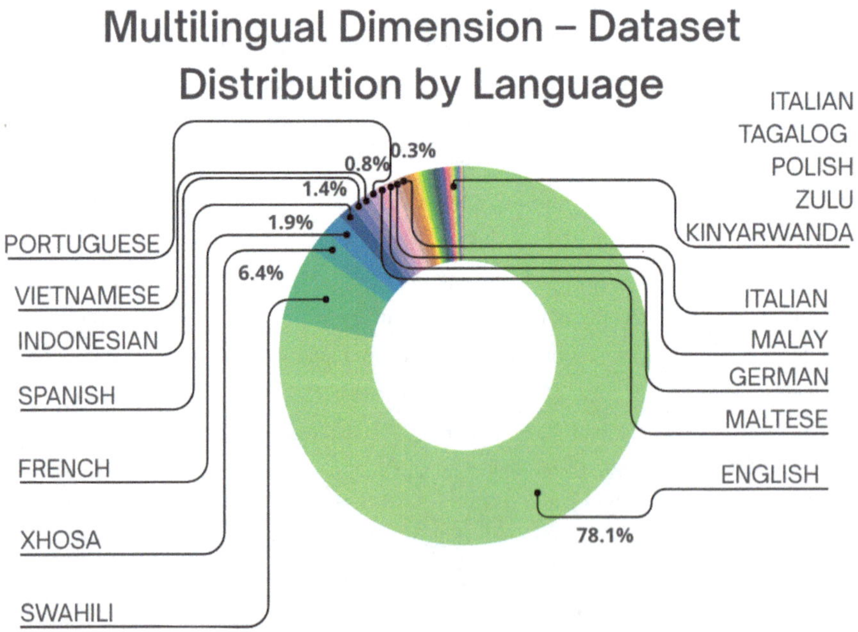

Fig. 4. Class distribution across models.

The first approach (fastText + LightGBM) leverages fastText embeddings, which were prioritized for their ability to capture subword information and handle morphological variation across multiple languages. Given that the collected dataset includes multilingual comments, this property makes fastText particularly suitable compared to static embeddings such as Word2Vec or GloVe, which are limited in cross-lingual adaptability and vocabulary coverage. Token-level embeddings were averaged to produce a single vector representation per comment.

These vectors were then standardized and used as input to a LightGBM classifier, chosen for its speed, interpretability, and strong performance on tabular and high-dimensional data. Alternative classifiers such as Random Forest or SVM were considered. However, LightGBM provides faster training, better handling of class imHowever, and built-in regularization mechanisms that reduce overfitting, making it more suitable for large-scale text-derived feature sets.

This combination (fastText + LightGBM) offers a lightweight yet robust baseline capable of modeling semantic information while maintaining computational efficiency.

The second approach employs DistilBERT, a distilled version of BERT, to generate contextualized embeddings that capture semantic and syntactic relationships within text. DistilBERT was preferred over larger transformer models (e.g., BERT, RoBERTa, or XLM-R) due to its reduced computational complexity and faster inference time, which make it more practical for real-world deployment without sacrificing much accuracy.

Also, a multi-layer perceptron (MLP) was adopted, as MLPs are well-suited for vectorized representations like sentence embeddings and provide flexibility in modeling complex feature interactions.

The combination of the 1st and the 2nd approach was selected to balance efficiency, interpretability, and semantic depth. The first pipeline emphasizes scalability and multilingual robustness, whereas the second focuses on contextual understanding and semantic modeling.

4.4 Training and Evaluation

A stratified train–test split was employed to preserve class distribution, allocating 80% of the data for training and 20% for testing. Model performance was assessed using standard evaluation metrics, including Accuracy, Precision, Recall, and F1-score, ensuring a comprehensive evaluation of both predictive capability and class-wise balance.

After initial model training, a second optimization phase was conducted to address class imbalance within the dataset. To enhance the reliability of minority class detection, the Synthetic Minority Over-sampling Technique (SMOTE) was applied, generating synthetic examples for underrepresented classes. This resampling strategy helped to mitigate bias toward the majority class and improve overall model generalization.

5 Results and Discussion

The performance of the proposed models was evaluated on the aggregated author-level dataset using multiple metrics, including Accuracy and AUC. Two modeling paradigms were compared.

5.1 Initial Performance

In the initial experiments, both models demonstrated strong predictive capabilities. For fastText + LightGBM, the classification report indicated high precision and recall, particularly for the majority class, yielding an overall accuracy of 0.97 and F1-scores ranging from 0.94 to 0.98 across classes. Similarly, the DistilBERT + MLP model achieved comparable results, confirming the effectiveness of transformer-based embeddings in capturing nuanced textual and behavioral patterns, especially within a moderately sized multilingual dataset (see Table 3).

Table 3. Comparison of Model Performance Metrics before optimization.

Model	Class	Precision	Recall	F1-Score
FastText + LightGBM	0	0.98	0.99	0.98
	1	0.94	0.91	0.93
DistilBERT + MLP	0	0.98	0.99	0.98
	1	0.94	0.91	0.92

5.2 Model Optimization

To further enhance performance and robustness, several targeted optimization strategies were applied. Preprocessing was refined to retain contextual elements such as URLs, mentions, and hashtags, ensuring richer semantic representation. The feature space was expanded with sentiment-based and author-level behavioral indicators, offering deeper insight into user intent. To address class imbalance, SMOTE oversampling and class weighting were implemented, improving minority class detection. Finally, hyperparameter tuning (including early stopping for MLP and parameter optimization for LightGBM) and optimal threshold selection contributed to a more balanced and accurate classification.

5.3 Optimized Performance

After optimization, both models maintained high predictive performance. DistilBERT + MLP achieved an AUC of 0.95. fastText + LightGBM achieved a slightly higher AUC of 0.96 (see Table 4). Visual analyses, including confusion matrices and probability distributions, confirmed that the transformer-based model better captures semantic nuances in multilingual comments, while fastText provides a lighter, faster alternative suitable for large-scale deployment (see Fig. 5).

Table 4. Final Comparison of Model Performance Metrics after Optimization

Model	Accuracy	Precision	Recall	F1-Score	AUC
DistilBERT + MLP	0.9164	0.8414	0.8246	0.8329	0.9569
FastText + LightGBM	0.9139	0.8507	0.7995	0.8243	0.9677

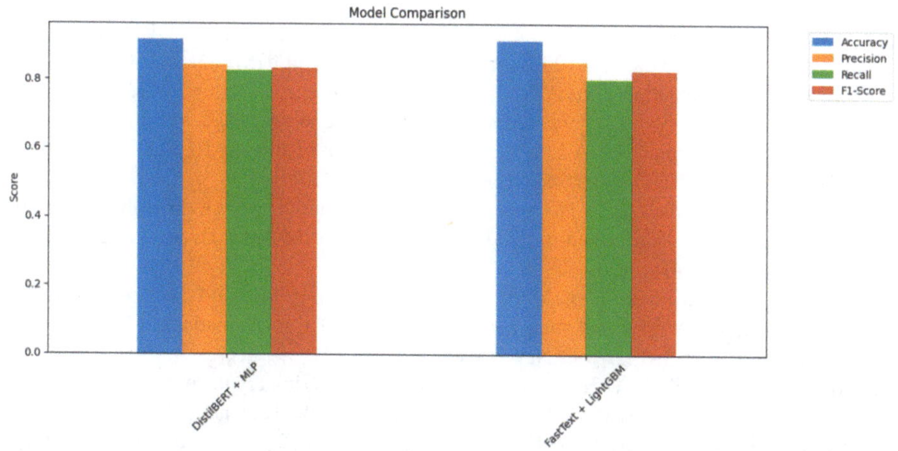

Fig. 5. Comparison of performance metrics between the two approaches after optimization.

6 Conclusion

This study introduced a comprehensive framework for detecting suspicious accounts within cryptocurrency-focused social media communities. By integrating behavioral features with textual embeddings, it combines the strengths of classical machine learning and transformer-based learning to effectively uncover automated or malicious activities. The experimental results demonstrate strong predictive performance across both modeling paradigms. The fastText-based approach offers a lightweight solution, while the DistilBERT architecture, through its attention mechanism, captures nuanced semantic and multilingual cues, enhancing the detection of subtle or context-dependent bot behaviors. Feature analysis further revealed that the fusion of behavioral and textual indicators is essential for reliable author-level classification, surpassing the limitations of single comment evaluation. Beyond its technical contributions, the framework facilitates proactive monitoring and contributes to building safer, more trustworthy digital ecosystems within cryptocurrency discussions. By constraining the influence of fraudulent or manipulative actors, it supports the integrity of online information flows. Looking ahead, expanding the model to larger and more diverse datasets could improve its generalizability. Incorporating additional account level attributes would also enrich feature representations. Finally, extending the framework to cross-platform contexts, such as Twitter or Telegram, would allow for the detection of coordinated and multi-platform bot operations.

Acknowledgments. Declaration of Conflicting Interests The author(s) declared no potential conflicts of interest with respect to the research, authorship, and/or publication of this article.

Data Availability Statement The dataset used in this study is available from the corresponding author upon reasonable request.

References

1. Lipschultz, J.H.: Social Media Communication: Concepts, Practices, Data, Law and Ethics, 4th edn. Routledge, New York (2023). https://doi.org/10.4324/9781003281924
2. Akhtar, M.M., Masood, R., Ikram, M., Kanhere, S.S.: False information, bots and malicious campaigns: demystifying elements of social media manipulations. In: Proceedings of the 19th International Conference on Availability, Reliability and Security (ARES 2024), pp. 1–10. ACM, New York (2024). https://doi.org/10.48550/arXiv.2308.12497
3. Ja'afaru, S.G., Asemah, E.S.: How social media shape public opinion through propaganda and the spread of disinformation. In: Communication and Media Dynamics, pp. 37–50. ResearchGate (2024)
4. Mirtaheri, M., Abu-El-Haija, S., Morstatter, F., et al.: Identifying and Analyzing cryptocurrency manipulations in social media. In: Proceedings of the International Conference on Social Systems, pp. 1–12. IEEE, New York (2021)
5. Krishnan, L.P., Vakilinia, I., Reddivari, S., Ahuja, S.: Analyzing cryptocurrency social media for price forecasting and scam detection. In: Proceedings of the 12th International Symposium on Digital Forensics and Security (ISDFS 2024), pp. 1–10. IEEE, San Antonio (2024). https://doi.org/10.1109/ISDFS60797.2024.10527330
6. Aljabri, M., Zagrouba, R., Shaahid, A., Alnasser, F., Saleh, A., Alomari, D.M.: Machine learning-based social media bot detection: a comprehensive literature review. In: Social Network Analysis and Mining. Springer, Cham (2023). https://doi.org/10.1007/s13278-022-01000-2
7. Kovács, G., Alonso, P., Saini, R.: Challenges of hate speech detection in social media: data scarcity and leveraging external resources. SN Comput. Sci. **2**, 95 (2021). https://doi.org/10.1007/s42979-021-00457-3
8. Heidari, M., James Jr., H., Uzuner, O.: An empirical study of machine learning algorithms for social media bot detection. In: Proceedings of the 2021 IEEE International IoT, Electronics and Mechatronics Conference (IEMTRONICS 2021), pp. 1–8. IEEE, Toronto (2021). https://doi.org/10.1109/IEMTRONICS52119.2021.9422605
9. Jalal, N., Ghafoor, K.Z.: Machine learning algorithms for detecting and Analyzing social bots using a novel dataset. ARO – Scienti. J. Koya University. **10**(2), 11–21 (2022). https://doi.org/10.14500/aro.11032
10. Cantini, R., Marozzo, F., Talia, D., Trunfio, P.: Analyzing political polarization on social media by deleting bot spamming. Big Data Cogn. Comput. **6**(1), 3 (2022). https://doi.org/10.3390/bdcc6010003
11. Santhosh Kumar, A.V., Suresh Kumar, N., Kanniga Devi, R., Muthukannan, M.: Analysis of deep learning-based approaches for spam bots and cyberbullying detection in online social networks. In: AI-Centric Modeling and Analytics, 1st edn, pp. 38–55. CRC Press, Boca Raton (2023)
12. Ellaky, Z., Benabbou, F., Matrane, Y., Qaqa, S.: A hybrid deep learning architecture for social media bots detection based on BiGRU-LSTM and GloVe word embedding. IEEE Access. **12**, 100278–100294 (2024)

13. Ellaky, Z., Benabbou, F., Ouahabi, S., Sael, N.: A survey of spam bots detection in online social networks. In: Proceedings of the 2021 IEEE International Conference on Data Analytics (ICDATA 2021), pp. 1–6. IEEE, Marrakech (2021). https://doi.org/10.1109/ICDATA52997.2021.00021
14. Liubchenko, N., Podorozhniak, A., Oliinyk, V.: Research of antispam bot algorithms for social networks. In: COLINS 2021, pp. 1–8. Academia.edu, Kharkiv (2021)

Predictive Modeling and Forecasting

Climate Change Forecasting: A Review of Machine Learning and Generative AI Models

Mohamed EL Yahyaoui[1(✉)], Hassan Oukhouya[2,3(✉)], Saida Amine[1], Abdelkrim Haqiq[4], and Salah-Eddine El Adlouni[5]

[1] Laboratory of Mathematics and Applications, Faculty of Science and Technologies of Mohammedia, Hassan II University of Casablanca, BP 146, 20650 Casablanca, Morocco
med.elyahyaoui19@gmail.com

[2] LaMSD, MSASE, Department of Economics, FSJES, Mohammed First University of Oujda, BV Mohammed VI B.P. 724, 60000 Oujda, Morocco
oukhouya.hassan@ump.ac.ma

[3] MAEGE, Department of Statistics and Applied Mathematics, FSJES Ain Sebaâ, Hassan II University of Casablanca, BP: 2634, Route des Chaux et Ciments Beausite, 20254 Casablanca, Morocco

[4] IR2MLab, Faculty of Science and Techniques, Hassan First University of Settat, B.P.: 577 Route de Casablanca, Settat, Morocco

[5] Department of Mathematics and Statistics, Faculty of Science, Université de Moncton, 18 Av Antonine-Maillet, Moncton NB E1A 3E9, Canada

Abstract. Nowadays, climate change represents a real challenge to the sustainability of human societies and ecosystems. This study analyzes recent research on generative approaches and machine learning (ML) algorithms in artificial intelligence (AI), focusing on their role in improving climate change predictions. To accomplish this, we conduct a systematic review of the existing literature. Then, we perform a bibliometric analysis using specialized software to explore the relationships between co-citation and emerging research themes. Finally, the analysis shows that AI and ML are effective tools that integrate ethical considerations into technological innovation to provide rigorous and morally acceptable solutions to the global climate crisis.

Keywords: Climate Change · Forecasting · Machine Learning · Generative AI · Systematic Review

1 Introduction

Climate change is one of the most pressing global issues of our time due to its profound and complex impacts on ecosystems, human societies, and economic systems [1]. Most scientists agree that the leading cause of global warming, or a sustained rise in the Earth's average surface temperature, is human activity, specifically the emission of greenhouse gases into the atmosphere, which

A. E. Hassanien et al. (Eds.): AISDS 2025, LNDECT 287, pp. 59–67, 2027.
https://doi.org/10.1007/978-3-032-18927-1_6

intensifies the natural greenhouse effect. Predictive modeling is necessary for integrative analytical approaches that use environmental assessments to address complex climate change issues. Recent advances in machine learning (ML) and artificial intelligence (AI) have yielded effective methods to enhance our ability to predict future climate dynamics and assess the environmental impacts of various human activities. Finding intricate, nonlinear patterns in climate data has been made much easier by deep learning (DL) techniques [2]. Another explanation is the high accuracy of ensemble approaches, such as XGBoost, in various environmental applications [3]. Furthermore, AI-based techniques are increasingly being accepted as a legitimate approach to addressing significant ecological issues in climate policy and governance [4]. In fact, climate change has been recognized to pose severe threats to both environmental and human systems; it impacts public health, water resources, and food security [5]. Climate change refers to the transformation of the regional or global climate patterns of the Earth, particularly as manifested through long-term temperature changes, precipitation, and atmospheric circulation systems [6]. Scientific consensus indicates that human activities, especially fossil fuel burning, deforestation, and industrial emissions, are the primary cause of current global warming [7]. All these activities of humanity have caused sea levels to rise, increased the frequency and intensity of storms, and led many plant and animal species to struggle for survival. Predictive modeling forms the basis for simulating and projecting future climate scenarios [8]. They develop complex climate models by combining multiple information sources, including historical climate records, satellite data, and atmospheric data. They help develop evidence-based policies and action plans for climate action [9]. AI and ML have developed innovative modeling methodologies that enable the identification of complex, multidimensional data and nonlinear correlations that cannot be interpreted by traditional models [10]. More specific predictions are made for certain climatic phenomena, such as the increase in the frequency of severe weather events, temperature rises, and variations in precipitation rates. AI is increasingly driving climate science decision-making through environmental predictions and analysis [11].

This article reviews the use of generative AI and machine learning models in environmental forecasting and decision support. It also highlights the primary challenges of applying AI to climate forecasting and modeling. The paper also outlines the main issues related to the use of AI in climate modeling and forecasting. The research is structured to address these issues, with a particular focus on the dynamic and fundamentally chaotic characteristics of climate systems. This research is structured as follows: Sect. 2 reviews the literature, Sect. 3 describes recent advances in machine learning and generative AI, Sect. 4 describes the exploration of potential research areas and provides practical conclusions.

2 Background and Related Work

Climate models are based on research to explain and predict changes in the Earth's climate system [12]. The models aim to measure and integrate the key exchanges of greenhouse gases, solar radiation, land and ocean surface circulation [13]. The development of these models is vital to climate research, as it has inspired many of the essential drivers of climate change [14]. The two types of traditional climate models are statistical models and physical models. Climate variables are analyzed in statistical models using regressions, autoregressive approaches, and time series decompositions [2]. Although these models efficiently utilize computational resources, they still face challenges in accurately capturing many complex, nonlinear relationships. In contrast, regional climate models (RCMs), general circulation models (GCMs), and Earth system models (ESMs) rely on the fundamental equations of physics and thermodynamics [15]. Although these complex models simulate energy, momentum, and mass flows, characterize thermodynamic phenomena, and offer mechanistic insights, they are nevertheless costly in terms of computing and energy resources. More recently, procedures such as downscaling, ensemble methods, and other hybrid methods have enhanced the reliability of climate models and their associated estimates of uncertainty, leading to a more direct application of ML and AI in climate models. Numerous ML and generative AI models have been developed in recent years to advance the forecasting of climate change and extreme weather events. We summarize some of these models and their key contributions in Table 1.

3 ML and Generative AI Models for Climate Change Forecasting

This review employs a systematic approach to locate, assess, and identify relevant studies on the application of generative AI and ML models in climate change research. In Fig. 1, a total of 549,954 scientific publications were identified using the terms "machine learning," "climate forecasting," and "generative AI" in databases such as Web of Science (WoS) (538,141) and Scopus (11,813). We carefully evaluated the studies' identifiers based on methodological rigidity, data reliability, reproducibility, and precision of results. This approach ensured the scientific reliability of the review, thereby providing a solid foundation for analyzing the primary ML and generative AI models applied to climate change forecasting.

3.1 Adoption of ML and Generative AI in Climate Change Forecasting

The integration of nonlinear, multidimensional, and spatiotemporal models into climate data, ML, and generative AI has given rise to a data-driven methodology that outperforms conventional models. Figure 1 shows the PRISMA flow diagram outlining the process of identifying, screening, and selecting studies from Scopus and WoS for inclusion in the review on ML and generative AI models in climate change forecasting. ML has been integrated into predicting climate change. Support vector machines (SVMs) can classify patterns and make predictions about precipitation and droughts. Random forests (RF) can capture

Table 1. The literature review on recent ML and generative AI models for climate change forecasting (2019–2025).

Ref.	Approaches	Summary of the Results
[16]	AI-Based Yield Prediction and Smart Irrigation	AI techniques improved crop yield forecasting, irrigation efficiency, and policy planning.
[17]	AI reconstructs missing climate information	GANs enhanced spatial detail in climate projections while preserving physical consistency.
[18]	A Physics Informed ML Framework	Physics-informed ML improved efficiency, generalization, and physical consistency in climate modeling.
[19]	Transformer-based Climate Models	Transformer-based models (FourCastNet) achieved accurate medium-range weather forecasts by capturing long-range dependencies.
[20]	Diffusion Models for Climate Simulation	Diffusion models outperformed classical solvers in turbulent flow simulation with stable, probabilistic predictions.
[21]	Hybrid Physics–AI Models	Hybrid physics–AI models reduced bias in seasonal precipitation forecasts.
[22]	Generative diffusion model for probabilistic downscaling	Generative diffusion models provided superior probabilistic downscaling and uncertainty quantification.
[23]	Spatio-temporal conditional GAN (cGAN) for probabilistic downscaling of ERA5 precipitation data	Spatio-temporal cGANs generated high-resolution precipitation data while preserving extremes and spatial patterns.
[24]	Multi-stage generative diffusion model for km-scale global climate emulation	Multi-stage generative diffusion (cBottle) emulated global climate at km-scale efficiently and consistently.
[25]	High-resolution coupled climate/ocean model	High-resolution coupled ocean–climate models showed mesoscale heat exchanges are critical for accurate energy projections.

temperature changes and climate variability [26]. Convolutional neural networks (CNNs) and DL models used for satellite imagery analysis, along with long short-term memory (LSTM) for time series data, can forecast temperature changes, elevated sea levels, and even extreme weather phenomena [27]. Applications such as climate scenario simulation, hydrology modeling, cyclone detection, precipitation forecasting, bias correction, integrated data processing, and downscaling are all facilitated by this. Although these techniques are helpful, several challenges remain, including poor data quality, overfitting, and the "black box" problem associated with extreme event prediction. ML must still be integrated with physical knowledge to increase the generalization, robustness, and reliability of climate risk assessment and decision-making [4]. Co-citation analysis based on WoS reveals distinct groups of influential authors who are key players at the intersection of climate change and AI research (see Fig. 2(a)). Similarly, Scopus mapping reveals important authors and collaboration networks, as well as the evolution of co-citation relations over the past few years (see Fig. 2(b)). As illustrated in Fig. 2(c), the mapping of co-occurrence links between author keywords (based on Scopus data) highlights the main research themes and their interconnections.

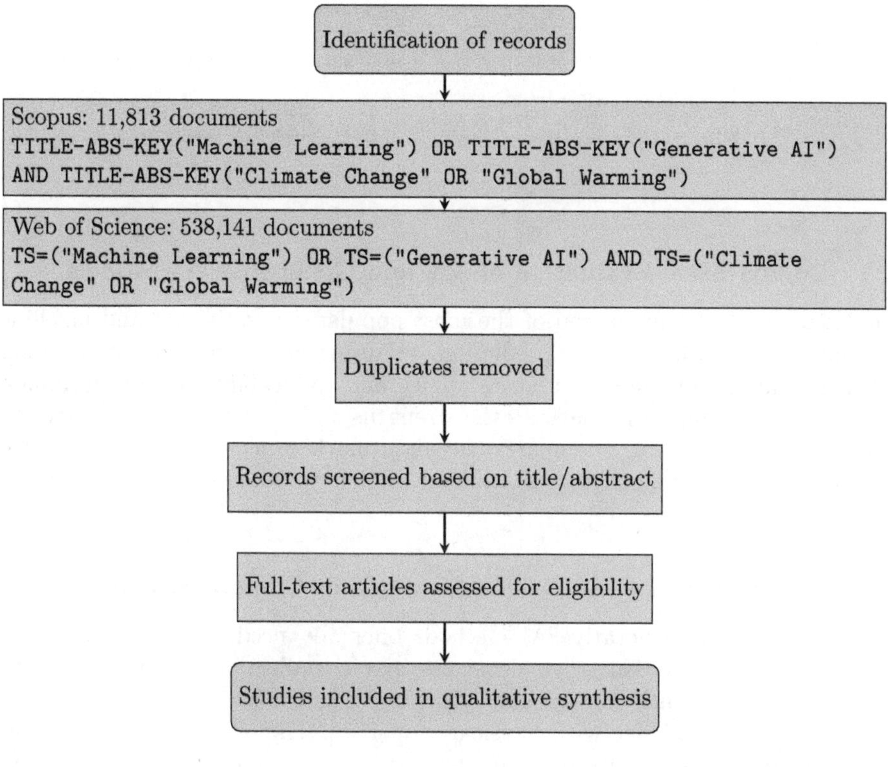

Fig. 1. PRISMA flow diagram from Scopus and WoS for the topic 'ML and Generative AI Models in Climate Change Forecasting.'

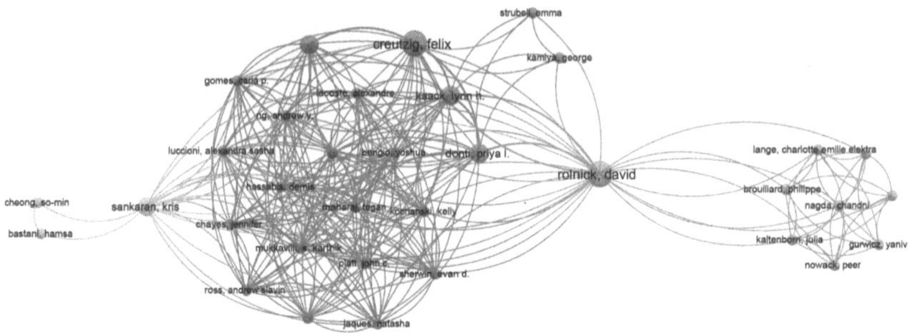

(a) Mapping co-citation links between authors (WoS).

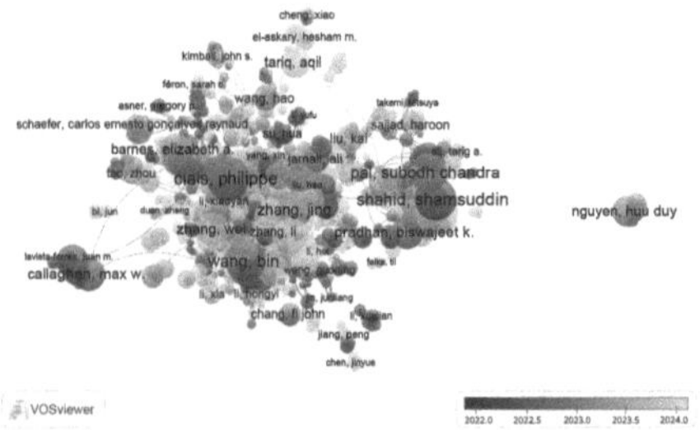

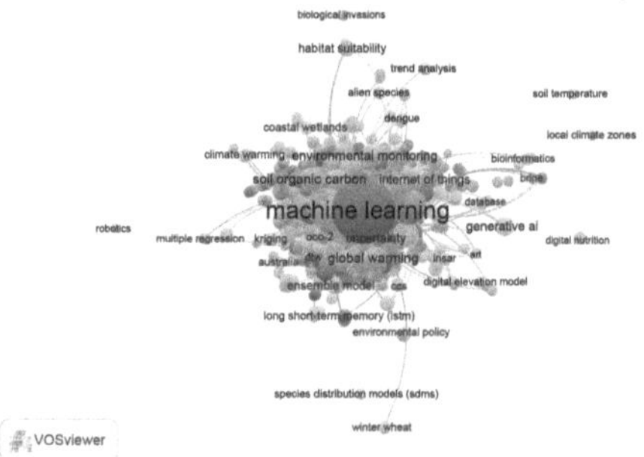

(b) Mapping co-citation links between authors (Scopus).

(c) Mapping co-occurence links between author keywords (Scopus).

Fig. 2. Comparative mapping of bibliometric networks.

3.2 Comparative Evaluation of ML and Generative AI Models

Table 2 presents a comparison of the most popular generative AI and machine learning methods for climate prediction. In terms of accuracy, computational efficiency, data requirements, interpretability, and applicability to certain climate variables, this comparison outlines the strengths and weaknesses of each model. This table systematically compares the main machine learning and generative AI techniques based on their performance, efficiency, and application to climate prediction.

3.3 Limitations of Data-Driven Methods in Climate Science

Although ML and generative AI methods offer advanced tools for generating accurate climate forecasts, their database-driven applications have certain limitations. The difficulties associated with applying these methods lie in their potential non-compliance with physical properties. The results of in-depth statistical models may deviate from fundamental concepts such as mass or energy conservation. They often struggle to derive meaning, showing limited reliability in extreme or unexpected situations, and only perform consistently within

Table 2. Comparative evaluation of major ML and Generative AI approaches in climate forecasting

Models	Accuracy	Timing	Data	Interpretability	Variables
RF	High	Moderate	Moderate	Good	Surface temperature, precipitation
SVM	Moderate	High	Low	Excellent	Precipitation and drought prediction
LSTM	Very High	Low	High	Low	Projected CO_2 concentration, temporal
CNN	Very High	Low	High	Low	Identification of spatial climatic patterns
GAN	High	Low	Very High	Low	Generation of synthetic climate data
Transformers	Very High	Moderate	Very High	Moderate	Multi-variable climate forecasting

the boundaries of recorded data. The "black box" problem related to advanced constructs like transformers and deep neural networks, as well as other complex systems, lies in understanding the fundamental principles that underpin the system's predictions. This understanding is essential, as a lack of clarity may lead to disillusioned users who fail to recognize the value of climate decision-making. To enhance the accuracy, reliability, and trustworthiness of predictions from climate models, hybrid methodologies that integrate data-driven and physics-based approaches should be employed.

4 Conclusion, and Research Directions

In conclusion, innovations in ML and AI are significantly improving our ability to understand, predict, and mitigate the impacts of climate change [28]. Advancements in algorithms, increased processing power, and the expanding availability of climate data are driving the development of more sophisticated climate models [29]. Additionally, the integration of AI and ML with remote sensing, the Internet of Things (IoT), and advanced computing facilitates real-time monitoring of ecological systems. It enhances predictive analysis to address extreme geophysical events [30]. In order to optimize these developments, it is important to encourage interdisciplinary cooperation among ethicists, political decision-makers, and experts in climate science, computer science, and data science. This initiative should also involve designing educational programs that bridge the gap between data science and environmental science. Current challenges that must be addressed include data standardization, model interpretability, ethics, governance, and policy integration [31]. AI and ML technologies can support evidence-based policy development, improve assessments of climate-related risks, and optimize planning for reforestation and renewable energy projects [32]. By emphasizing transparency, flexibility, and strong ethical standards, these technologies can foster trust and collaboration. Ultimately, AI and machine learning are effective approaches that integrate ethical aspects into technological development to produce strong and morally accurate responses to climate change [33].

Funding. The study was conducted without any external financial support.

Conflict of Interest. The authors declare no competing interests or personal relationships influencing this work.

References

1. Zhang, L., Xu, M., Chen, H., Li, Y., Chen, S.: Globalization, green economy and environmental challenges: state of the art review for practical implications. Front. Environ. Sci. **10**, 870271 (2022)
2. Goodfellow, I., Bengio, Y., Courville, A., Bengio, Y.: Deep Learning, vol. 1. MIT press, Cambridge (2016)
3. Chen, T., Guestrin, C.: XGBoost: a scalable tree boosting system. In: Proceedings of the 22nd ACM SIGKDD International Conference on Knowledge Discovery and Data Mining, pp. 785–794 (2016)
4. Rolnick, D., et al.: Tackling climate change with machine learning. ACM Comput. Surv. (CSUR) **55**(2), 1–96 (2022)
5. Gomez-Zavaglia, A., Mejuto, J.C., Simal-Gandara, J.: Mitigation of emerging implications of climate change on food production systems. Food Res. Int. **134**, 109256 (2020)
6. Mashwani, Z.u.R.: Environment, Climate, Plant and Vegetation Growth, pp. 473–501. Springer, Cham (2020)
7. Kumari, B., Solanki, H., Kumar, A.: Climate change: a burning issue for the world. Medicine **35**, 501–507 (2020)
8. Yerlikaya, B.A., Ömezli, S., Aydoğan, N.: Environment, Climate, Plant and Vegetation Growth, pp. 109–122. Springer, Cham (2020)
9. Bibri, S.E., Krogstie, J., Kaboli, A., Alahi, A.: Smarter eco-cities and their leading-edge artificial intelligence of things solutions for environmental sustainability: a comprehensive systematic review. Environ. Sci. Ecotechnol. **19**, 100330 (2024)
10. Yadav, S., Yadav, S.P., Raj, P., Tiwari, P., Albuquerque, V.H.C.D.: Novel AI Applications for Advancing Earth Sciences, pp. 186–203. IGI Global Scientific Publishing (2024)
11. Galaz, V., et al.: Artificial intelligence, systemic risks, and sustainability. Technol. Soc. **67**, 101741 (2021)
12. Qiu, H., Li, T., Zhang, B.: The impact of climate change on the earth system and its simulation predictions: progress, challenges, and future directions. Geograph. Res. Bull. **3**, 231–246 (2024)
13. Bonan, G.: Climate Change and Terrestrial Ecosystem Modeling. Cambridge University Press (2019)
14. Bonan, G.B., Doney, S.C.: Climate, ecosystems, and planetary futures: the challenge to predict life in earth system models. Science **359**(6375), eaam8328 (2018)
15. Eliseev, A.: Hierarchy of the models of the earth climate system. Radiophys. Quantum Electron. **67**(7), 493–507 (2024)
16. Sinwar, D., Dhaka, V.S., Sharma, M.K., Rani, G.: Internet of Things and Analytics for Agriculture, vol. 2, pp. 155–180. Springer, Singapore (2019)
17. Kadow, C., Hall, D.M., Ulbrich, U.: Artificial intelligence reconstructs missing climate information. Nat. Geosci. **13**(6), 408–413 (2020)
18. Kashinath, K., et al.: Physics-informed machine learning: case studies for weather and climate modelling. Phil. Trans. R. Soc. A **379**(2194), 20200093 (2021)
19. Pathak, J., et al.: FourcastNet: a global data-driven high-resolution weather model using adaptive Fourier neural operators. arXiv preprint arXiv:2202.11214 (2022)

20. Kohl, G., Chen, L.W., Thuerey, N.: Benchmarking autoregressive conditional diffusion models for turbulent flow simulation. arXiv preprint arXiv:2309.01745 (2023)
21. Nguyen, T., Brandstetter, J., Kapoor, A., Gupta, J.K., Grover, A.: Climax: a foundation model for weather and climate. arXiv preprint arXiv:2301.10343 (2023)
22. Watt, R.A., Mansfield, L.A.: Generative diffusion-based downscaling for climate. arXiv preprint arXiv:2404.17752 (2024)
23. Glawion, L., Polz, J., Kunstmann, H., Fersch, B., Chwala, C.: Global spatio-temporal era5 precipitation downscaling to km and sub-hourly scale using generative AI. npj Clim. Atmos. Sci. **8**(1), 219 (2025)
24. Brenowitz, N.D., et al.: Climate in a bottle: towards a generative foundation model for the kilometer-scale global atmosphere. arXiv preprint arXiv:2505.06474 (2025)
25. Li, M., Cai, J., Yang, H., Chen, Z., Wu, L.: Oceanic eddy kinetic energy in the spectral space regulated by mesoscale air-sea heat exchange in the Kuroshio extension. In: Advances in Atmospheric Sciences, pp. 1–11 (2025)
26. Smola, A.J., Schölkopf, B.: A tutorial on support vector regression. Stat. Comput. **14**(3), 199–222 (2004)
27. LeCun, Y., Bengio, Y., Hinton, G.: Deep learning. Nature **521**(7553), 436–444 (2015)
28. Singh, S., Goyal, M.K.: Enhancing climate resilience in businesses: the role of artificial intelligence. J. Clean. Prod. **418**, 138228 (2023)
29. MacEachern, S.J., Forkert, N.D.: Machine learning for precision medicine. Genome **64**(4), 416–425 (2021)
30. Ataei Kachouei, M., Kaushik, A., Ali, M.A.: Internet of things-enabled food and plant sensors to empower sustainability. Adv. Intell. Syst. **5**(12), 2300321 (2023)
31. ElZarrad, M.K., Lee, A.Y., Purcell, R., Steele, S.J.: Advancing an agile regulatory ecosystem to respond to the rapid development of innovative technologies. Clin. Transl. Sci. **15**(6), 1332–1339 (2022)
32. Sharifi, A., Allam, Z., Bibri, S.E., Khavarian-Garmsir, A.R.: Smart cities and sustainable development goals (SDGs): a systematic literature review of co-benefits and trade-offs. Cities **146**, 104659 (2024)
33. Bachmann, N., Tripathi, S., Brunner, M., Jodlbauer, H.: The contribution of data-driven technologies in achieving the sustainable development goals. Sustainability **14**(5), 2497 (2022)

Predicting Fish Catches Along Moroccan Coasts Using Ocean Remote Sensing Data and Machine Learning

Younes Oulad Sayad(✉)

Innovation in Mathematics and Intelligent Systems Research Laboratory, Faculty of Applied Sciences Ait Melloul, Ibn Zohr University, BP 6146 Cité Azrou, Agadir, Morocco
y.ouladsayad@uiz.ac.ma

Abstract. Sustainable fisheries management increasingly relies on predictive insights, which can help anticipate changes in marine resources. Although remote sensing technologies now offer continuous monitoring of ocean conditions, the complex relationships between environmental variability and fish population dynamics remain insufficiently captured by current models, especially in regions with contrasting marine ecosystems. In this study, we introduced an integrated predictive framework that combines satellite-derived environmental parameters with machine learning algorithms to predict fish catch across Moroccan waters. The model was validated using official data from the Moroccan Ministry of Fisheries and Maritime Affairs. To achieve this, we implemented a comprehensive analytical pipeline for the period 2005–2023 using MODIS-Aqua parameters: sea surface temperature (SST), chlorophyll-a (CHL-A), particulate backscattering (BBP), diffuse attenuation coefficient (KD490), photosynthetically active radiation (PAR), particulate organic carbon (POC), and normalized fluorescence line height (NFLH). Eight regression models were trained and optimized separately for the Atlantic and Mediterranean regions, including both linear baselines and tree-based algorithms. The analysis revealed that ensemble models, trained on feature-selected parameters, predicted fish catch better than linear approaches. Random Forest achieved the best performance in the Atlantic ($R^2 = 0.599$, RMSE = 23,813, MAE = 17,158), while Gradient Boosting outperformed in the Mediterranean ($R^2 = 0.568$, RMSE = 698, MAE = 547). The Atlantic's dynamic conditions favor nonlinear ensemble learning, whereas the more stable Mediterranean ecosystem benefits from gradient-based optimization. Overall, this research demonstrates the strong potential of combining remote sensing with machine learning to improve marine resource assessment.

Keywords: Machine learning · Remote sensing · MODIS-Aqua · Regression Models · Sustainable fisheries management

1 Introduction

Fisheries play a vital role in global food security, economic development, and marine ecosystem health. However, environmental variability and habitat degradation threaten these resources, making sustainable fish management necessary.

A. E. Hassanien et al. (Eds.): AISDS 2025, LNDECT 287, pp. 68–78, 2027.
https://doi.org/10.1007/978-3-032-18927-1_7

Predicting fish catches is a challenging task due to the complex interactions among biological parameters, oceanographic conditions, and environmental variability. Traditional methods, which rely primarily on historical data, often struggle to capture nonlinear patterns, particularly in heterogeneous coastal ecosystems. Moreover, fish production is influenced not only by oceanographic factors and light penetration, but also by broader climatic, ecological, and socio-economic drivers.

Satellite remote sensing (e.g., MODIS-Aqua SST, CHL-A, BBP) provides high-resolution environmental parameters that reveal spatiotemporal patterns affecting fish distribution. Coupled with machine learning, these datasets enable predictive models to capture complex, nonlinear relationships and improve forecasting accuracy.

Despite this potential, most studies focus on single variables or a limited set of models, with few applications in regions with strong environmental gradients. A comprehensive, multi-parameter approach integrating remote sensing and machine learning is therefore needed to enhance fisheries prediction.

This research developed an integrated predictive framework using multi-parameter MODIS-Aqua satellite data and machine learning algorithms to predict fish catch volumes along Morocco's Atlantic and Mediterranean coasts from 2005 to 2023. The framework employed regression-based machine learning algorithms to identify the most influential environmental drivers, and validate outputs against official landing reports from the Moroccan Ministry of Fisheries and Maritime Affairs.

This research is guided by the following questions:

1. Which environmental variables most strongly influence fish catches along the Moroccan coasts?
2. How does model performance differ between the dynamic Atlantic and the stable Mediterranean regions?
3. What is the practical efficacy of a satellite-based machine learning framework as a decision-support tool for promoting sustainable fisheries management?

The remainder of this paper is structured as follows. Section 2 describes the methodological workflow. This is followed by an Experiment conducted in the selected study area. Finally, the paper concludes with a summary of the main findings, a discussion of practical implications, and suggestions for future research.

2 Research Methodology

This study implements a comprehensive and well-structured methodology to predict fish catch volumes by integrating long-term satellite data with advanced machine learning algorithms. The integrated workflow is composed of five sequential steps:

1. **The identification of the study area and temporal framework**. Capturing contrasting marine environments and long-term environmental variability.
2. **Data acquisition and processing.** Collects oceanographic parameters from remote satellites, which are preprocessed through cropping, temporal aggregation, interpolation, and normalization to ensure consistency.

3. **Data diagnostic, correlation, and feature selection.** Integrate these data with official fish catch records to visualize relationships, assess correlations, and select the most influential features.
4. **Data prediction.** This is then performed using supervised regression algorithms, trained for each region from the study area to account for environmental heterogeneity. Then, Model performance was evaluated using hyper-parameter tuning, cross-validation, and regression metrics such as R^2, RMSE, and MAE to ensure robust performance.
5. **Data visualization.** Translates predictions into interpretable spatial patterns, providing actionable insights for sustainable fisheries management.

This workflow builds upon our previous PFZ mapping methodology [1].

2.1 Study Area and Temporal Framework

The first phase of the methodology sets the spatial and temporal scope of the study. This involves specifying the geographic region for analysis as well as the time period over which data will be collected.

The spatial framework defines the spatial domain of the study by selecting a specific, bounded geographic region where fishing activities occur and oceanographic data are available. The area must be ecologically meaningful, and the boundaries should be defined by coordinates to ensure accurate satellite data extraction [2].

The temporal framework defines the period for which both real fish catch data and oceanographic variables are available. The selection of this period is critical, as it must be long enough to capture the system's full range of natural variability [3].

The result of this step is a defined historical period, long enough to capture climatic cycles and provide sufficient data for reliable machine learning analysis.

2.2 Data Acquisition and Processing

The objective of this step is to collect raw satellite data, then transform it into a clean, analysis-ready dataset that can be integrated with the target variable (fish catch). The result is a unified dataset in which each record pairs environmental conditions with the observed outcome for a given time period.

The processing pipeline consisted of several automated steps:

1. **Data download.** Download the chosen oceanographic parameters for the entire defined historical period from a satellite data portal like NASA's OceanColor Web. The data will typically be in NetCDF format [4].
2. **Format conversion.** Convert the downloaded NetCDF files into a GIS-compatible format, such as GeoTIFF.
3. **Spatial cropping.** Crop each GeoTIFF image using the specific shapefile that defines the boundaries of the study area [5].
4. **Temporal gap-filling.** Fill missing values in the image time series (caused by clouds or sensor errors) using a temporal interpolation method, such as linear interpolation, to create a continuous data record [6].

5. **Feature extraction.** For each processed image (representing one parameter on one date), calculate the average value across the entire study area.
6. **Dataset assembly and target integration.** Compile the averaged environmental values into a table with dates as rows and parameters as columns. Then, join this table with official fish catch records.

The final output is a complete, integrated dataset (CSV file or DataFrame). This is the primary input for supervised learning. Its structure is shown in Table 1 below:

Table 1. Structure of the final integrated dataset for supervised learning.

Date	SST	CHL-A	POC	NFLH	BBP	....	....	Fish catch
2005–01	19.5	0.8	210	0.17	1.38	....	....	550
2005–02	18.9	1.1	305	0.16	1.50	....	....	720
....	....	....	....	....	....	....	...	....

2.3 Data Diagnostic, Correlation, and Feature Selection

This phase explores the dataset patterns through visual analysis and statistical correlations between environmental variables and fish catch. Feature selection is applied to identify the most influential parameters, used as inputs for machine learning models.

This phase consists of three steps that uncover data patterns and relationships:

1. **Visual diagnostics.** Create time series and seasonal plots for each variable as well as the real fish catch, to identify differences between regions [7].
2. **Statistical correlation.** Calculate multiple correlation coefficients (Pearson, Spearman, Kendall) to evaluate the relationships between environmental parameters and fish catch, visualized through comprehensive charts [8].
3. **Feature selection.** Identify the most influential ocean parameters for each region using a unified feature selection; these selected parameters were subsequently used as optimized inputs for all machine learning models.

After exploring and analyzing the data, we identified the main features influencing fish catch, which will guide the development of the prediction model.

2.4 Data Prediction

This phase involves building, training, and evaluating machine learning models to predict fish catch based on environmental variables through several key steps:

1. **Feature engineering.** Convert raw data into informative inputs by creating seasonal features and interaction terms to reveal ecological relationships [8].
2. **Model selection and training.** Apply a range of regression algorithms, including Linear Models, Ensemble Methods, and advanced boosting [9].

3. **Hyperparameter optimization.** Optimize key hyper-parameters for each model through a tuning process using cross-validation and regularization, identifying the optimal configuration without over-fitting [10].
4. **Model evaluation and validation.** Use a comprehensive set of metrics, including R-squared (R^2), Root Mean Squared Error (RMSE) and Mean Absolute Error (MAE) are used to quantify accuracy and precision.

The output is a fully optimized model with performance benchmarks, providing validated, reliable tools for fisheries forecasting.

2.5 Data Visualization

The final step of the workflow focuses on presenting model evaluation results in a clear and interpretable way. Visualization is used to transform numerical performance metrics and predicted outputs into meaningful insights.

This includes: Plots of predicted versus observed values to assess alignment with actual catch data, feature importance rankings to highlight key environmental drivers, error distribution plots to evaluate biases and variability. Accuracy tolerance plots for intuitive assessment of reliability, and temporal plots comparing observed and predicted catches to examine stability over time.

3 Experiment

This section details the practical implementation of the methodology outlined in Sect. 2. Here, we apply the analytical workflow to the selected study area and period, process the satellite data, integrate it with official fish catch records, and build predictive models. The aim is to demonstrate the actual execution of the framework, validate model performance, and generate actionable insights for fisheries management.

3.1 Study Area and Temporal Framework

3.1.1 Study Area

The study area covers the Moroccan Atlantic and Mediterranean coasts, which display contrasting oceanographic conditions. The Atlantic features dynamic, upwelling systems and variable environmental conditions, while the Mediterranean is more stable, with lower primary productivity and moderate seasonal changes. A shapefile-based map was created to delineate these regions. Figure 1 shows the study area boundaries, highlighting fishing zones and locations of satellite-derived environmental data.

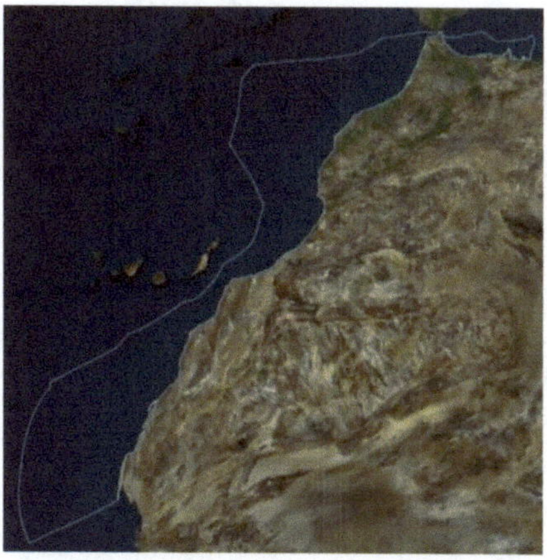

Fig. 1. Study area: Moroccan Atlantic and Mediterranean coasts.

3.1.2 Temporal Framework

For this study, the period from January 2005 to December 2023 was selected. This 19-year span was chosen because fish catch records for this period are available from the Moroccan Ministry of Fisheries and Maritime Affairs [11]. This timeframe captures seasonal cycles, inter-annual variability, and key climatic events that influence fish populations, while providing sufficient data for robust machine learning analysis.

3.2 Data Acquisition and Processing

In this phase, the methodology is applied to the study area and period defined in Sect. 3.1. This includes selecting relevant environmental parameters, acquiring satellite data, processing it, and integrating it with official fish catch records to build the experimental dataset for modeling.

3.2.1 Parameter Selection

Seven oceanographic parameters were selected based on their ecological relevance to fish catch dynamics. The selected parameters, their description, range, and known relationships with fish catch are summarized in Table 2.

Table 2. Selected oceanographic parameters and their relationship with fish catch.

Parameter	Description	Range	Relation with fish catch
SST	Sea Surface Temperature	15–30 °C	Influences fish distribution and metabolism
CHL-A	Chlorophyll-a concentration	0.1–5 mg/m^3	Indicator of primary productivity and food availability

(*continued*)

Table 2. (*continued*)

Parameter	Description	Range	Relation with fish catch
POC	Particulate Organic Carbon	50–500 mg/m^3	Reflects organic matter availability in the water column
PAR	Photosynthetically Active Radiation	0–2000 $Einstein/m^2/day$	Affects primary production
KD490	Diffuse attenuation coefficient	0.05–0.5 m^{-1}	Indicates water clarity, influences light penetration
BBP	Particulate backscattering	0.01–0.2 m^{-1}	Related to suspended particles, can affect habitat quality
NFLH	Normalized Fluorescence Line Height	0–0.06 $W\ m^{-2}\ \mu m^{-1}\ sr^{-1}$	Proxy for phytoplankton health and productivity

The selected parameters were downloaded from the NASA OceanColor Web portal (MODIS-Aqua) [8] for the period 2005–2023. Level 3 (4 km) monthly Standard Mapped Images (SMI) were obtained in NetCDF format covering both Atlantic and Mediterranean study zones.

The downloaded NetCDF files were converted to GeoTIFF format. Each image was cropped using the study area shapefile, isolating the Atlantic and Mediterranean zones for subsequent analysis

Missing values due to clouds or sensor errors were filled using linear interpolation over time, ensuring a continuous dataset for each parameter as shown in Fig. 2.

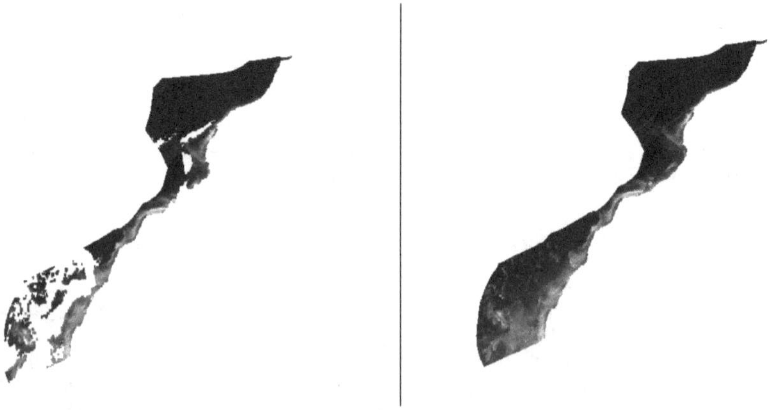

Fig. 2. GeoTIFF conversion and temporal gap-filling of satellite data.

After temporal correction, each satellite-derived parameter was scaled using its calibration formula to convert raw pixels into meaningful environmental units. This ensures the dataset reflects real-world measurements and is suitable for integration with fish catch data. Table 3 summarizes the scaling formulas used for each product.

Table 3. Calibration and scaling of satellite-derived oceanographic parameters.

Product	Scaling formula	Unit	Remarks
SST	SST = raw * 0.005	°C	Offset = 0
CHL-A	CHL-A = raw * 1.0	mg/m^3	No offset, direct value
POC	POC = raw * 0.2 + 6400	mg/m^3	Correction of the baseline value
NFLH	NFLH = raw * 8.5e-05 + 2.25	$W\ m^{-2}\ \mu m^{-1}\ sr^{-1}$	Ensures positive values
PAR	PAR = raw * 0.002 + 65.5	$Einstein/m^2/day$	Corrects the sensor baseline
KD490	KD490 = raw * 0.0002	m^{-1}	Offset = 0
BBP	BBP = raw * 1.0	m^{-1}	No offset, direct value

After scaling, the mean value of each parameter was computed for each image, producing a single representative value. These environmental features were combined with official monthly fish catch data from the Moroccan Ministry of Fisheries and Maritime Affairs [11]. The final dataset contains one record per zone each month, linking all environmental parameters with actual fish catch, as shown in Table 4.

Table 4. Experimental dataset of monthly environmental parameters and fish catch.

Date	Zone	CHL-A	SST	POC	NFLH	PAR	KD490	BBP	Fish catch
2005–01	Atlantic	1.06	17.95	306.86	0.175	30.21	0.11	1.38	23853
2005–01	Mediterranean	0.54	15.49	127.12	0.160	23.17	0.07	1.50	3102
…..	…..	…..	…..	…..	…..	…..	…..	…..	…..
2023–11	Atlantic	1.49	22.02	163.12	0.15	32.65	0.16	1.54	140725
2023–12	Atlantic	1.02	20.81	144.69	0.14	28.38	0.11	1.53	96915

This step resulted in a complete dataset linking oceanographic parameters with fish catches, comprising 456 rows of data, ready for the experimental modeling phase.

3.3 Data Diagnostic, Correlation and Feature Selection

Figure 3 summarizes the results of correlation and feature selection analysis for both the Atlantic and Mediterranean regions, across the selected oceanographic parameters.

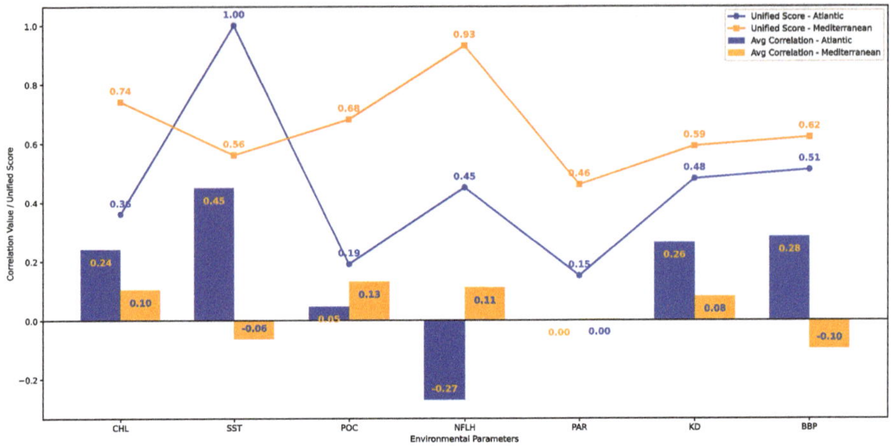

Fig. 3. Parameter correlations and feature selection linked to fish catch in the study area.

From the figure, several patterns emerge:

In the **Atlantic region**, Environmental parameters show strong, significant correlations with fish catch. SST has the highest positive correlation (Pearson: 0.51, Spearman: 0.50, Kendall: 0.34), with an average correlation of 0.45. While BBP (0.28) and KD (0.26) also contribute positively, NFLH exhibits a notable negative correlation, reflecting complex ecological interactions.

Unified feature selection confirmed SST as the dominant predictor (score: 1.00), followed by BBP (0.51) and KD (0.48), validating the correlation findings while incorporating nonlinear relationships.

In the **Mediterranean region**, Correlations are weaker. CHL (0.10), POC (0.13), and NFLH (0.11) show slight positive correlations, whereas SST, PAR and BBP are near zero or negative, indicating limited influence on fish catch.

Unified feature selection revealed stronger nonlinear dependencies, identifying NFLH (0.93), CHL (0.74), and POC (0.68) as the primary drivers.

3.4 Data Prediction

This step evaluates the predictive performance of the trained machine learning models using the selected environmental features.

In the **Atlantic region**, Linear Regression performed poorly ($R^2 = -0.267$, RMSE = 42,331), while Ridge ($R^2 = 0.519$, RMSE = 26,075) and Lasso ($R^2 = 0.414$, RMSE = 28,794) showed moderate results. Random Forest achieved the best performance ($R^2 = 0.599$, RMSE = 23,813), capturing nonlinear relationships.

In the **Mediterranean region**, Linear Regression and regularized models offered moderate accuracy ($R^2 \approx 0.40$), whereas ensemble models surpassed linear approaches. Gradient Boosting led ($R^2 = 0.568$, RMSE = 698), followed by LightGBM ($R^2 = 0.565$) and Random Forest ($R^2 = 0.518$), while SVM performed poorly ($R^2 = 0.276$).

Overall, ensemble methods outperformed linear models, showing their effectiveness for fish catch prediction. Performance varies by region: Random Forest excelled in the Atlantic, and Gradient Boosting slightly outperformed in the Mediterranean.

3.5 Data Visualization

The scatter plots of predicted versus actual values clearly illustrate these results, where predictions align more closely with the 1:1 reference line in the Atlantic than in the Mediterranean (Fig. 4). Overall, ensemble methods consistently outperformed linear models in both zones, confirming their suitability for fish catch prediction and potential fishing zone identification.

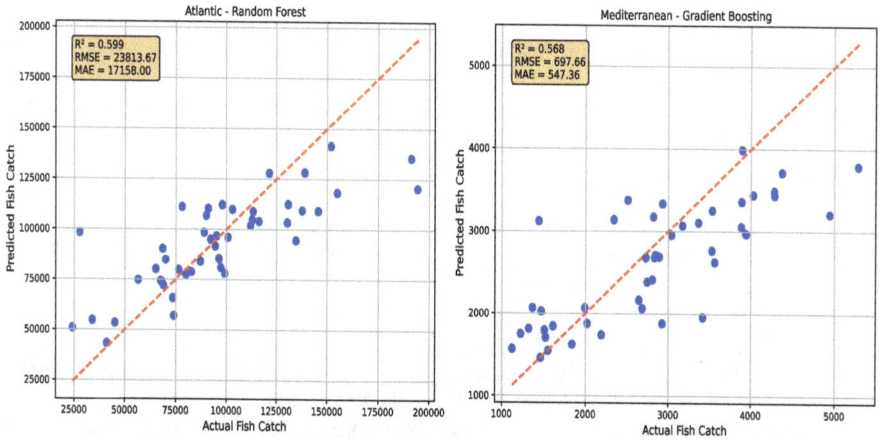

Fig. 4. Predicted vs. Actual fish catch for Atlantic and Mediterranean zones.

4 Conclusion and Future Works

This study developed and validated an integrated framework combining multi-parameter MODIS-Aqua satellite data with machine learning to predict fish catch along the heterogeneous Moroccan coasts.

Model performance varied across regions, with feature selection significantly enhancing predictive accuracy by identifying region-specific environmental drivers. In the Atlantic, Random Forest trained on selected parameters (SST, BBP, KD) explained 60% of fish catch variance ($R^2 = 0.599$), with errors of 17,158 tons (MAE) and 23,813 tons (RMSE). In the Mediterranean, Gradient Boosting using key biological parameters (NFLH, CHL, POC) accounted for 56.8% of variance ($R^2 = 0.568$), with lower errors of 547 tons (MAE) and 698 tons (RMSE).

The observed performance difference between the Atlantic and the Mediterranean results can be explained by the environmental variability of each region. The Atlantic coast, characterized by dynamic upwelling systems and high spatial–temporal variability, benefits from the Random Forest's ability to handle nonlinear relationships and complex feature interactions. In contrast, the Mediterranean coast presents more stable and homogeneous environmental conditions, where Gradient Boosting stepwise learning approach allows it to uncover the gradual patterns characterizing this zone. This regional difference confirms that environmental dynamics directly influence model behavior and predictive performance.

Future work will focus on feature engineering to enhance model performance by creating more informative features. We will also pursue collaborations with fisheries institutions to incorporate crucial non-environmental parameters, including fishing effort, fish migration behavior, and socioeconomic factors. By combining technical refinement with real-world contextual data, we aim to develop more accurate and practical decision-support tools for sustainable fisheries management in Morocco.

References

1. Oulad Sayad, Y.: Mapping potential fishing zones using remote sensing data and GIS: a case study of Moroccan waters. Multimedia Res. **6**(2), 04009 (2023)
2. Paradinas, I., Giménez, J., Conesa, D., López-Quílez, A., Pennino, M.G.: Evidence for spatiotemporal shift in demersal fishery management priority areas in the western Mediterranean. Can. J. Fish. Aquat. Sci. **79**(10), 1641–1654 (2022)
3. Morfin, M., Fromentin, J.M., Jadaud, A., Bez, N.: Spatio-temporal patterns of key exploited marine species in the Northwestern Mediterranean Sea. PLoS One. **7**(5), e37907 (2012)
4. NASA.: OceanColor Web. *Ocean Color Data Access.* NASA Goddard Space Flight Center. Available at: https://oceancolor.gsfc.nasa.gov/
5. Bivand, R.S., Pebesma, E., Gómez-Rubio, V.: Applied Spatial Data Analysis with R. Springer (2013)
6. Beckers, J.M., Rixen, M.: Calculations and data filling from incomplete oceanographic datasets. J. Atmos. Ocean. Technol. **20**(12), 1839–1856 (2003)
7. Mendiolar, M., Filar, J.A., Yang, W.H., Leahy, S., Courtney, A.J.: Capturing episodic impacts of environmental signals. Environ. Model Softw. **170**, 105837 (2023)
8. Sampaio, N.A.S., Mazza, F.C., Siqueira, S.S.S., Miranda Junior, J.E., Moutinho, J.V.S., Pacífico, L.O.: Applications of correlation analysis in environmental problems. Revista de Gestão Social e Ambiental. **18**(3), 1–16 (2024)
9. Tanaka, T., Tahara, Y., Ohsuga, A., Sei, Y.: Fish catch prediction by combining fishing, weather and tidal data using XGBoost. In: Proceedings of the 2025 International Conference on Environmental Data Science, pp. 1–13. Springer, Heidelberg (2025)
10. Jelovica, B., Kovačič, M., Kovačič, M.: Atlantic Salmon habitat-abundance modeling using machine learning algorithms. Environ. Model Softw. **160**, 105515 (2024)
11. Ministry of Agriculture, Maritime Fisheries, Rural Development, and Water and Forests. Available at: https://www.mpm.gov.ma/

Prediction of Structural Deformation Using Adaptive Neuro-Fuzzy Inference Systems and Simulation-Based Soft Computing

Aya Elhozayen(✉) and Bahy Y. Elgendy

The British University in Egypt, Cairo 11837, Egypt
aya.elhozayen@bue.edu.eg

Abstract. The widespread use of artificial intelligence (AI) and machine learning (ML) has provided effective substitutes for compute-intensive simulations. This research applies ML to model nonlinear deformation undergoing dynamic impact loading, specifically referring to auxetic sandwich panels. The dataset was generated through finite element (FE) simulations with different drop weights and heights representing different environmental conditions. To (partially) mitigate shortcomings of tradtional FE analysis, an Adaptive Neuro-Fuzzy Inference System (ANFIS) model was effectively employed, bringing together neural learning and intepretable fuzzy models. The predictive model was successfully trained for a unique purpose with high accuracy ($R^2 > 0.93$) and low error (RMSE $\approx$ 3.3 mm), reflecting both accuracy and generalizability. Validation using the residuals, parity plots, and response surfaces demonstrated robustness. The findings in this study confirm and set forth ANFIS modelling as an effective, interpretable, and scalable alternative to black-box ML models and an avenue for real-time prediction, design optimization, and intelligent monitoring. This research provides a reliable framework for the use of ML as a bridge between physics-based simulations and ML-driven surrogates for future intelligent land and infrastructure issues.

Keywords: Auxetic sandwich panel · surrogate model · impact loading · machine learning · composite structures

1 Introduction

One of the great challenges of modern engineering is reliably predicting nonlinear deformation of sophisticated structural systems subjected to extreme loading. While classical Finite Element Analysis (FEA) provides accurate solutions, it is too computationally intensive and is therefore not valid for situations that either require rapid decisions or where parametric studies can involve thousands of models [1, 2].

Auxetic sandwich panels have received attention, exhibiting a unique negative Poisson's ratio and extraordinary energy absorption. They represent promising structurally systems for use in impact resistant solutions [3, 4]. Their highly nonlinear responses under dynamic loading can make prediction difficult, frustrating and expensive [5].

A. E. Hassanien et al. (Eds.): AISDS 2025, LNDECT 287, pp. 79–88, 2027.
https://doi.org/10.1007/978-3-032-18927-1_8

This paper highlights the use of an Adaptive Neuro-Fuzzy Inference System (ANFIS) as a predictive surrogate for auxetic sandwich panels under impact loading. The intention of the proposed framework is to provide the flexibility of neural networks and the transparency of fuzzy logic, while minimizing the computational expense and maintaining transparency and accuracy.

The specific objectives of this research are:

- Develop and train ANFIS computational model that explicitly captures nonlinear deformation behavior
- Validate the ANFIS predictive model to consider the accuracy and generalization capacity against the FEA data
- Evaluate the potential of ANFIS as interpretable, scalable surrogate for general design, optimization and structural health monitoring

These objectives will be critical for improved potential of intelligent surrogates to help fasttrack design, provide resilience, and provide a data-driven decision-making capability in the next generation of structural systems.

2 Literature Review

For predicting structural deformation, traditionally, Finite Element Analysis (FEA) has been used, which is useful for detailed information about stress–strain, but is computationally expensive and very sensitive to assumptions [1, 6]. Although FEA is appropriate for case-specific analysis, there is no question that it is impractical for large parametric studies, or when quick decisions are necessary; hence the need for data-driven surrogates [7].

Machine learning (ML) methods have been investigated more in relation to structural engineering problems [8, 9]. Artificial Neural Networks (ANN) have produced impressive results when predicting buckling, strength, and deformation, but due to their "black box" nature, they limit the knowledge distilled from their use in safety-critical scenarios [10, 11]. Fuzzy inference systems provide an understandable and rule-based approach but do not offer the capacity to learn highly nonlinear dynamics as stand-alone systems [2, 12]. The Adaptive Neuro-Fuzzy Inference System (ANFIS) was designed as a hybrid model that merges the adaptability of neural networks with the explanatory power of fuzzy logic, identifying a path toward balance between accuracy and interpretability. The prior studies clearly identified ANFIS as a viable surrogate model for foundation settlement prediction, buckling load prediction of beams and vibration mode identification of composite structures, and consistently outperformed ML or fuzzy models [13–15].

In addition to ANFIS, hybridized models and physics informed models have emerged like Physics-Informed Neural Networks (PINNs) and deep learning – FEA couplings [16]. These approaches allow engineers to develop surrogate models with high accuracy and/or physical consistency, however, many times they also introduce complexity, high training cost and lack of interpretability, which can limit their overall adoption for practice [11].

With respect to auxetic sandwich panels, literature related to their performance has predominantly focused on laboratory-based testing or simulated FEA [17, 18]. These testing methods provide relyable assessments of the negative Poisson's ratio but track of negative microstructural geometry and greater impact resistance. What's lacking, is frameworks that are computatively inexpensive (compared to FEA) but are more interpretable [19, 20].

This gap serves as motivation for the current work, which proposes an ANFIS-based predictive model of auxetic sandwich panels (SPs) under impact loading. Unlike black-box ML models, the proposed approach provides both accuracy and explainability, suggesting a scalable pathway to the use of intelligent surrogates in engineering design and structural health monitoring.

3 Methodology

The approach for this research followed two intertwined stages, using impact simulations as a way to carry out finite-element analysis (FEA) using ANSYS and then predictive modeling using Adaptive Neuro-Fuzzy Inference System (ANFIS). The complete approach is summarized in Fig. 1 which outlines the steps of the predefined work process from simulations, data preprocessing, to developing the ANFIS model and evaluating performance.

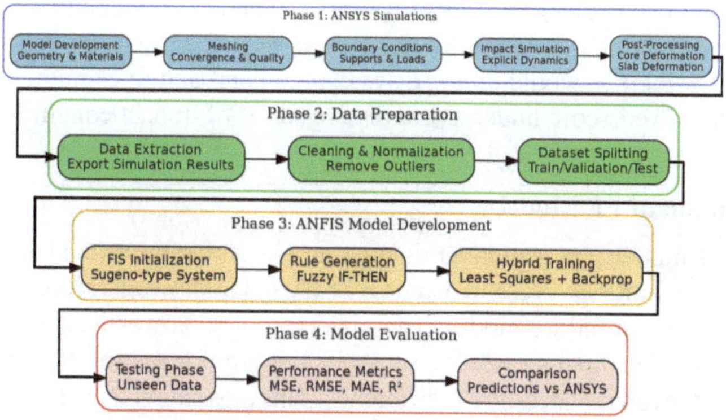

Fig. 1. Methodological workflow of the study.

3.1 Finite Element Modeling

3.1.1 Geometry, Boundary Conditions, and Contact Modeling

All simulations were run on ANSYS Explicit Dynamics with the goal of replicating the impact response of sandwich panels with aluminum auxetic cores test by Qi et al. [21]. All symmetry conditions allowed for only a representative section to be modelled, which resulted in reduced computational cost. The RC slab base was fixed and uni-form pressure of 0.328 MPa was applied to the upper face. The assembly was an RC slab with metallic face sheets and auxetic core, the face sheets and core midsurfaces were modelled in shell elements, and the slab and core solids were modelled in 3D solid elements. Working with an initial mesh size of 5 mm, determined by an appropriate sensitivity analysis and which provided an appropriate balance between accuracy and efficiency. The core-face sheet interfaces were assumed perfectly bonded, and contact was modeled using hard contact (normal) and a penalty method (tangential, with a coefficient of friction 0.3).

3.1.2 Material Characterstics and Modeling

The auxetic core and face sheets were modeled with AA6061 aluminum alloy, characterized by isotropic elasticity and plasticity using the Steinberg–Guinan model. Key properties were density 2.71×10^{-6} kg/mm^3, Young's modulus 40.3 MPa, Poisson's ratio 0.33, and yield stress 113 MPa. Nonlinear hardening with strain failure was used to simulate the progressive material response under loading.

3.1.3 Impact Load Modeling

Dynamic impact loading was modeled using a drop-weight impactor test, where the load is defined both by the equivalent pressure of the weight of the impactor and the equivalent velocity which was generated by the drop height from a free fall. A baseline 580 kg drop weight (0.328 MPa) was used, then step-loaded weights from 600 kg–10,000 kg were modeled, converted to equivalent pressures. All performed six drop heights (1–6 m) to generate drop velocities, where it was determined all velocities fell within the range of 4.43–10.85 m/s. Performing these drop weight caused the measurements of the equivalent velocities would create a systematic approach to examine the structural response of the auxetic core under a progressive series of impact conditions.

3.1.4 Validation of FE Model

The numerical model was validated against an experimental reference model under similar impact conditions presented by Qi et al. [21], showing strong correlation in peak deformation and failure modes. The main comparison was peak displacement of the auxetic core at the final simulation time: 99.574 mm for the validated model versus 95.365 mm for the reference, a 4.41% difference. This close agreement (Fig. 2) confirmed the simulation setup's accuracy and its suitability for parametric analysis and machine learning applications.

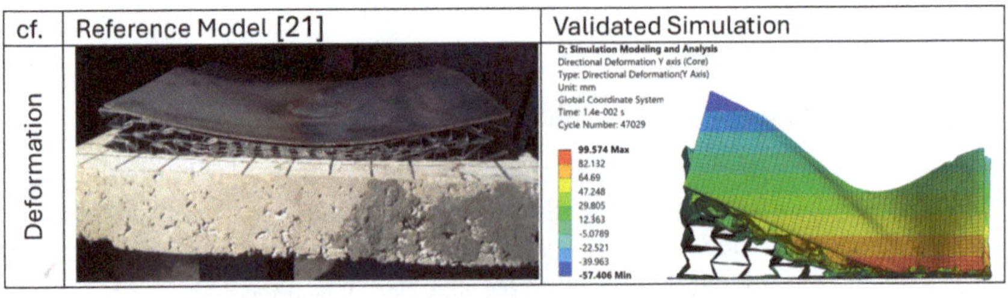

Fig. 2. Validation of the finite element model.

3.2 Adaptive Neuro-Fuzzy Inference System (ANFIS) Modeling

As a supplement to the finite element simulations, a MATLAB-based Adaptive Neuro-Fuzzy Inference System (ANFIS) was developed to predict the nonlinear structural response to impact loading conditions. ANFIS is composed of fuzzy logic and neural networks, where fuzzy if–then rules are created and membership functions are then adapted through a hybrid learning mechanism. The input variables were defined as impact pressure (MPa) and velocity (m/s), and the output variable is the core deformation (mm). The complete workflow begins with the input data, which are fuzzified using Gaussian membership functions. There were three membership functions for pressure and two for velocity, which sufficiently divided the input space. These inputs are then mapped through a rule base of six fuzzy rules (Fig. 3), which cover the entire input domain. Defuzzification was performed using a Sugeno-type fuzzy inference method to determine the predicted deformation as the output.

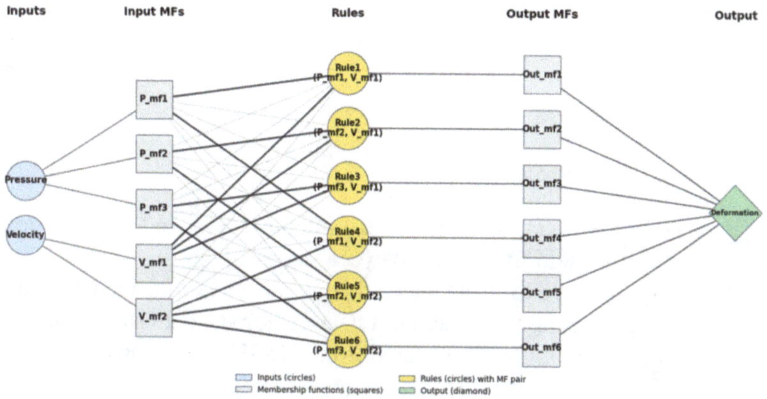

Fig. 3. Architecture of the ANFIS model.

The ANFIS components shown in (Fig. 3) illustrate the mapping of inputs to membership functions before the two membership functions were aggregated together while utilizing the fuzzy rule layer before the output connected to the ANFIS architecture. The membership plot of Fig. 4 provided solid and smooth coverage across the input domains while the rule schematic depicts how the fuzzy rules combine input terms responsible for the governing behavior of the system. Training was accomplished using a hybrid algorithm taking least-squares (for the consequent parameters) estimation along with backpropagation (for antecedent parameters). The training data set was obtained from explicit dynamic simulations and divided into a 70–15–15 training/validation/testing split in order to generalize the ANFIS model.

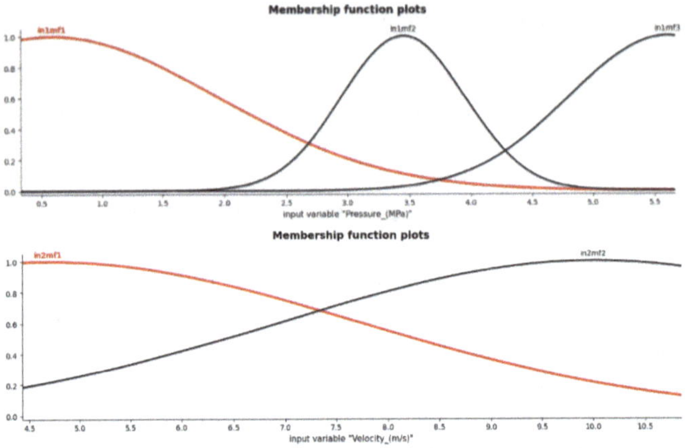

Fig. 4. Membership function plots for the ANFIS model.

Overall, the developed ANFIS successfully captured the complex mapping between impact conditions and structural deformation, providing a computationally efficient surrogate for the more resource-intensive finite element simulations.

4 Results and Discussion

The performance of the Adaptive Neuro-Fuzzy Inference System (ANFIS) model was evaluated to predict core deformation (mm) over a variety of input conditions of pressure (MPa) and velocity (m/s). To evaluate the predictive capacity of the model, results were examined for the training, validation and test datasets for standard statistical measures: mean squared error (MSE), root mean squared error (RMSE), mean absolute error (MAE) and coefficient of determination (R2).

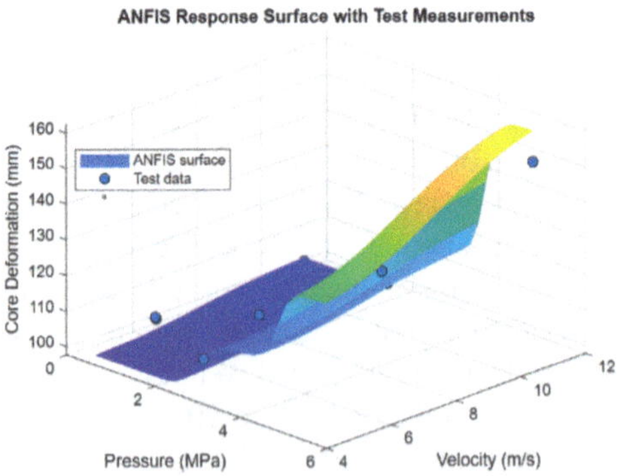

Fig. 5. Three-dimensional ANFIS response surface.

The design and overall workflow of the ANFIS structure guaranteed that the input data stream was mapped into fuzzy membership values, rules were evaluated and defuzzified to provide deformation estimates that matched experimental values. Figure 5 shows the three-dimensional response and surface produced by the trained ANFIS model alongside experimental test data points overlaid onto it. The smoothness of the surface and closeness of the test measurements to the predicted manifold shows that the model was able to suitably map nonlinear input-output relations.

The parity plots (Fig. 6) for individual splits create a better understanding of the performance consistency. Analyzing the training set, the predictions cluster closely about the 1:1 reference line with $R^2 = 0.932$, RMSE = 4.11 mm, and MAE = 2.70 mm. The validation set yielded a similar degree of accuracy with $R^2 = 0.944$, while also showing somewhat larger deviations (RMSE = 4.95 mm MAE = 4.05 mm), indicating that the trained model generalized well and did not overfit the training data. The test set showed the degree of predictive reliability that can be obtained, with R^2 value = 0.952 and the smallest error measures (RMSE = 3.31 mm, MAE = 2.43 mm), which confirms the robustness that the training data had given the trained ANFIS model.

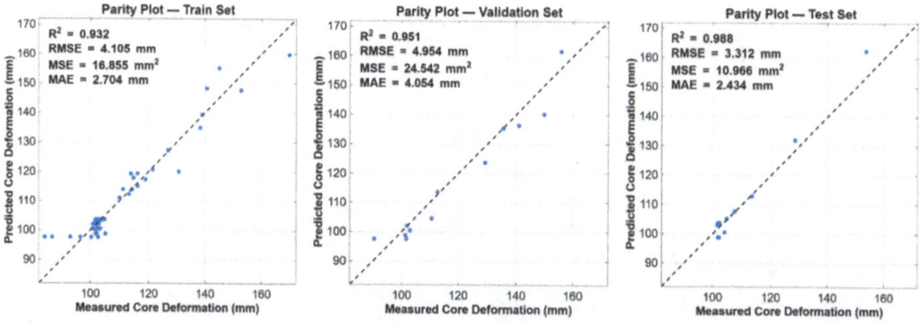

Fig. 6. Parity plot for train, validation, and test sets, show close alignment with the 1:1 line.

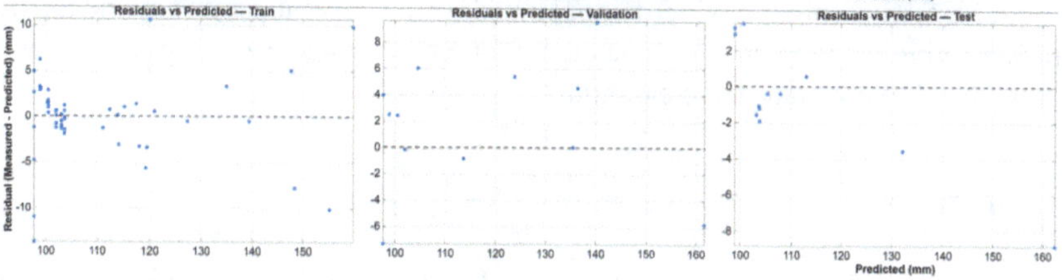

Fig. 7. (top) Residual distribution histograms for training, validation, and test sets. (bottom) Residuals versus predicted plots showing random scatter without bias.

Error distribution histograms as shown in the top of Fig. 7 also support these conclusions with a almost symmetric spread of residuals around zero, particularly for the test data, that suggests that no systematic bias exists in the predictions. The residual-versus-predicted plots as predicted in the bottom of Fig. 7 conclude the case with seemingly random scatter that is void of any trend, validating that errors are homoscedastic over the predicted range. Overall, the slightly less variable spread of residuals in the test set compared to the training and validation sets may also account for the lower computed error indices.

When visually comparing the measured data versus the predicted deformation in the series of plots (Fig. 8), the ANFIS model is adequate at tracing changes from sample to sample, and this occurred even in areas with quickly changing deformation values. The fact that this consistency was observed in all three splits suggests that the fuzzy rule base and Gaussian membership functions mapped the input space sufficiently enough to capture regional nonlinearities.

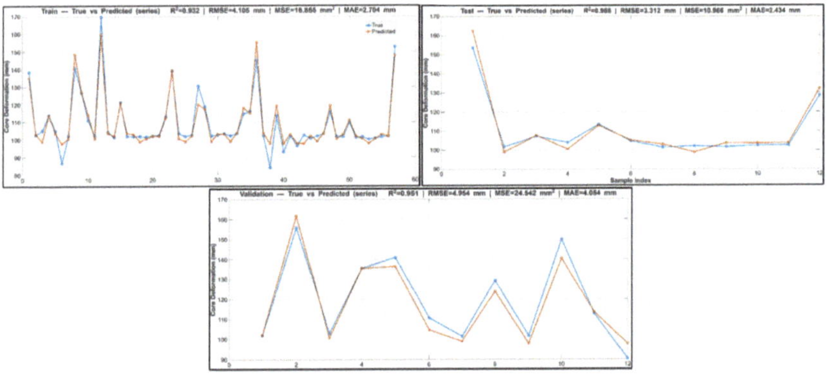

Fig. 8. Series plot of measured vs. predicted deformation for training, validation, and test sets.

The ANFIS model delivered consistent performance across training, validation, and test splits, with predictions closely matching measured values and the test set achieving the highest accuracy. Across all splits, the model achieved R^2 values above 0.93 with very low errors on the order of a few millimeters, confirming its accuracy and interpretability in capturing complex deformation behavior. By combining adaptive learning with rule-based reasoning, ANFIS demonstrated strong generalization beyond the training data, making it a reliable surrogate model for engineering design applications where robustness and extrapolation are essential.

5 Conclusion

The research developed and validated an Adaptive Neuro-Fuzzy Inference System (ANFIS) as a surrogate prediction model of the nonlinear impact response of auxetic sandwich panels; by combining finite element simulations with a data-driven soft computing framework, we have achieved a compromise between accuracy, interpretability, and efficiency – all important distinctions for safety-critical work. The data outcomes suggest ANFIS can be a viable substitute for resource-heavy finite element analysis (FEA), providing generalized responses to a range of impact conditions in a transparent way to support decision-making. Thus providing a scalable framework for incorporating AI-driven predictive analytics into engineering practices for future generations.

Key findings and impact:

- The ANFIS predictive model achieved $R^2 > 0.93$ across all datasets, with RMSE's as low as 3.3 mm, confirming its reliability.
- An error analysis (parity plots, histograms, residuals) confirmed strong generalisation with no systematic errors.
- The framework provided a computationally efficient surrogate to FEA, with demonstrated reductions in analysis time whilst maintaining good accuracy.
- The interpretability of the model as it used fuzzy rule bases provides a reasoning process that is important in applications within safety of structures.
- The study has identified a scalable framework for real-time design evaluation, optimisation, and structural health monitoring of advanced materials.

Future research should build on this framework in several important aspects. First, towards multi-scale modeling, which could explicitly connect the micro-structural geometry of the auxetic core with the macro impact response of the panel. Second, the creation of physics-informed fuzzy hybrids could ensure robustness by introducing governing physical equations as constraints into the ANFIS model. Finally, this model could be an ideal candidate for real-time implementation in structural health monitoring, such as embedding the trained surrogate on in-situ sensor nodes to provide real-time deformation analysis and warning signs of damage immediately following the impact event.

References

1. Kudela, J., Matousek, R.: Recent advances and applications of surrogate models for finite element method computations: a review. Soft. Comput. **26**, 13709–13733 (2022). https://doi.org/10.1007/s00500-022-07362-8
2. Karaboga, D., Kaya, E.: Adaptive network based fuzzy inference system (ANFIS) training approaches: a comprehensive survey. Artif. Intell. Rev. **52**, 2263–2293 (2019). https://doi.org/10.1007/s10462-017-9610-2
3. Bohara, R.P., Linforth, S., Nguyen, T., Ghazlan, A., Ngo, T.: Anti-blast and -impact performances of auxetic structures: a review of structures, materials, methods, and fabrications. Eng. Struct. **276**, 115377 (2023). https://doi.org/10.1016/j.engstruct.2022.115377
4. Xiao, P., Bin, L., Vescovini, R., Zheng, S.: Optimal design of composite sandwich panel with auxetic reentrant honeycomb using asymptotic equivalent model and PSO algorithm. Compos. Struct. **328**, 117761 (2024). https://doi.org/10.1016/j.compstruct.2023.117761
5. Elgendy, B.Y., Elhozayen, A.: Auxetic metamaterials for seismic wave mitigation: a state-of-the-art review of recent literature. IOP Conf. Ser.: Earth Environ. Sci. **1530**, 012031 (2025). https://doi.org/10.1088/1755-1315/1530/1/012031
6. Astroza, R., Alessandri, A.: Effects of model uncertainty in nonlinear structural finite element model updating by numerical simulation of building structures. Struct. Control. Health Monit. **26**, e2297 (2019). https://doi.org/10.1002/stc.2297

7. Hou, C.K.J., Behdinan, K.: Dimensionality reduction in surrogate modeling: a review of combined methods. Data Sci. Eng. **7**, 402–427 (2022). https://doi.org/10.1007/s41019-022-00193-5
8. Etim, B., Al-Ghosoun, A., Renno, J., Seaid, M., Mohamed, M.S.: Machine learning-based modeling for structural engineering: a comprehensive survey and applications overview. Buildings. **14**, 3515 (2024). https://doi.org/10.3390/buildings14113515
9. Malekloo, A., Ozer, E., AlHamaydeh, M., Girolami, M.: Machine learning and structural health monitoring overview with emerging technology and high-dimensional data source highlights. Struct. Health Monit. **21**, 1906–1955 (2022). https://doi.org/10.1177/14759217211036880
10. Ly, H.-B., et al.: Development of hybrid machine learning models for predicting the critical buckling load of I-shaped cellular beams. Appl. Sci. **9**, 5458 (2019). https://doi.org/10.3390/app9245458
11. Rudin, C.: Stop explaining black box machine learning models for high stakes decisions and use interpretable models instead. Nat. Mach. Intell. **1**, 206–215 (2019). https://doi.org/10.1038/s42256-019-0048-x
12. Cao, J., et al.: Fuzzy inference system with interpretable fuzzy rules: advancing explainable artificial intelligence for disease diagnosis—a comprehensive review. Inf. Sci. **662**, 120212 (2024). https://doi.org/10.1016/j.ins.2024.120212
13. Özbayrak, A., Ali, M.K., Çıtakoğlu, H.: Buckling load estimation using multiple linear regression analysis and multigene genetic programming method in cantilever beams with transverse stiffeners. Arab. J. Sci. Eng. **48**, 5347–5370 (2023). https://doi.org/10.1007/s13369-022-07445-6
14. Jakkamputi, L., et al.: Experimental and computational vibration analysis for diagnosing the defects in high performance composite structures using machine learning approach. Appl. Sci. **12**, 12100 (2022). https://doi.org/10.3390/app122312100
15. Sahu, D.P., Das, R., Prusty, J.K., Mohanty, S.C.: Frequency analysis of skew sandwich plates with polyethylene terephthalate foam-core and carbon/basalt fiber-reinforced hybrid face layers using ANFIS model and experimental validation. Mech. Adv. Mater. Struct., 1–23 (2024). https://doi.org/10.1080/15376494.2024.2438910
16. Herrmann, L., Kollmannsberger, S.: Deep learning in computational mechanics: a review. Comput. Mech. **74**, 281–331 (2024). https://doi.org/10.1007/s00466-023-02434-4
17. Galea Mifsud, R., et al.: Auxetics and FEA: modern materials driven by modern simulation methods. Materials. **17**, 1506 (2024). https://doi.org/10.3390/ma17071506
18. Indreş, A.I., et al.: Particularities on the low-velocity impact behavior of 3D-printed sandwich panels with re-entrant and honeycomb core topologies. J. Compos. Sci. **8**, 426 (2024). https://doi.org/10.3390/jcs8100426
19. Garg, A., Sharma, A., Zheng, W., Li, L.: A review on artificial intelligence-enabled mechanical analysis of 3D printed and FEM-modelled auxetic metamaterials. Virtual Phys. Prototyp. **20**, e2445712 (2025). https://doi.org/10.1080/17452759.2024.2445712
20. Danesh, H., Di Lorenzo, D., Chinesta, F., Reese, S., Brepols, T.: FFT-based surrogate modeling of auxetic metamaterials with real-time prediction of effective elastic properties and swift inverse design. Mater. Des. **248**, 113491 (2024). https://doi.org/10.1016/j.matdes.2024.113491
21. Qi, C., Remennikov, A., Pei, L.-Z., Yang, S., Yu, Z.-H., Ngo, T.D.: Impact and close-in blast response of auxetic honeycomb-cored sandwich panels: experimental tests and numerical simulations. Compos. Struct. **180**, 161–178 (2017). https://doi.org/10.1016/j.compstruct.2017.08.020

Towards Early Crop Classification Using Deep Learning

Salma Oumoussa[1,2](✉) and Diego H. Peluffo-Ordoñez[1]

[1] College of Computing (COLCOM), Mohammed VI Polytechnic University, Lot 660, Hay Moulay Rachid Ben Guerir, 43150 Ben Guerir, Morocco
salma.oumoussa@um6p.ma, diego.peluffo@sdas-group.com
[2] National School of Applied Sciences (ENSA), University Hassan I, ENSA Campus, 10587 Berrechid, Morocco

Abstract. Early crop classification enables time-sensitive agricultural decisions months before harvest. We adapt EarlyRNN, a dual-head LSTM with adaptive stopping, for Sentinel-2 imagery. Using 70 time steps (versus 147 for full-season models), we achieve accuracy 86% (BavarianCrops) and 80% (BreizhCrops), only 3% points below the best full-season ensemble (LITE: 89%), with predictions 60–90 days before harvest. The adaptive stopping mechanism learned crop-specific phenological patterns, enabling earlier predictions for phenologically distinctive classes while deferring decisions for ambiguous ones. Confidence-based filtering enables operational risk calibration, achieving 96.5% accuracy on 28.7% of high-confidence cases. The results demonstrate that reliable early crop classification is feasible with favorable accuracy-earliness trade-offs for operational precision agriculture.

Keywords: Early crop classification · LSTM · Sentinel-2 · Adaptive stopping · Precision agriculture

1 Introduction

Early crop classification is essential for precision agriculture, helping with timely resource allocation, yield forecasting, and risk management months before harvest. Traditional methods rely on complete growing-season data, delaying actionable insights when decisions must be made. While that Sentinel-2 provides frequent global coverage [1,2], exploiting partial-season observations remain challenging due to spectral noise, cloud gaps, and region-specific phenology.

Recent deep learning advances, particularly LSTM networks, can capture temporal dynamics from satellite sequences [3]. However, existing approaches provide limited guidance on hyperparameter optimization, class imbalance handling, and cross-region transferability, restricting practical deployment.

D. H. Peluffo-Ordoñez—Contributing author.

S. Oumoussa and D. H. Peluffo-Ordoñez—These authors contributed equally to this work.

A. E. Hassanien et al. (Eds.): AISDS 2025, LNDECT 287, pp. 89–97, 2027.
https://doi.org/10.1007/978-3-032-18927-1_9

This work adapts EarlyRNN [4], a dual-head LSTM with adaptive stopping, to Sentinel-2 crop classification. We contribute:

(1) an architecture enabling predictions 60–90 days before harvest, (2) preprocessing strategies for sequence variability and class imbalance, (3) comprehensive evaluation on two benchmarks (BavarianCrops, BreizhCrops) with behavioral analysis of stopping mechanisms and confidence-earliness relationships, and (4) cross-region generalization diagnostics demonstrating favorable accuracy-earliness trade-offs for operational monitoring.

Section 2 describes materials and methods, Sect. 3 presents results, and Sect. 4 concludes with future directions.

2 Materials and Methods

This study focuses on developing and evaluating a deep learning-based crop classification pipeline using Sentinel-2 satellite imagery. The methodology includes data preprocessing, model architecture, baseline comparisons, and evaluation protocols designed to assess both classification accuracy and prediction timeliness.

2.1 Datasets

We evaluate our approach on two public benchmark datasets providing parcel-level Sentinel-2 time series with validated crop labels. BavarianCrops (Bavaria, Germany) contains 16,600 parcels across 7 crop classes and serves as the primary training dataset. BreizhCrops (Brittany, France) contains 122,614 parcels across 9 classes and provides a more challenging generalization test due to smaller field sizes, higher cloud cover from maritime climate, and more severe class imbalance.

Both datasets exhibit class imbalance: Meadow comprises 40% of BavarianCrops and 55% of BreizhCrops, while minority classes such as Winter Triticale represent less than 4% of samples. Raw sequence lengths vary from 71 to 147 observations depending on cloud cover and acquisition frequency. Both datasets use an 80/20 train-test split with stratified sampling.

For notation, let $\mathbf{X} \in \mathbb{R}^{N \times T \times D}$ denote the input data, where N is the number of parcels, $T = 70$ is the sequence length after preprocessing, and $D = 13$ is the number of Sentinel-2 spectral bands: visible (B1–B4), red-edge (B5–B7), near-infrared (B8, B8A, B9), cirrus (B10), and shortwave infrared (B11, B12).

2.2 Preprocessing

Sequence Standardization

Raw sequences varied from 71 to 147 observations due to cloud cover. All sequences were standardized to $T = 70$ timesteps through zero-padding (for shorter sequences) or truncation to early-season observations (for longer ones). Missing values were linearly interpolated to maintain temporal continuity.

Spectral Features

We used the 13 original Sentinel-2 spectral bands as input features without deriving vegetation indices (NDVI, EVI, NDWI). This allows the LSTM to learn spectral-temporal relationships directly from raw reflectance data.

Class Imbalance Mitigation

To address class imbalance, we applied two strategies: (i) inverse-frequency class weighting in the cross-entropy loss ($w_c = N/n_c$, where n_c is the sample count for class c), and (ii) random oversampling of minority classes to ensure at least 5% representation per epoch. This combination was chosen for its simplicity and computational efficiency over alternatives like SMOTE.

2.3 Model Architecture

We adapt EarlyRNN [4], a dual-head LSTM with adaptive stopping. Input sequences of shape 70 × 13 (time steps × spectral bands) are first normalized across features and then projected into a 64-dimensional hidden space through a linear transformation. A two-layer LSTM with 64 hidden units per layer processes the sequence to capture both short and long-term temporal dependencies.

The network has two parallel output heads. The classification head applies a softmax activation to produce class probabilities at each time step, while the stopping head outputs a sigmoid-valued probability of halting. During inference, the model monitors the stopping probability and issues its prediction once this exceeds a learned threshold, allowing predictions to adaptively occur at different points in the growing season.

2.4 Baselines

We compare EarlyRNN against baselines spanning traditional machine learning to state-of-the-art deep architectures.

Simple Baselines

- LightGBM: Gradient boosting on flattened sequences (70 × 13 features).
- MLP: Two-layer feedforward network (512 and 256 hidden units) with ReLU activation and dropout.
- CNN-1D: Temporal convolutions with two blocks (convolution, pooling, dropout) followed by global average pooling.

Deep Temporal Baselines

- FCN [5]: Fully convolutional network for time series.
- InceptionFCN [6]: Multi-scale temporal feature extraction via inception modules.
- LITE [7]: Lightweight temporal encoder with attention.
- LITE Ensemble: Majority voting across five independently trained LITE models.

2.5 Evaluation Protocol

A key methodological consideration concerns temporal information availability. EarlyRNN was trained and evaluated on sequences truncated to $T = 70$ timesteps, corresponding to partial-season observations typically available by late May or early June. In contrast, all baseline models (LightGBM, MLP, CNN-1D, FCN, InceptionFCN, LITE, and LITE ensemble) were trained and evaluated using complete time series extending through harvest (up to 147 observations per parcel).

This asymmetry is intentional and reflects the operational objective of early classification. We assess whether competitive accuracy can be achieved with substantially less temporal information. The comparison evaluates the accuracy-earliness trade-off rather than architectural superiority under equivalent conditions. A model achieving 86% accuracy with 70 timesteps versus 89% with 147 timesteps demonstrates that early prediction is feasible, where the modest accuracy reduction is the operational cost of months of predictive lead time.

Performance was assessed using: (i) overall accuracy and Cohen's κ, (ii) macro and weighted F1-scores, (iii) per-class precision-recall, (iv) earliness metrics (fraction observed and stopping DOY), and (v) prediction confidence. Bootstrap resampling (1,000 replicates) provided confidence intervals.

3 Results and Discussion

We evaluate model performance on both datasets and analyze the adaptive stopping mechanism's behavior. Subsections 3.2–3.4 focus on BreizhCrops due to its larger size (122,614 parcels) and more challenging characteristics, enabling robust statistical inference and stringent generalization testing.

3.1 Overall Performance and Baseline Comparison

Table 1 reports overall accuracy (OA), Cohen's κ, and F1 scores across both benchmark datasets. EarlyRNN achieves 86% overall accuracy on BavarianCrops and 80% on BreizhCrops, with balanced macro-F1 scores of 0.77 and 0.70, respectively. These results demonstrate competitive performance against traditional baselines (LightGBM, MLP, CNN-1D) while enabling adaptive early prediction through the dual-head stopping mechanism.

Table 1. Overall performance of EarlyRNN on both benchmark datasets (mean ± std over 3 runs).

Model	Dataset	OA	κ	Macro-F1	Weighted-F1
EarlyRNN	BavarianCrops	0.860 ± 0.003	0.77 ± 0.01	0.77 ± 0.01	0.86 ± 0.00
	BreizhCrops	0.800 ± 0.004	0.71 ± 0.01	0.70 ± 0.01	0.79 ± 0.01

The 6% point performance gap between datasets reflects BreizhCrops' more challenging characteristics: higher class imbalance (55% Meadow vs. 40%), smaller mean field sizes, and frequent cloud cover due to the maritime climate. This cross-region evaluation demonstrates model robustness across diverse agro-climatic conditions, though performance degradation under severe imbalance remains a limitation addressed in Sect. 3.4 (Fig. 1).

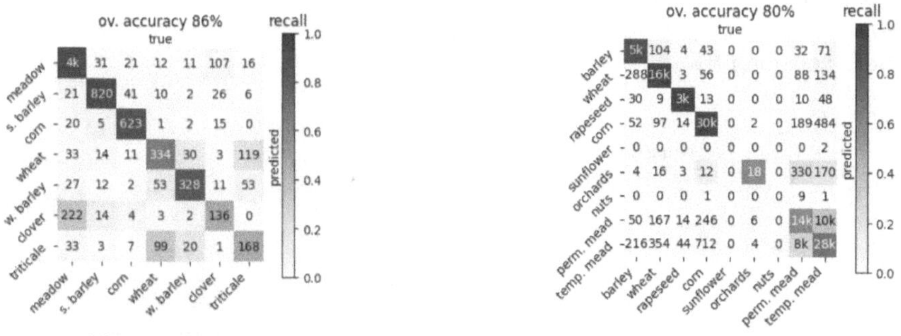

(a) BavarianCrops (b) BreizhCrops

Fig. 1. Recall-normalized confusion matrices showing classification patterns. Errors concentrate among winter cereals with similar phenology; Meadow, Corn, and Sunflower exhibit strong separability.

Table 2 compares EarlyRNN against representative state-of-the-art methods on BavarianCrops, including fully convolutional networks (FCN, InceptionFCN) and lightweight temporal encoders (LITE). The ensemble of five LITE models achieves the highest overall accuracy (89% OA, 0.79 macro-F1), followed by individual InceptionFCN and tuned LITE models (both 88% OA). EarlyRNN attains 86% overall accuracy with 0.74 macro-F1.

Table 2. State-of-the-art comparison on BavarianCrops. Baseline models use complete time series (147 observations); EarlyRNN uses T=70 for early prediction.

Method	Macro-F1	Weighted-F1	OA
FCN [5]	0.66	0.81	0.80
InceptionFCN [6]	0.78	0.88	0.88
LITE [7]	0.73	0.86	0.86
LITE (tuned)	0.77	0.88	0.88
LITE Ensemble (5 models)	**0.79**	**0.89**	**0.89**
EarlyRNN (ours, T = 70)	0.74	0.87	0.86

As described in Sect. 2.5, EarlyRNN achieves competitive performance using partial-season data (T = 70), demonstrating that the modest accuracy reduction is operationally justified by months of lead time for agricultural planning.

3.2 Adaptive Stopping Behavior

Figure 2 shows stopping time distributions for both datasets. BavarianCrops exhibits mean DOY 144 $\pm$ 65.6 with substantial variability, while BreizhCrops shows mean DOY 152 $\pm$ 48.3 with tighter synchronization due to maritime climate constraints.

(a) BavarianCrops (b) BreizhCrops

Fig. 2. Stopping DOY distributions for both datasets. Predictions typically occur in late May to early July.

Per-class analysis on BreizhCrops (Fig. 3) reveals systematic patterns driven by phenological distinctiveness. Sunflower, Barley, and Wheat exhibit tight stopping distributions (median DOY 142–160), while Meadows and Orchards show wide distributions (IQR > 50 days) due to difficulty distinguishing perennial vegetation without extended observations. Corn displays intermediate variability (IQR $\approx$ 40 days) from diverse regional planting dates.

Importantly, per-class accuracy does not correlate with stopping time (Spearman $\rho = -0.23, p = 0.08$). This indicates the model stops early for phenologically distinctive crops without sacrificing accuracy, demonstrating adaptive temporal attention where stopping decisions reflect crop-specific identification difficulty.

3.3 Confidence-Earliness Relationship

To assess prediction reliability across stopping times, we analyzed the relationship between stopping DOY and classification confidence (maximum softmax probability). A moderate negative correlation emerges (Pearson $r = -0.326$, $p < 0.001$), indicating that earlier predictions exhibit higher confidence (Fig. 4).

Stratified analysis reveals the following:

- Early predictions (DOY $\leq$ 142): mean confidence 0.780, accuracy 91.9%
- Late predictions (DOY $\geq$ 180): mean confidence 0.637, accuracy 72.1%

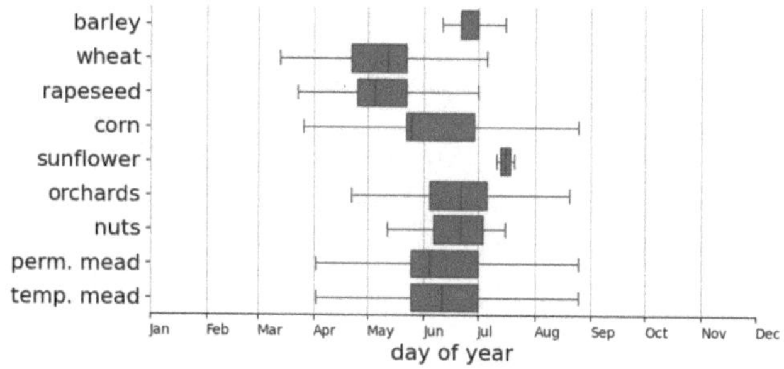

Fig. 3. Per-class stopping DOY distributions (BreizhCrops). Early-identifiable crops (Sunflower, Wheat, Barley) trigger predictions earlier than ambiguous classes (Meadows, Orchards).

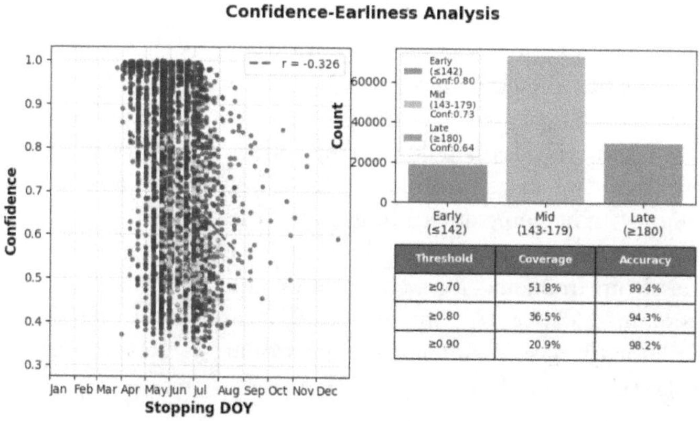

Fig. 4. Confidence-earliness relationship (BreizhCrops). Left: Scatter plot showing negative correlation ($r = -0.326$) between stopping DOY and prediction confidence. Top right: Distribution of predictions across temporal categories. Bottom right: Confidence threshold trade-offs for operational filtering.

This pattern suggests the adaptive stopping mechanism has learned to trigger early predictions for phenologically distinctive crops (e.g., Sunflower, Wheat) where high confidence aligns with high accuracy. Conversely, the model waits longer for spectrally ambiguous classes (e.g., Meadows), though extended observation does not fully resolve classification difficulty for inherently challenging crops.

For operational deployment, confidence-based filtering enables risk calibration (Table 3). At a 0.85 threshold, 28.7% of parcels receive high-confidence predictions (96.5% accuracy) at mean DOY 146, while lower thresholds trade accuracy for coverage.

Table 3. Comprehensive confidence threshold analysis (BreizhCrops). Higher confidence thresholds yield lower coverage but substantially higher accuracy and earlier predictions, enabling risk-calibrated operational filtering.

Confidence ≥	Coverage	Accuracy	Mean DOY
0.70	51.8%	89.4%	152
0.75	44.0%	91.9%	150
0.80	36.5%	94.3%	148
0.85	**28.7%**	**96.5%**	**146**
0.90	20.9%	98.2%	143
0.95	14.2%	98.9%	139

3.4 Cross-Region Generalization Analysis

The 6-point accuracy gap reflects three interacting factors. First, BreizhCrops' more severe class imbalance creates stronger majority-class bias despite mitigation strategies (Sect. 2), with the dominant Meadow class inherently challenging due to lack of structured phenology, amplifying errors. Second, smaller field sizes increase mixed-pixel contamination from boundaries and hedgerows while exhibiting higher management variability (planting dates, irrigation), reducing within-class spectral consistency. Third, maritime climate produces sparser, noisier time series from frequent cloud cover and atmospheric variability. Weighted-F1 (0.79) remained closer to Bavaria's (0.86) than macro-F1 (0.70 vs. 0.77), confirming minority classes suffer disproportionately. Nevertheless, 80% accuracy demonstrates reasonable generalization.

4 Final Remarks

Adaptive temporal modeling enables reliable early crop classification months before harvest, reaching 86% (BavarianCrops) and 80% (BreizhCrops) while using <50% of full-season temporal data. A dual-head LSTM learned crop-specific phenologies-stopping early for distinctive classes and waiting for ambiguous ones producing an emergent temporal-attention effect (confidence–earliness $r = -0.326$).

The 3-percentage-point accuracy reduction approach versus full-season (86% vs 89%) is an acceptable operational trade-off: months of lead time for procurement, hedging, and policy decisions outweigh small end-season gains. Confidence-based filtering lets stakeholders tune coverage–accuracy risk.

Despite its limitations (uses raw spectral bands only, requires multiple horizons to build accuracy–earliness curves, its geographic scope is temperate Europe), our approach has demonstrated that early crop classification is operationally viable and provides actionable lead-time intelligence. As future work, continental scalability and SAR fusion are to be explored.

Declarations

This work was supported by the College of Computing (UM6P) and the SDAS Research Group, with no external funding. The authors declare no conflict of interest. Ethics approval, consent to participate, and consent for publication are not applicable. The BavarianCrops and BreizhCrops datasets are publicly available via the ElectS repository. The code is based on the open source ElectS framework, with additional scripts available upon request. Author contributions: S. Oumoussa (conceptualization, experiments, draft preparation) and D. Peluffo (methodology, review and editing).

References

1. Adhinata, F.D., Sumiharto, R.: A comprehensive survey on weed and crop classification using machine learning and deep learning. Artif. Intell. Agric. (2024)
2. Bouguettaya, A., Zarzour, H., Kechida, A., Taberkit, A.M.: Deep learning techniques to classify agricultural crops through UAV imagery: a review. Neural Comput. Appl., 1–26 (2022)
3. Shahid, M.F., Khanzada, T.J., Aslam, M.A., Hussain, S., Baowidan, S.A., Ashari, R.B.: An ensemble deep learning models approach using image analysis for cotton crop classification in AI-enabled smart agriculture. Plant Methods **20**(1), 104 (2024)
4. Rußwurm, M., Courty, N., Emonet, R., Lefévre, S., Tuia, D., Tavenard, R.: Endtoend learned early classification of time series for in-season crop type mapping. ISPRS J. Photogramm. Remote. Sens. **196**, 445–456 (2023)
5. Fawaz, H.I., Forestier, G., Weber, J., Idoumghar, L., Muller, P.-A.: Time series classification from scratch with deep neural networks: a strong baseline. In: Proceedings of the International Joint Conference on Neural Networks (IJCNN), pp. 1578–1584 (2017)
6. Ismail Fawaz, H., Forestier, G., Weber, J., Idoumghar, L., Muller, P.-A.: Time series classification with inceptionFCN. In: Proceedings of the International Joint Conference on Neural Networks (IJCNN), pp. 1–7 (2019)
7. Fawaz, H.I., et al.: Look into the lite in deep learning for time series classification. arXiv preprint arXiv:2409.02869 (2024)

Markov Chain for Behavior Prediction in Road Safety

Mohammed Laouadi(✉), Omar El Beggar, and Adil El Bouziri

LIM Laboratory, Faculty of Sciences and Techniques of Mohammedia, University Hassan II, BP 146, 28806 Mohammedia, Morocco
mohammedlaouadi5@gmail.com, {omar.elbeggar,adil.elbouziri}@fstm.ac.ma

Abstract. Human behavior, the primary factor in most road accidents, is inherently aleatory and resists deterministic prediction. This paper presents a *Markov chain* framework integrated with the *Gamified Transtheoretical Model (GTTM)* for probabilistic behavior prediction in road safety. The proposed approach leverages transition matrices to model behavioral evolution through five distinct stages while incorporating GTTM's fundamental rules including time constraints, progression thresholds, and regression conditions. By formalizing these psychological principles into a structured Markov matrix, we enable weekly probabilistic predictions of user behavior evolution. The framework serves as a computational testbed for intervention design, allowing researchers to simulate behavioral trajectories and optimize road safety strategies through matrix-based multi-step predictions. Simulation results demonstrate the model's capability to capture complex behavioral dynamics and predict long-term intervention outcomes.

Keywords: Markov Chain · Behavioral Prediction · Road Safety · Gamified Transtheoretical Model · Probabilistic Modeling · Transition Matrix

1 Introduction

Road traffic accidents remain a critical global challenge, with human behavior identified as the primary contributing factor in approximately 90% of incidents [1]. The complex interplay between psychological readiness, environmental factors, and behavioral consistency necessitates advanced modeling approaches that can capture the inherent uncertainty of human actions in road environments.

Traditional road safety interventions have predominantly focused on infrastructural and legislative measures, often overlooking the dynamic nature of human behavior. The fundamental challenge lies in the aleatory nature of human actions—characterized by inherent randomness and influenced by numerous psychological and contextual variables. This stochastic reality necessitates a

O. El Beggar—Contributing author.

A. E. Hassanien et al. (Eds.): AISDS 2025, LNDECT 287, pp. 98–106, 2027.
https://doi.org/10.1007/978-3-032-18927-1_10

paradigm shift from deterministic approaches to probabilistic modeling frameworks capable of capturing behavioral evolution over time.

Markov chains provide the mathematical foundation for modeling such stochastic processes, enabling prediction of future states based on current conditions through transition probability matrices [2]. This paper integrates *Markov chain* theory with the *Gamified Transtheoretical Model (GTTM)*, creating a robust framework that respects the psychological principles of behavioral change while providing mathematical rigor for prediction.

Our approach focuses on leveraging the structural properties of transition matrices to encode GTTM rules and compute multi-step behavioral predictions. This integration bridges psychological theory with computational mathematics, enabling effective analysis of intervention strategies before real-world implementation through sophisticated simulation techniques.

The remainder of this paper is organized as follows. Section 2 presents the theoretical background and related work. Section 3 details our methodological framework for integrating *Markov chains* with the GTTM. Section 4 describes the simulation process and experimental analysis. Finally, we discuss our findings and conclude with directions for future work.

2 Background and Theoretical Framework

To effectively influence behavior, an intervention model must be both psychologically sound and engaging. The *GTTM* is a composite framework designed to meet these criteria by integrating two powerful concepts.

First, its psychological foundation is the *Transtheoretical Model (TTM)*, a cornerstone of health behavior change. The *TTM* postulates that individuals move through five distinct stages of readiness: *Precontemplation (PC)*, *Contemplation (C)*, *Preparation (P)*, *Action (A)*, and *Maintenance (M)* [3]. Its application to transport and road safety provides a structured map for interventions [4], but it does not inherently prescribe the mechanisms for transitioning between stages. Recent studies have demonstrated the effectiveness of TTM-based approaches in promoting sustainable transportation behaviors [5].

Second, its motivational engine is Gamification, the use of game design elements in non-game contexts [6]. Gamification has emerged as a powerful tool to foster engagement and has shown promise in promoting safer driving behaviors [7,8]. Research indicates that gamification services can significantly enhance user motivation and participation in behavior change programs [9].

The *GTTM* uses the *TTM* stages as a roadmap and applies tailored gamification mechanics at each stage to motivate users to progress. Its logic is divided into two distinct phases. The initial "*Off-Road*" phase (*PC*, *C*, *P*) is a protected educational environment where users can only progress by reaching score thresholds within a time limit; regression is disabled. The subsequent "*On-Road*" phase (*A*, *M*) incorporates real-world behavior, exposing the user to regression risk. In the *Action* stage, progression to *Maintenance* is possible, but so is regression to any preceding stage, triggered by an accumulation of unsafe driving events (e.g., measured by Time-to-Zebra/Time-to-Collision [10]). The *Maintenance* stage is the final goal, where progression is no longer possible.

2.1 Markov Chain Foundations for Behavioral Prediction

Markov chains model systems that transition between discrete states with probabilities dependent only on the current state [2]. They are widely used in health, education, and transportation to simulate stage-based processes [11,12]. In road safety, they have modeled route choice, mode switching, and driver state transitions [13]. A key challenge, especially in novel applications, is the estimation of the transition probability matrix. When empirical data is sparse or non-existent, as in our case, modelers must derive probabilities from mechanistic models or theoretical principles [14]. Our work contributes to this literature by proposing a method to formalize transition probabilities directly from the logical rules of a psychological model, filling a critical gap between qualitative behavioral theory and quantitative, simulatable models.

As illustrated in Fig. 1, our proposed framework operates through a two-level architecture that integrates user interactions with probabilistic behavioral predictions.

3 Methodological Framework

3.1 Markov Chain Formalization and GTTM Rules Integration

We define the five distinct stages of the *GTTM* as the states of our Markov Chain: *Precontemplation (PC)*, *Contemplation (C)*, *Preparation (P)*, *Action (A)*, and *Maintenance (M)*. The state space is denoted as $S = \{\text{PC, C, P, A, M}\}$, representing the complete behavioral change continuum from initial awareness to sustained safe behavior.

The transition probability matrix P structurally encodes the *GTTM* rules, constructed as follows:

$$P = \begin{bmatrix} p_{\text{PC}\to\text{PC}} & p_{\text{PC}\to\text{C}} & 0 & 0 & 0 \\ 0 & p_{\text{C}\to\text{C}} & p_{\text{C}\to\text{P}} & 0 & 0 \\ 0 & 0 & p_{\text{P}\to\text{P}} & p_{\text{P}\to\text{A}} & 0 \\ p_{\text{A}\to\text{PC}} & p_{\text{A}\to\text{C}} & p_{\text{A}\to\text{P}} & p_{\text{A}\to\text{A}} & p_{\text{A}\to\text{M}} \\ p_{\text{M}\to\text{PC}} & p_{\text{M}\to\text{G}} & p_{\text{M}\to\text{P}} & 0 & p_{\text{M}\to\text{M}} \end{bmatrix} \tag{1}$$

The matrix structure embodies the core *GTTM* principles through systematic constraint implementation:

Off-Road **Stages:** The upper triangular block (*PC*, *C*, *P*) enforces sequential forward progression only, reflecting the educational building-block approach where users must master foundational concepts before advancing. The zero elements represent psychologically impossible transitions, such as regression during early learning stages or skipping developmental steps.

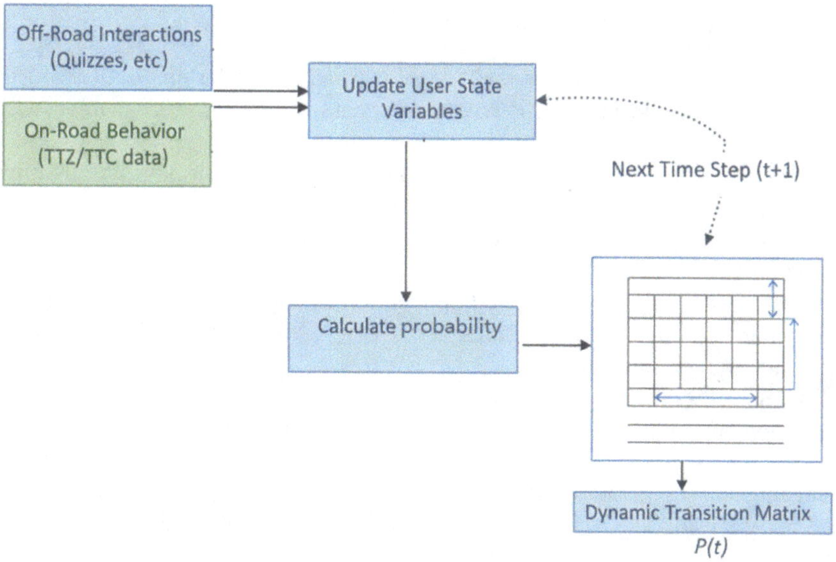

Fig. 1. Architecture of the GTTM-Markov Model. At the first level, user interactions update state variables based on *Off-Road* and *On-Road* activities. At the second level, these updated variables inform the calculation of weekly transition probabilities between TTM stages.

***On-Road* Stages:** The *Action* and *Maintenance* rows incorporate bidirectional transitions, allowing regression based on real-world behavioral performance. The specific regression pathways reflect severity-graded consequences, with more significant behavioral lapses triggering deeper regression to earlier stages.

Gamification Modulation: All transition probabilities are influenced by engagement factors, with gamification elements amplifying progression tendencies and providing regression resistance for consistently engaged users.

The transitions occur over defined weekly time steps, with each transition representing user movement between stages based on their engagement with gamified elements and demonstrated behavioral performance.

3.2 Probability Calculation and Prediction

The probability calculation framework operates through a two-level architecture where micro-level behavioral updates inform macro-level transition computations. Probabilities are not static. They update weekly based on:

- Current stage duration (time pressure intensifies near deadlines),
- Score relative to threshold (higher scores increase forward transition likelihood),
- Recent unsafe events (weighted more heavily than older ones),

- Gamification engagement (e.g., badge collection, streaks).

The n-step transition matrix $P^{(n)} = P^n$ provides probabilistic distributions across future time horizons, allowing researchers to answer critical intervention questions.

This forecasting capability enables proactive intervention planning by identifying future risk periods and optimal intervention timing. Researchers can simulate how different parameter settings affect long-term behavioral outcomes, optimizing time constraints, progression thresholds, and gamification intensity before real-world implementation. The framework supports cohort-level analysis by tracking population distributions across stages over time, identifying bottlenecks and progression patterns.

The model serves as a computational testbed for intervention strategy evaluation, allowing comparison of alternative approaches through simulation rather than costly real-world trials. This predictive capability represents a significant advancement over traditional post-hoc analysis methods, enabling evidence-based intervention design grounded in both psychological theory and mathematical rigor.

4 Simulation Process and Experimental Analysis

4.1 Simulation Process

The simulation operates through a discrete-time iterative process where each cycle (weekly iteration) updates user states based on the Markov transition matrix. At each temporal iteration, the model systematically updates the user state components (e.g., scores, time spent in current stages), then calculates the corresponding transition probabilities. These updates follow a rigorous methodology derived from qualitative and quantitative behavioral analysis.

The simulation process incorporates multiple behavioral dimensions: observed overall performance, time spent in the current stage, action consistency, and adherence to temporal criteria (e.g., six-month minimum for *Precontemplation*). Each stage associates with a potential score that reflects behavior, serving as the foundation for transition evaluations between stages.

To illustrate the simulation mechanics, consider the following exemplary transition matrix derived from *GTTM* rules and behavioral patterns:

$$P_{\text{example}} = \begin{bmatrix} 0.85 & 0.15 & 0 & 0 & 0 \\ 0 & 0.80 & 0.20 & 0 & 0 \\ 0 & 0 & 0.75 & 0.25 & 0 \\ 0.10 & 0.15 & 0.20 & 0.40 & 0.15 \\ 0.05 & 0 & 0.10 & 0.15 & 0.70 \end{bmatrix} \tag{2}$$

In this matrix representation:

- $p_{11} = 0.85$: High probability of remaining in *Precontemplation*, reflecting initial resistance to change

- $p_{12} = 0.15$: Moderate progression probability to *Contemplation* for engaged users
- $p_{45} = 0.15$: *Action* to *Maintenance* transition probability for consistent performers
- $p_{41} = 0.10$: Regression probability to *Precontemplation* for severe behavioral lapses
- $p_{55} = 0.70$: High persistence in *Maintenance* for successful behavior internalization

These probabilities are not arbitrary but emerge from systematic analysis of behavioral quality, temporal consistency, and threshold adherence. The matrix serves as the computational engine for multi-step predictions through repeated matrix multiplication.

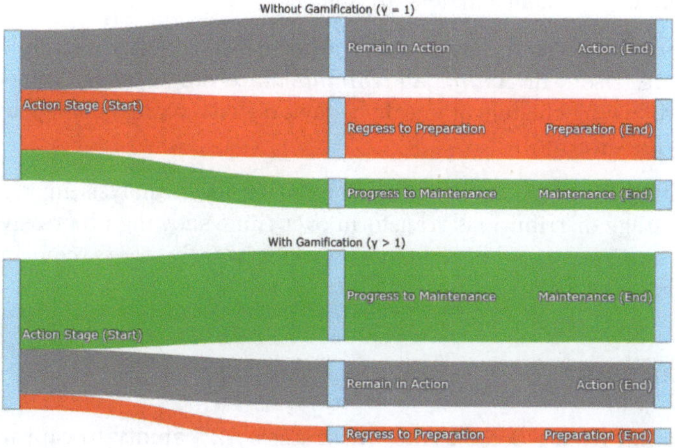

Fig. 2. Sankey diagram illustrating cohort flow through GTTM stages over one year, highlighting progression pathways, regression loops, and timeout recycling effects.

4.2 Prediction and Behavioral Trajectories

The predictive power of the framework emerges through matrix operations that project behavioral distributions across future time horizons. The n-step transition matrix $P^{(n)} = P^n$ enables forecasting of stage distributions after n weeks, providing valuable insights for intervention planning.

For the example matrix P_{example}, the 4-week projection $P^{(4)}$ reveals emerging behavioral patterns:

$$P_{\text{example}}^{(4)} = \begin{bmatrix} 0.522 & 0.285 & 0.120 & 0.055 & 0.018 \\ 0.008 & 0.410 & 0.328 & 0.185 & 0.069 \\ 0.015 & 0.024 & 0.316 & 0.391 & 0.254 \\ 0.092 & 0.128 & 0.185 & 0.312 & 0.283 \\ 0.064 & 0.085 & 0.142 & 0.248 & 0.461 \end{bmatrix} \quad (3)$$

This projection demonstrates how initial stage distributions evolve over time, with significant accumulation in later stages for engaged users and persistence in early stages for struggling users.

The simulation reveals several key behavioral patterns emerging from *GTTM* rule integration:

Progression Trajectories: Users demonstrate characteristic progression patterns based on engagement levels. Engaged learners show steady advancement through stages with minimal regression, while struggling users exhibit oscillation between stages and frequent timeout events.

Timeout Impact Analysis: The enforcement of time constraints significantly affects user progression. Approximately 27% of simulated users experience at least one timeout reset, particularly in *Precontemplation* and *Contemplation* stages where engagement barriers are highest.

Regression Pathways: The simulation identifies common regression triggers and patterns. Users in *Action* and *Maintenance* stages show vulnerability to regression following clusters of unsafe behaviors, with regression depth correlated with behavior severity.

Cohort Distribution Evolution: Population-level analysis in Fig. 2 reveals how initial stage distributions transform over time, showing progressive accumulation in *Maintenance* and persistent subgroups in early stages requiring targeted interventions.

5 Discussion

The simulation results demonstrate the framework's ability to capture complex behavioral dynamics inherent in road safety interventions. The integration of *GTTM* rules within the Markov structure provides several significant advantages for intervention design and analysis.

5.1 Behavioral Dynamics and Intervention Insights

The simulation reveals that behavioral change follows nonlinear patterns with critical transition points. Users often experience rapid progression once they overcome initial barriers in *Precontemplation* and *Contemplation* stages. However, the transition from *Action* to *Maintenance* presents a significant challenge, with many users oscillating between these stages before achieving stability.

The timeout mechanism, while necessary for ensuring stage mastery, creates distinct behavioral subgroups. Persistent users who eventually progress after resets demonstrate resilience and learning, while users repeatedly timing out may require alternative intervention approaches or modified time constraints.

The framework demonstrates strong predictive capabilities for behavioral evolution, with particular accuracy in identifying timeout risks 4–6 weeks in advance. The matrix-based approach enables efficient scenario analysis, allowing researchers to evaluate intervention strategies without costly real-world implementations.

The simulation accurately reproduces key behavioral phenomena observed in real-world interventions, including the S-shaped progression curve, regression clustering following stressful periods, and the positive impact of consistent engagement on long-term outcomes.

5.2 Practical Applications and Intervention Optimization

The framework serves as a computational testbed for optimizing *GTTM* parameters. Simulation analysis suggests that adaptive time constraints based on user engagement patterns could reduce timeout rates by 18–25% without compromising learning outcomes. Similarly, personalized progression thresholds based on initial assessment scores may improve overall success rates.

The multi-step prediction capability enables proactive intervention planning by identifying future risk periods and optimal intervention timing. This forward-looking approach represents a significant advancement over traditional post-hoc analysis methods.

6 Conclusion and Future Work

This research presents a comprehensive integration of *Markov chain* theory with the *Gamified Transtheoretical Model*, creating a robust framework for behavioral prediction in road safety contexts. The simulation framework demonstrates how psychological principles can be formalized into computational structures that capture the complexity of behavioral change processes.

The key contributions include the development of a structured transition matrix that encodes *GTTM* rules, the implementation of a temporal iteration process that updates behavioral probabilities based on performance metrics, and the demonstration of multi-step prediction capabilities through matrix operations.

The simulation results provide valuable insights for intervention design, highlighting the importance of time constraints, the impact of engagement on progression patterns, and the identification of critical transition points in the behavior change process.

For future work, we identify several promising directions:

- Integration with real-world intervention data for model calibration and validation
- Development of personalized transition matrices based on user characteristics
- Application to other behavioral domains beyond road safety

The Markov-GTTM integration represents a significant step toward more effective, evidence-based behavioral interventions, providing a rigorous computational foundation for understanding and predicting complex human behavior in safety-critical contexts.

References

1. Singh, S.: Critical reasons for crashes investigated in the National Motor Vehicle Crash Causation Survey. National Highway Traffic Safety Administration (2015)
2. Norris, J.R.: Markov Chains. Cambridge University Press (1998)
3. Prochaska, J.O., DiClemente, C.C.: Stages and processes of self-change of smoking: toward an integrative model of change. J. Consult. Clin. Psychol. **51**(3), 390 (1983)
4. Batool, T., et al.: Promoting sustainable transportation: a transtheoretical examination of active transport modes. Sustainability **16**(2), 472 (2023)
5. Friman, M., Huck, J., Olsson, L.E.: Transtheoretical model of change during travel behavior interventions: an integrative review. Int. J. Environ. Res. Public Health **14**(6), 581 (2017)
6. Deterding, S., Dixon, D., Khaled, R., Nacke, L.: From game design elements to gamefulness: defining gamification. In: Proceedings of the 15th International Academic MindTrek Conference, pp. 9–15 (2011)
7. Kirchner, K., Signer, B.: EcoGO: a gamified mobile application for encouraging eco-friendly driving behavior. In: CHI Conference Extended Abstracts, pp. 1779–1786 (2017)
8. Alonso, F.: DriveSafe: a gamified application for improving driving behavior. In: Advances in Human Factors and Systems Interaction, pp. 3–14. Springer (2020)
9. Hamari, J., Koivisto, J.: Why do people use gamification services? Int. J. Inf. Manage. **35**(4), 419–431 (2015)
10. Varhelyi, A.: Drivers speed behaviour at a zebra crossing: a case study. Accid. Anal. Prev. **30**(6), 731–743 (1998)
11. Ma, J., Chan, W., Tilley, B.C.: Continuous time Markov chain approaches for analysing transtheoretical models. Stat. Methods Med. Res. **27**(2), 593–607 (2018)
12. Johnson, M., Thompson, K.: Predictive modeling of driver behavior using stochastic processes. Transport. Res. F Traffic Psychol. Behav. **88**, 45–62 (2022)
13. Kitamura, R.: A Markov chain Monte Carlo approach to dynamic models of activity and travel behavior. Transp. Res. Part B Methodol. **32**(6), 387–400 (1998)
14. Mizutani, D., Lethanh, N., Adey, B.T., Kaito, K.: Improving the estimation of Markov transition probabilities using mechanistic-empirical models. Front. Built Environ. **3**, 58 (2017)

Sentiment Analysis and Text Mining

Enhancing Arabic Sentiment Analysis with Contextual Embeddings and Class Imbalance Handling Using Recurrent Neural Networks

Hamza Jakha[1(✉)], Souad El Houssaini[1], Mohammed-Alamine El Houssaini[2], and Souad Ajjaj[3]

[1] ELIITS Laboratory, Department of Computer Science, Faculty of Sciences, Chouaib Doukkali University, El Jadida, Morocco
{Jakha.h,elhoussaini.s}@ucd.ac.ma

[2] Higher School of Education and Training, Chouaib Doukkali University, El Jadida, Morocco
elhoussaini.m@ucd.ac.ma

[3] Department of Computer Science, Faculty of Sciences, Ibn Tofail University, Kenitra, Morocco
souad.ajjaj@uit.ac.ma

Abstract. The rapid digitalization of applications across various domains has generated vast amounts of user-generated content, making sentiment analysis essential, particularly in business intelligence for improving marketing strategies and decision-making. Building a high-performance sentiment analysis model requires careful attention to each stage of the process, as well as addressing the challenges encountered along the way. In this study, we focus on binary sentiment classification using two unbalanced benchmark datasets. The first step involves feature extraction, where we compare two transformer-based contextual embedding techniques, AraBERT and AraELECTRA. We then address the issue of class imbalance by evaluating three scenarios: the original unprocessed dataset, a version with class weighting, and a balanced version generated using the SMOTE oversampling technique applied to the training set, in order to assess the impact of balancing on model performance. For classification, we employ three recurrent neural architectures including RNN, LSTM, and GRU, known for their effectiveness in processing textual data. Evaluation is conducted using multiple metrics, including accuracy, recall, f1-score, cohen's kappa, and matthews correlation coefficient, along with runtime and embedding matrix size. Results show that combining contextual embeddings with SMOTE and recurrent networks achieves superior performance, reaching 97.84% accuracy for the HTL dataset and 93.10% for the RES dataset. Comparisons with existing studies on the same benchmarks demonstrate the effectiveness of the proposed approach.

Keywords: Arabic sentiment analysis · Contextual word embeddings · Imbalanced datasets · Recurrent neural networks · Business intelligence

A. E. Hassanien et al. (Eds.): AISDS 2025, LNDECT 287, pp. 109–119, 2027.
https://doi.org/10.1007/978-3-032-18927-1_11

1 Introduction

Large volumes of data are generated and circulated daily across social networks, websites, and online forums. This content, which may take textual, audio, or visual form, plays a central role in natural language processing (NLP) applications [1]. However, before use, this data requires preprocessing to ensure quality and usability. Sentiment analysis, a major NLP task, identifies and interprets opinions or emotions toward products, services, or topics, proving vital for business intelligence by guiding communication and marketing decisions [2]. This study focuses on the hotel and restaurant sector, emphasizing the Arabic language, which spoken by about 334 million people and marked by rich morphology which demands specialized processing distinct from widely studied languages like English.

The sentiment analysis process, from data collection to model evaluation, has evolved through multiple techniques and algorithms at each stage. In particular, the feature extraction phase has progressed from traditional statistical approaches, such as bag-of-words (BoW) and term frequency inverse document frequency (TF-IDF), to distributed word representations such as Word2Vec, global vectors for word representation (GloVe), and FastText, and more recently to contextual models based on transformers, including BERT, GPT and others pre-trained models, which capture both semantic and syntactic relationships between words and phrases. Another major challenge in sentiment analysis is class imbalance, which can bias predictions toward the majority class and reduce model reliability. Classifier selection is also crucial, requiring alignment between the algorithm, data type, and feature representation. Deep learning models, especially recurrent neural networks, are well suited for transformer-based embeddings and sequential data. However, in Arabic sentiment analysis, many approaches still inadequately address class imbalance and lack systematic evaluation combining contextual embeddings with advanced recurrent architectures.

Research on Arabic sentiment analysis has evolved from traditional feature-based approaches with machine learning classifiers [3] to word embedding techniques such as Word2Vec and FastText combined with deep learning models [4]. More recently, pre-trained transformers like Arabic-BERT and its variants have shown superior performance, achieving up to 94.34% accuracy on large Twitter datasets [5] and f1-scores above 88% in aspect-based sentiment analysis tasks [6]. To address class imbalance, ensemble methods with SMOTE [7] and class-weighting strategies have further improved robustness, confirming the effectiveness of contextual embeddings and balancing techniques in Arabic sentiment classification.

In light of these limitations, this study aims to evaluates two contextual embedding models, AraBERT and AraELECTRA, integrated with recurrent architectures such as RNN, LSTM, and GRU, while examining the impact of advanced class imbalance handling techniques on Arabic sentiment analysis. Three dataset versions are used: the original unbalanced, class-weighted, and SMOTE-balanced versions. Experiments are conducted on two benchmark Arabic sentiment datasets, HTL and RES datasets.

The paper is structured as follows: Sect. 2 reviews related work, Sect. 3 describes the methodology, Sect. 4 presents evaluation metrics, experiments, and results, and Sect. 5 concludes with future directions.

2 Methodology

The sentiment analysis process follows a standardized pipeline, where each stage critically influences model performance as shown in Fig. 1.

2.1 Datasets

In this study, we investigate the performance of AraBERT and AraELECTRA on two Arabic sentiment analysis datasets [8], both collected from hotel and restaurant review platforms.

Hotel reviews dataset (HTL): this dataset was scraped from TripAdvisor and includes user reviews from various hotels. It contains over 15,000 reviews, originally categorized into three sentiment labels: positive, negative, and neutral. Since the majority of the reviews fall into the positive and negative categories, we retained only these two classes to focus on binary sentiment classification. This binary setup is especially relevant for improving decision-making in the business intelligence domain.

Restaurant reviews dataset (RES): this dataset combines reviews from TripAdvisor and Qaym websites, comprising more than 3,000 user reviews spanning over 4,500 restaurants. As with the HTL dataset, the majority of the reviews fall into the positive and negative categories. To maintain consistency and improve the clarity of our comparison, we retained only these two sentiment classes and removed the neutral ones.

A preprocessing step was applied to both the HTL and RES datasets before proceeding to feature extraction. In this phase, several cleaning techniques were employed to enhance the quality of the textual data. We began by removing Arabic diacritics (Tashkil), elongation characters (Tatweel), and repeated letters. To normalize Arabic orthography while preserving meaning, we standardized certain letters, for example, converting "ى" and "ئ" to "ى". Additional preprocessing steps included the removal of Arabic punctuation, URLs, and HTML tags. We also eliminated stop words while deliberately retaining negation words and particles, as these elements are critical in determining sentiment polarity. Although Transformer-based models are typically trained on raw text, we opted to clean our Arabic corpora to reduce noise while avoiding techniques such as stemming or lemmatization. This decision was made to preserve the original morphological form of words, which is often important in Arabic sentiment analysis.

2.2 Data Representation

In this section, we introduce two advanced contextual word embedding techniques: AraBERT and AraELECTRA.

2.2.1 AraBERT

Based on the BERT architecture, AraBERT [9] is a transformer model pre-trained on over 70 million Arabic sentences from diverse domains and dialects. It mirrors the BERT-base structure with 12 encoder layers, 12 attention heads, a 768-dimensional hidden size, and around 110 million parameters. AraBERT generates contextual embeddings by combining token, position, and segment embeddings, capturing both semantic meaning and sentence order. Using the Masked Language Modeling (MLM) objective, it randomly masks 15% of tokens and predicts them from bidirectional context, allowing the model to learn rich semantic representations of Arabic text.

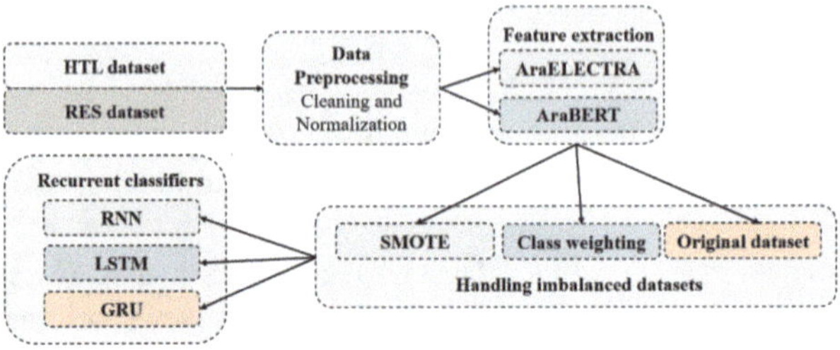

Fig. 1. The proposed methodology

2.2.2 AraELECTRA

AraELECTRA introduced by [10] is a Transformer-based language model adapted for the Arabic language and built upon the ELECTRA architecture. It was trained on the same Arabic corpora used for AraBERT, totaling approximately 77 GB of data from various domains. AraELECTRA follows the same core structural configuration as AraBERT in terms of Transformer components. Unlike AraBERT, which uses the MLM objective, AraELECTRA employs the Replaced Token Detection (RTD) objective with two components: a generator and a discriminator. The generator, a smaller MLM network, predicts and replaces masked tokens to create corrupted input, while the discriminator identifies whether each token is original or replaced, improving data efficiency and representation quality.

In AraBERT, only 15% of the tokens are masked and used in training loss, meaning that 85% of the tokens do not contribute to the learning process. In contrast, AraELECTRA evaluates every token in the input sequence during training, allowing all tokens to contribute to the model's optimization.

In this study to reduce computational cost and prevent overfitting on small datasets, the transformer embeddings were kept frozen during RNN training.

2.3 Resampling and Weighting Techniques

This section presents the techniques used to handle class imbalance, including the original dataset and versions balanced using SMOTE and class weighting.

2.3.1 SMOTE

Synthetic minority over-sampling technique (SMOTE) [11] is a widely used method for handling class imbalance. Instead of duplicating samples, it generates new minority instances by interpolating between neighboring points in the feature space, enriching class distribution and reducing learning bias. In this study, SMOTE is applied during preprocessing, immediately after feature extraction, and only on the training set.

2.3.2 Class Weighting

Class weighting [12] is applied during training to assign greater importance to the minority class by adjusting the loss function, ensuring that errors on underrepresented samples have a stronger influence on optimization. For example, in a dataset with 90% positive and 10% negative samples, the model may ignore the minority class; increasing its weight encourages better learning and improves performance on imbalanced data.

2.4 Classifiers Baselines

In this part, we define the recurrent neural network models used in our study.

2.4.1 Recurrent Neural Network

RNN is a deep learning model designed to process sequential data such as text, speech, or time series [13]. Unlike feedforward networks, it incorporates memory through recurrent connections to retain information from previous steps and learn temporal patterns. However, it struggles with long-term dependencies due to the vanishing gradient problem, limiting its ability to preserve information over long sequences.

2.4.2 Long Short-Term Memory

LSTM introduced by [14], is a variant of RNN architecture was developed to address the vanishing gradient problem that hinders long-term dependency learning in sequential data. It introduces a memory cell and gating mechanisms to control information flow: the input gate adds new information, the forget gate removes irrelevant past data, and the output gate determines what to pass as the hidden state, allowing effective preservation of long-term context.

2.4.3 Gated Recurrent Unit

GRU [15] another variant of RNN architecture and a simplified version of the LSTM. GRU enhances computational efficiency by using fewer parameters while retaining the ability to model long-term dependencies and mitigate the vanishing gradient problem. It employs two gates: the update gate, which controls how much of the previous state is retained, and the reset gate, which determines how much past information to discard.

3 Results and Discussion

In this study, we evaluate the performance of AraBERT and AraELECTRA embeddings when combined with RNN, LSTM, and GRU. The experimental setup involves comparing these recurrent models across three different resampling strategies.

3.1 Evaluation Metrics

We use multiple metrics to evaluate model performance and compare embeddings, while examining the effect of resampling on class balance. The metrics include accuracy, f1-score, cohen's kappa, and matthews correlation coefficient (MCC), with a focus on recall of the negative (minority) class to assess balancing techniques. We also measure execution time and embedding matrix size to evaluate computational efficiency.

3.2 Experiments Setup

The experiments were conducted on Google Colab using Python 3.11.13, with a T4 GPU for embedding generation and a TPU v2-8 for model training. The dataset was split into 80% training including 10% validation and 20% testing, using the Adam optimizer with a learning rate of 0.0005, batch size of 32, and 10 epochs.

4 Results and Discussion

This section presents a comprehensive analysis of the performance of each model in combination with its corresponding embedding technique, either AraBERT or AraELECTRA, across two Arabic datasets. For each dataset, considering different resampling strategies.

Table 1 presents the results of AraELECTRA combined with LSTM, GRU, and RNN, applied on HTL and RES datasets in their three versions.

For the HTL dataset, the GRU–AraELECTRA–SMOTE configuration achieves the highest scores, with an accuracy of 96.82%, an f1-score of 96.78%, cohen's kappa of 93.64%, and an MCC of 93.67%, indicating its strong ability to capture patterns in the balanced data. Similarly, the same model obtains the best results on the RES dataset, with an Accuracy of 90.75%, an f1-score of 90.84%, a kappa of 81.51%, and an MCC of 81.52%. In contrast, the original dataset generally yields lower kappa and MCC values, particularly for the RNN, which exhibits limited capability in learning from imbalanced data. The class weighting technique provides moderate improvements over the original dataset but does not reach the performance gains achieved with SMOTE.

The results presented in Table 2 concern the AraBERT embedding technique combined with LSTM, GRU, and RNN classifiers. The GRU–AraBERT–SMOTE configuration delivers the highest performance across all metrics for both datasets. For the HTL dataset, this model achieves an accuracy of 97.84%, a f1-score of 97.83%, cohen's kappa of 95.68%, and a MCC of 95.69%. Similarly, for the RES dataset, it reaches an accuracy of 93.06%, an f1-score of 93.10%, a kappa of 86.11%, and an MCC of 86.12%, demonstrating its robustness across different datasets. Conversely, the original dataset records lower performance compared to the two resampling and weighting techniques, particularly with the RNN classifier, which shows marked limitations compared to LSTM and GRU across all evaluation metrics. The results presented in Tables 1 and 2 show that combining AraBERT and AraELECTRA embeddings with the SMOTE technique significantly enhances recurrent classifiers performance across all metrics and datasets. AraBERT slightly outperforms on the HTL dataset, with the GRU model achieving

97.84% accuracy compared to 96.82% for AraELECTRA, while results on RES remain close to 93.06% vs. 90.75%. These results confirm that both embeddings effectively capture the semantic and contextual nuances of Arabic text. the comparable effectiveness of AraBERT and AraELECTRA, combined with the performance gains from SMOTE and advanced recurrent architectures, demonstrates that the integration of contextual embeddings with effective

balancing strategies such as SMOTE is a key factor for achieving high accuracy binary sentiment classification in Arabic language.

Table 1. Model performance with AraELECTRA across original, SMOTE-balanced, and class-weighted datasets

HTL dataset					RES dataset			
Recurrent classifiers	Acc	F1	K	MCC	Acc	F1	K	MCC
Original dataset								
LSTM	0.9423	0.9637	0.8227	0.8235	0.8613	0.9083	0.6236	0.6240
GRU	0.9382	0.9616	0.8035	0.8035	0.8519	0.9068	0.5517	0.5730
RNN	0.8916	0.9310	0.6788	0.6822	0.7688	0.8634	0.1709	0.2356
SMOTE technique								
LSTM	0.9559	0.9548	0.9118	0.9130	0.9044	0.9064	0.8088	0.8096
GRU	0.9682	0.9678	0.9364	0.9367	0.9075	0.9084	0.8151	0.8152
RNN	0.9371	0.9360	0.8742	0.8748	0.7905	0.7662	0.5809	0.5939
Class weight								
LSTM	0.9326	0.9578	0.7901	0.7903	0.8277	0.8790	0.5827	0.5928
GRU	0.9378	0.9610	0.8082	0.8088	0.8239	0.8745	0.5844	0.6006
RNN	0.8950	0.9326	0.6962	0.7029	0.7142	0.7771	0.4091	0.4581

Table 2. Model performance with AraBERT across original, SMOTE-balanced, and class-weighted datasets

HTL dataset					RES dataset			
Recurrent classifiers	Acc	F1	K	MCC	Acc	F1	K	MCC
Original dataset								
LSTM	0.9508	0.9696	0.8405	0.8414	0.8697	0.9115	0.6647	0.6664
GRU	0.9460	0.9661	0.8342	0.8350	0.8762	0.9173	0.6713	0.6713
RNN	0.9270	0.9558	0.7473	0.7571	0.8491	0.9035	0.5606	0.5703
SMOTE technique								
LSTM	0.9773	0.9773	0.9545	0.9545	0.9237	0.9256	0.8474	0.8486
GRU	0.9784	0.9783	0.9568	0.9569	0.9306	0.9310	0.8611	0.8612

(continued)

Table 2. (*continued*)

HTL dataset					RES dataset			
Recurrent classifiers	Acc	F1	K	MCC	Acc	F1	K	MCC
RNN	0.9582	0.9575	0.9165	0.9170	0.8904	0.8848	0.7808	0.7845
Class weight								
LSTM	0.9490	0.9680	0.8420	0.8424	0.8113	0.8601	0.5812	0.6171
GRU	0.9423	0.9635	0.8254	0.8273	0.8678	0.9107	0.6566	0.6575
RNN	0.9259	0.9532	0.7752	0.7768	0.8234	0.8734	0.5872	0.6059

To assess the effect of resampling, we examined recall for the minority class in both HTL and RES datasets as shown in Figs. 2 and 3. With AraBERT embeddings, recall in the original versions was relatively low (70% for HTL and 57% for RES with RNN), but improved with class weighting, reaching 87% for RNN and remaining above 90% for GRU and LSTM in HTL, while in RES the RNN attained 86%. Applying SMOTE produced the best results, with recall reaching 95% for RNN and 98% for both GRU and LSTM in HTL, and over 90% for all classifiers in RES. With AraELECTRA, recall was also weak in the original sets, particularly for RNN in RES (15%), though GRU and LSTM remained above 83% in HTL. Class weighting brought moderate gains, but SMOTE again delivered the highest recall, with GRU and LSTM at 98% in HTL and all models exceeding 88% in RES.

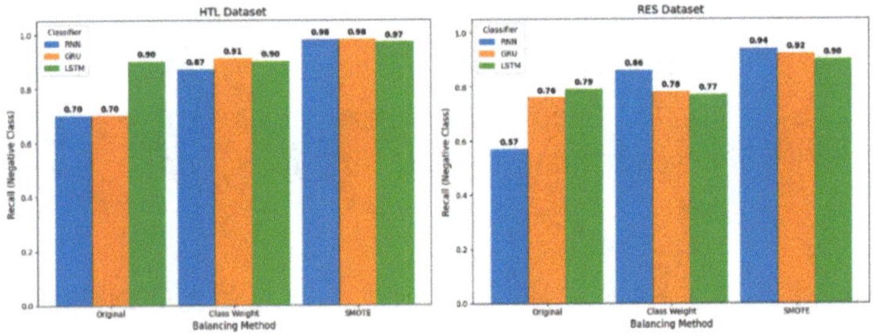

Fig. 2. AraBERT Negative Class Recall by Balancing Methods and Classifier

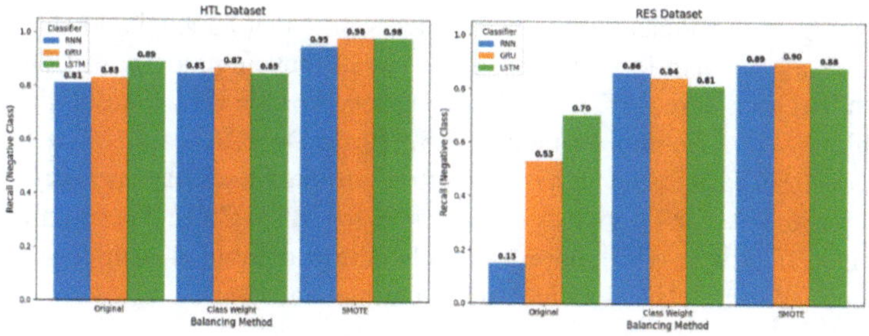

Fig. 3. AraELECTRA Negative Class Recall by Balancing Methods and Classifier

Overall, these findings confirm that without balancing, models tend to overlook the minority class in favor of the majority. Although class weighting slightly improves negative class recall, SMOTE consistently outperforms across datasets, classifiers, and embeddings, proving more effective in detecting minority class instances

Table 3 provides an additional evaluation of the two embedding techniques used in this study, comparing execution time and embedding size across the HTL and RES datasets. The experiments were conducted on the Google Colab platform using a GPU T4 processor. Both AraBERT and AraELECTRA produce large embedding matrices exceeding 1.5 GB, reflecting the computational complexity of transformer-based architectures. For the HTL dataset, AraELECTRA generated embeddings in 2310.80 s, outperforming AraBERT by 596 s. In contrast, for the RES dataset, AraBERT achieved faster processing with a time advantage of 191 s over AraELECTRA. These findings suggest that while AraELECTRA tends to be more efficient for larger datasets, performance differences may vary depending on dataset size and characteristics.

Comparing our approach with other related works that adopted a binary sentiment classification, as shown in Table 4, the results indicate that our model consistently outperforms both baselines across HTL and RES datasets. For the HTL dataset, the proposed model achieves a relative improvement of 3.52% over DeepASA and 3.98% over MCE-CNN. For the RES dataset, the gain is even more substantial, with an accuracy of 93.10% compared to 86.84% for DeepASA and 84.08% for MCE-CNN. These improvements highlight the effectiveness of contextual embeddings combined with class imbalance handling and recurrent classifiers in Arabic sentiment analysis.

Table 3. Execution time and embedding size comparison of AraBERT and AraELECTRA across HTL and RES datasets

Datasets	Environment	Architecture	Dataset Size	AraBERT embeddings		AraELECTRA embeddings	
				Time of execution	Embedding size	Time of execution	Embedding size
HTL	Google Colab	GPU T4	13422	2907.74 s	1.92 Go	2310.80 s	1.92 Go
RES	Google Colab	GPU T4	10705	547.15 s	1.53 Go	738.73 s	1.53 Go

Table 4. Comparison with state-of-the-art approaches on HTL and RES Datasets

Ref	Method	Data Presentation	Dataset	Results
[16]	Deep learning for Arabic Sentiment Analysis using GRU and LSTM	FastText Skip-gram	HTL RES	Accuracy = 94.32% Accuracy = 86.84%
[17]	Multi-Channel Embedding Convolutional Neural Network	Word2Vec + GloVe+CBOW	HTL RES	Accuracy = 93.86% Accuracy = 84.08%
proposed model	GRU-AraBERT-SMOTE	AraBERT	HTL RES	Accuracy = 97.84% Accuracy = 93.10%

5 Conclusion

Several limitations in the sentiment analysis process, such as the traditional feature extraction, reliance on conventional machine learning classifiers, and inadequate dataset balancing, make this task challenging. This study addressed these issues by comparing

AraBERT and AraELECTRA as contextual embedding that preserve semantic relationships, combined with deep learning classifiers such as RNN, GRU, and LSTM, which are well-suited for processing sequential data. Experiments on two benchmark datasets, HTL and RES, each prepared in three versions. The combination of AraBERT with GRU and SMOTE achieved the best performance (97.84%) on HTL and 93.10% on RES. These results confirm that combining contextual embeddings, effective balancing and recurrent networks improves Arabic sentiment classification. Future work could extend this approach to larger datasets, investigate the use of GPT models for either feature extraction or data augmentation, explore GPT models for feature extraction or data augmentation, develop hybrid deep learning–lexicon models, and evaluate end-to-end fine-tuning for improved contextual representation.

References

1. Jakha, H.: Optimizing sentiment analysis in multilingual balanced datasets: a new comparative approach to enhancing feature extraction performance with ML and DL classifiers. ASI. **8**, 104 (2025)
2. Jakha, H., El Houssaini, S., El Houssaini, M.-A., Ajjaj, S.: Graph embeddings vs. contextual embeddings: an efficient approach to Arabic sentiment analysis for improving decision making in business intelligence. In: 2025 5th International Conference on Innovative Research in Applied Science, Engineering and Technology, pp. 1–8. IEEE, Fez (2025)
3. Hadwan, M.: Arabic sentiment analysis of users' opinions of governmental mobile applications. Comput. Mater. Continua. **72**, 4675–4689 (2022)
4. Abdelhady, N.: Stacked-CNN-BiLSTM-COVID: an effective stacked ensemble deep learning framework for sentiment analysis of Arabic COVID-19 tweets. J. Cloud Comp. **13**, 85 (2024)
5. Mourad, A.: A large-scale analysis of COVID-19 tweets in the Arab region. Soc. Netw. Anal. Min. **12**, 71 (2022)
6. Bensoltane, R.: Towards Arabic aspect-based sentiment analysis: a transfer learning-based approach. Soc. Netw. Anal. Min. **12**, 7 (2022)
7. Habbat, N.: Sentiment analysis of imbalanced datasets using BERT and ensemble stacking for deep learning. Eng. Appl. Artif. Intell. **126**, 106999 (2023)
8. ElSahar, H., El-Beltagy, S.R.: Building large Arabic multi-domain resources for sentiment analysis. In: Gelbukh, A. (ed.) Computational Linguistics and Intelligent Text Processing, vol. 9042, pp. 23–34. Springer International Publishing, Cham (2015)
9. Antoun, W., Baly, F., Hajj, H.: AraBERT: transformer-based model for Arabic language understanding. In: Proceedings of the 4th Workshop on Open-Source Arabic Corpora and Processing Tools, with a Shared Task on Offensive Language Detection, pp. 9–15 (2020)
10. Antoun, W., Baly, F., Hajj, H.: AraELECTRA: pre-training text discriminators for Arabic language understanding. Preprint at https://doi.org/10.48550/arXiv.2012.15516 (2021).
11. Chawla, N.: SMOTE: synthetic minority over-sampling technique. jair. **16**, 321–357 (2002)
12. Bakirarar, B.: Class weighting technique to deal with imbalanced class problem in machine learning: methodological research. Turkiye Klinikleri J Biostat. **15**, 19–29 (2023)
13. Elman, J.L.: Finding structure in time. Cogn. Sci. **14**, 179–211 (1990)
14. Hochreiter, S.: Long short-term memory. Neural Comput. **9**, 1735–1780 (1997)
15. Chung, J: Empirical evaluation of gated recurrent neural networks on sequence modeling. Preprint at https://doi.org/10.48550/arXiv.1412.3555 (2014).
16. Alharbi, A.: Arabic sentiment analysis using deep learning and ensemble methods. Arab. J. Sci. Eng. **46**, 8913–8923 (2021)
17. Dahou, A.: Multi-Channel embedding convolutional neural network model for Arabic sentiment classification. ACM Trans. Asian Low-Resour. Lang. Inf. Process. **18**, 1–23 (2019)

Sentiment Analysis of Patient-Generated Social Media Data on Chronic Diseases: A Systematic Exploration of Applications

Khadija Arradi(✉), Khadija Tlemçani, Kebira Azbeg, and Said Jai Andaloussi

Laboratory of Data Engineering and Intelligent Systems (LIDSI), Department of Mathematics and Computer Sciences, Faculty of Sciences Ain Chock, Hassan II University, Casablanca, Morocco
{khadija.arradi-etu,khadija.tlemcani-etu}@etu.univh2c.ma, {Kebira.azbeg, Said.jaiandaloussi}@univh2c.ma

Abstract. Sentiment Analysis, an influential technique in Natural Language Processing, offers valuable insights into emotions expressed in Patient-Generated Content shared across online platforms. Exploring this data provides opportunities to extract knowledge that is relevant for informed decision-making in healthcare. In this research work, we synthesized a total of 23 studies into a comprehensive comparison table. Sentiment analysis demonstrates important promise for healthcare stakeholders in improving the quality of life for individuals with chronic diseases. The findings show that Sentiment Analysis is mainly used to detect adverse drug reactions, monitor emotional well-being, and identify patients' daily challenges. Yet, most studies focus on diabetes, with limited exploration of other chronic diseases; particularly in regions lacking patient feedback systems. The scarcity of studies from Arab and African regions shows the difficulties faced when applying Sentiment mining to low-resource languages. This work explores the applications, limitations and future trends of applying SA to social media content in the context of chronic disease management.

Keywords: Sentiment Analysis · Social Media data · AI in healthcare · Chronic Diseases · Patient Generated Content

1 Introduction

In 1943, Warren McCulloch and Walter Pitts introduced the first mathematical model of a neural network, the "McCulloch-Pitts neuron" [1]. Inspired by the base biological structure of a "Cortical Neuron", they showed that a network of these simple, binary-threshold units could perform any logical or arithmetic function. This work was a cornerstone in the early birth of Artificial Intelligence (AI), as it provided a theoretical base modeling and simulating thoughts through numerical computation. Their work paved the way for the development of modern neural networks and AI. Nowadays, AI has emerged

A. E. Hassanien et al. (Eds.): AISDS 2025, LNDECT 287, pp. 120–128, 2027.
https://doi.org/10.1007/978-3-032-18927-1_12

as a transformative power across all global sectors, including finance, transportation, manufacturing, human resource management, and entertainment [2]. While AI's critical role in Smart Cities, such as reducing traffic accidents through road analysis which is crucial for public safety, its application in healthcare is even more vital and impactful. Beyond streamlining redundant tasks, the rapid adoption of AI in healthcare highlights its revolutionary potential in saving lives. Healthcare service delivery is being fundamentally transformed through the adoption of AI technologies: Machine Learning, Natural Language Processing, Deep Learning , and Robotic Process Automation which are promoting innovation in both clinical practices and administrative operations [3]. In the era of technological shifts, digital platforms and social media have driven substantial growth in unstructured textual data. Platforms such as Reddit, Quora, Facebook, X (formerly Twitter), Instagram, and LinkedIn generate massive volumes of user-generated content, often containing critical insights within chronic diseases patient communities. Therefore, unlocking this knowledge requires the deployment of advanced Natural Language Processing (NLP) techniques and algorithms. Since the 2000s, due to the rise in computing power resources and the availability of large datasets, language models have become efficient, leading to significant performance improvements. Sentiment analysis (SA) is among the most adopted NLP approaches, designed to identify affective states, mine subjective opinions, and evaluate judgments conveyed through textual data. Beyond clinical records, SA is equally applied to diverse sources of Patient-Generated Content (PGC), refers to information that patients share on digital platforms, thus enriching our understanding of the patient's dimension in medical care. Current literature lacks reviews on applying Sentiment Analysis to chronic disease patients' social media posts. Filling this gap is crucial as these analyses can uncover patients' everyday struggles, leading to key insights that can improve their quality of life. This systematic exploration explores the applicability of SA methodologies in extracting clinically meaningful insights from online PGCs. We underline their potential to advance emotion-conscious healthcare, highlighting how these computational methods can uncover unmet needs, detect emotional distress, and identify NCDs concerns that may otherwise be hidden in physical clinical appointments. By integrating SA into healthcare analytics, clinicians and researchers can reach a more understanding of patient experiences across chronic and infectious disease contexts, thereby supporting precision care and improved healthcare delivery.

2 Related Works

Several systematic reviews have examined sentiment analysis applied to user-generated content, particularly from social media, yet most remain limited to broad scopes and technical SA approaches. For instance, some reviews focused on vaccine hesitancy during the Coronavirus pandemic [4], while others concentrated on methodological aspects, prioritizing computational techniques over healthcare applications [5]. Broader investigations have surveyed SA within public health contexts [6], whereas few domain-specific efforts, such as in cardiology

[7], explored techniques for analyzing patient generated content across different platforms and data sources. Although conducted from distinct perspectives, these reviews consistently reported critical gaps, including insufficient handling of multimodal content (e.g., emojis, images, videos) [4]; limited multilingual and cross-cultural coverage [4,6]; weak integration of explainable and privacy-preserving AI and a lack of scalable methods [4,6]; poor model generalization across contexts and inadequate support for low resource languages [6]; imprecise treatment of contextuality, subjectivity, and polarity/neutrality definition particularly in domain-specific settings such as cardiology [7]; and insufficient ethical and regulatory guidance for the use of public online data [6,7]. Taken together, these gaps point to the limited attention given in prior reviews to the healthcare applications of SA. In particular, the application of SA to social media data for capturing the lived experiences of patients with chronic diseases remains insufficiently investigated. This work seeks to address this gap by systematically reviewing healthcare applications of SA in relation to chronic conditions that have not previously been the subject of dedicated reviews, thereby offering a contribution to the existing body of knowledge. (see Table 1)

Table 1. Selected existing reviews related to the proposed work

Study	Year	Scope of Review	Healthcare Domain	SM	Other Data Sources	Short search Query
[4]	2025	public debate on Covid19 vaccines	vaccination	yes	None	Social Media and vaccination and Covid19
[5]	2025	SA in social media using big data	not focused on healthcare	yes	None	not specified
[6]	2025	SA in public health	public health	yes	health sentiment datasets, literature, survey data, news media, clinical data, and wearable data	SA and Public Health and LLMs or Policy Making
[7]	2022	SA technique in cardiology	cardiology	yes	news media, blogs, SMS, and app reviews	SA and Cardiovascular diseases or Coronary disease

3 Methodology

This review was conducted following systematic review practices inspired by the PRISMA guidelines. Although no formal protocol was registered, clear eligibility criteria were defined, restricting the scope to peer-reviewed journal articles published in English. No time restriction was applied to the search, as research in this thematic area remains limited. Conference proceedings, review papers, and studies not directly addressing social media or sentiment analysis in healthcare were excluded. The primary sources of information were Scopus and Web of Science (WoS), where a Boolean-based search strategy was applied using combinations of keywords: "Cognitive Computing," "NLP", "AI", "SA" and "SM data". The search retrieved 29 records from Scopus and 40 from WoS. After deduplication and a three-phase screening and given that queries related to chronic diseases yielded no relevant results, the final dataset from these databases was refined to 10 journal articles. To complement this corpus, an AI-assisted search was conducted using *Liner*, which identified 13 additional relevant publications not captured by Scopus or WoS. Data extraction focused on key study characteristics (authors, publication year, methodology, datasets ...). The overall selection and exclusion process is summarized in a PRISMA-inspired flow diagram to ensure transparency and reproducibility(see Fig. 1), also no formal quality appraisal or bias assessment was conducted; as this review aimed to provide an exploratory overview of existing studies.

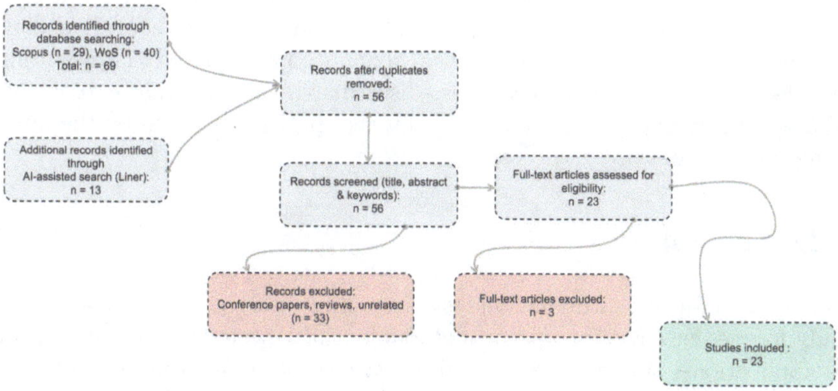

Fig. 1. PRISMA-inspired Flow Diagram of Study Selection Process

4 Previous Application of Sentiment Analysis

Within the in-depth exploration, non-communicable diseases constitute the primary focus of SA applications on patient-generated social media data. According to the World Health Organization (WHO) states that genetics, physiology, environment, and behavior all interact in complex ways to cause NCDs, and they encompass a wide spectrum of chronic conditions. In this body of work, diabetes stands out as the most extensively studied chronic disease, with studies analyzing Twitter and Reddit data to assess patient concerns and public attitudes, for instance, [8] explored general sentiment in diabetes tweets, in this study,

[9] researchers applied an aspect-level approach, while [10,11] provided region-specific insights into Moroccan diabetic populations, and [12] Reddit study captured changes in type 2 diabetes management during the coronavirus pandemic. Renal NCDs have received less attention: [13] used Zhihu posts to identify risk factors associated with kidney stones, whereas [14] examined Twitter/X discussions among dialysis patients to highlight care burdens in end stage kidney disease (EKD). Earlier studies have explored other long-term diseases; [15] investigated Crohn's disease patient experiences across Facebook and Twitter, while [16,17] applied sentiment and thematic analyses to cancer-related social media content. Pediatric Multiple long-term conditions (MLTCs) are rarely addressed, with [18] demonstrating how SA can uncover the emotional challenges of kids with MLTCs. Mental health disorders, which WHO also recognizes as chronic NCDs given their relapsing and persistent nature, have been studied through the lens of social media: [19] employed explainable AI to interpret depressive and suicidal behaviors, while the Boamente project [20] advanced digital phenotyping for continuous monitoring of suicidal ideation. Beyond these, sentiment analysis has also been applied to palliative care [21], maternal health [22], and vaccine hesitancy during the pandemic [23], reflecting how online discourse captures patient perceptions and health system challenges during global crises (see Table 2). Collectively, these applications demonstrate that while diabetes research has achieved relative maturity, other NCDS like Crohn's disease, cancer, pediatric comorbidities, and renal chronic diseases remain underexplored. This gap highlights the need for more balanced, cross-disease investigations that exploit social media sentiment to capture the lived realities of patients affected by the burden of long terms diseases.

5 Discussion

Previous reviews of sentiment analysis on online user-generated content have mainly focused on broad scopes or methodological advances rather than healthcare applications. While some focused on vaccine hesitancy during coronavirus pandemic [4] or technical architectures [5], few examined domain-specific healthcare contexts [6,7]. Commonly reported gaps include limited multimodal analysis, poor multilingual and cross-cultural coverage, lack of support for low-resource languages, and insufficient ethical guidance. The use of SA on social media to mine and capture the lived experiences of patients with chronic diseases remains underexplored. This research work addresses this gap by concentrating on chronic conditions not previously covered in dedicated reviews. The literature is dominated by diabetes [8–12,24], while other NCDs such as Crohn's disease [15], cancer [16,17], and renal chronic diseases [13,14] remain overlooked. Research is especially scarce in Arab and African regions, raising questions about data availability and institutional support. Another contextual observation concerns the Covid19 pandemic: several studies collected data during this period, focusing on vaccine hesitancy [4], maternal health [22], or diabetes management [12]; however, these remain short term analyses and lack longitudinal evidence

Table 2. Overview of studies applying sentiment analysis to SM data related to NCDs

Study	Year	NCD focus	Pandemic context	Analytical Focus	Social Network(s)
[19]	2024	Mental health disorders (possibly chronic)	No	–Text classification for mental health detection –XAI techniques –Topic analysis –Sentiment recognition (implicit in topics)	Not specified
[8]	2019	Diabetes (T1 and T2)	pre-covid19	–Sentiment analysis software	Twitter
[10]	2024	Diabetes	No	–Text classification: imbalanced sentiment analysis –Large imbalanced tweet datasets	Twitter
[24]	2022	Diabetes	No	–Machine learning-based SA –Gaussian distribution for estimating the probability of developing diabetes	No (PIDD dataset)
[9]	2017	Diabetes	pre-covid19	–Aspect-level SA –N-gram methods –They used ontologies	Twitter
[11]	2024	Diabetes	No	–SA of tweets	Twitter
[25]	2023	Chronic health conditions	No	–Stress-Level Classification	No
[13]	2025	Urinary disease (Kidney stones)	No	–Prompting LLMs for identifying risk factors	Zhihu (a Chinese website)
[14]	2025	End-stage kidney disease (dialysis)	No	–Statistical SA –Topic modeling	X (formerly Twitter)
[15]	2017	Crohn's disease	pre-covid	–Sentiment analysis –Ethnographic review	Facebook, Twitter
[16]	2023	Laryngeal cancer	No	–Sentiment analysis –Thematic content analysis	Twitter
[17]	2023	Cancer	No	–Sentiment analysis –Topic modelling	Tweets, blogs, communities
[18]	2025	Pediatric MLTCs	post-covid19	–Sentiment analysis –Emotion categorization	Care Opinion (platform)
[21]	2025	Palliative care	Yes (data were generated during the pandemic)	–Clustering comments for identifying themes –Sentiment analysis	Chinese social media
[12]	2025	Diabetes (T2/chronic)	Yes	–Text classification –Topic modelling –Sentiment analysis	Reddit
[20]	2022	Mental health conditions (possibly chronic)	Yes (data was generated during the pandemic)	–Detection of suicidal ideation from tweets	Twitter

of how disruptions caused by pandemic shaped chronic disease experiences over time. Finally, the existence of patient feedback platforms such as Care Opinion (UK, Australia and Canada) and Haodaifu (China) [26] can facilitate data collection, encourage more opinion-mining studies on PGC, and allow systematic and reproducible research. In contrast, countries such as Morocco lack similar infrastructures, which restricts data curation and limits research opportunities. This presents both a challenge and an opportunity. Most of the reviewed studies (see Table 2) relied mostly on Twitter (now X) [8–11,14–17,20], with fewer works used platforms such as Facebook (Meta) [15], Reddit [12], blogs [17], Care Opinion [18], or Zhihu [13], and one study employed the Pima Indians Diabetes Database (PIDD), a structured numerical dataset rather than a social network, where sentiment analysis was applied in the context of diabetes diagnosis [24]. This dominance of Twitter underscores both its easy accessibility and the limited exploration of alternative or local social platforms.

5.1 Future Directions

Future research works should also combine SA with topic modeling to capture both polarity and thematic concerns such as treatment commitment and medication use. Multimodal analysis remain largely overlooked but could improve insights into patient behavior and experience. Plus, most pandemic-era studies provide only short term evidence, emphasizing the need for longitudinal perspectives. The creation of feedback platforms in underserved regions could further enable systematic, privacy-preserving research.

5.2 Limitations

This review excluded conference papers, which may have resulted in the omission of relevant contributions. Furthermore, the evidence base remains limited by the lack of patient-feedback infrastructures in many LMICs, including Arab and African contexts. This absence of dedicated platforms restricts systematic data curation and hinders reproducible research. Another limitation arises from the scarcity of longitudinal studies; most Covid19 related works reviewed are short-term analyses, offering only limited insights into how disruptions, caused by pandemic, affected chronic disease patients experiences over time.

6 Conclusion

This research reviewed studies applying SA to patient-generated SM data on chronic non-communicable diseases. While diabetes dominates the current literature, other chronic conditions remain underexplored, particularly in Low- and Middle-Income Countries and environments where patient feedback infrastructures are absent. The scarcity of studies from the Arab and African regions also reflects the broader challenge of working with low-resource languages, where suitable datasets and tools remain limited. Across the reviewed works, SA is often combined with topic modeling, yet advanced directions such as multimodality and federated learning are still insufficiently developed. This work provides a base for extending disease coverage to diverse chronic conditions.

References

1. Piccinini, G.: The first computational theory of mind and brain: a close look at Mcculloch and Pitts's "logical calculus of ideas immanent in nervous activity." Synthese **141**(2), 175–215 (2004)
2. Padmaja, C., Narayana, S., Anga, G., Bhansali, P.: The rise of artificial intelligence: a concise review. IAES Int. J. Artif. Intell **13**, 2224–2233 (2024)
3. Haddiya, I., Ramdani, S.: Artificial intelligence in healthcare: a focus on the best practices. In: ITM Web of Conferences, vol. 69, p. 02010 (2024). EDP Sciences
4. Langbecker, A., Catalan-Matamoros, D.: The public debate of COVID-19 vaccines on social media: a systematic review. Heliyon **11**(13) (2025)
5. Patil, S.S., Suryawanshi, V.P., Patil, S.M., Girase, S.P., Bhagat, D.A.: Review of sentiment analysis in social media using big data: techniques, tools, and frameworks
6. Villanueva-Miranda, I., Xie, Y., Xiao, G.: Sentiment analysis in public health: a systematic review of the current state, challenges, and future directions. Front. Public Health **13**, 1609749 (2025)
7. Brezulianu, A., Burlacu, A., Popa, I.V., Arif, M., Geman, O.: "Not by our feeling, but by other's seeing": sentiment analysis technique in cardiology–an exploratory review. Front. Public Health **10**, 880207 (2022)

8. Gabarron, E., Dorronzoro, E., Rivera-Romero, O., Wynn, R.: Diabetes on twitter: a sentiment analysis. J. Diabetes Sci. Technol. **13**(3), 439–444 (2019)
9. Salas-Zárate, M.d.P., Medina-Moreira, J., Lagos-Ortiz, K., Luna-Aveiga, H., Rodriguez-Garcia, M.A., Valencia-Garcia, R.: Sentiment analysis on tweets about diabetes: an aspect-level approach. Comput. Math. Methods Med. **2017**(1), 5140631 (2017)
10. Mohammed, R., et al.: Twitter-sentiment analysis of Moroccan diabetic: a comparison study. In: International Conference on Big Data and Internet of Things, pp. 880–896. Springer (2024)
11. Roudani, M., Elkari, B., El Moutaouakil, K., Ourabah, L., Hicham, B., Chellak, S.: Twitter-sentiment analysis of Moroccan diabetic using fuzzy c-means smote and deep neural network. Math. Model. Comput. **11**, 835–847 (2024)
12. Nagpal, M., Jalali, N., Sherifali, D., Morita, P., Cafazzo, J.A.: Investigating reddit data on type 2 diabetes management during the COVID-19 pandemic using latent Dirichlet allocation topic modeling and valence aware dictionary for sentiment reasoning analysis: Content analysis. JMIR Form. Res. **9**, 51154 (2025)
13. Mao, C., Li, J., Pang, P.C.-I., Zhu, Q., Chen, R.: Identifying kidney stone risk factors through patient experiences with a large language model: text analysis and empirical study. J. Med. Internet Res. **27**, 66365 (2025)
14. Shankar, R., Xu, Q., Bundele, A.: Patient voices in dialysis care: sentiment analysis and topic modeling study of social media discourse. J. Med. Internet Res. **27**, 70128 (2025)
15. Roccetti, M., et al.: Attitudes of Crohn's disease patients: infodemiology case study and sentiment analysis of Facebook and twitter posts. JMIR Public Health Surveill. **3**(3), 7004 (2017)
16. Rao, D., Singh, R., Prakashini, K., Vijayananda, J.: Investigating public sentiment on laryngeal cancer in 2022 using machine learning. Indian J. Otolaryngol. Head Neck Surg. **75**(3), 2084–2090 (2023)
17. Edara, D.C., Vanukuri, L.P., Sistla, V., Kolli, V.K.K.: Sentiment analysis and text categorization of cancer medical records with LSTM. J. Ambient. Intell. Humaniz. Comput. **14**(5), 5309–5325 (2023)
18. Oluwalade, T.I., Ahmadi, H., Huo, L., Sharpe, R., Zhou, S.-M.: Comorbidities and emotions-unpacking the sentiments of pediatric patients with multiple longterm conditions through social media feedback: a large language model-driven study. J. Affect. Disord., 119752 (2025)
19. Malhotra, A., Jindal, R.: XAI transformer based approach for interpreting depressed and suicidal user behavior on online social networks. Cogn. Syst. Res. **84**, 101186 (2024)
20. Diniz, E.J., et al.: Boamente: a natural language processing-based digital phenotyping tool for smart monitoring of suicidal ideation. In: Healthcare, vol. 10, p. 698 (2022). MDPI
21. Zhou, Q., Lei, Y., Tian, L., Ai, S., Yang, Y., Zhu, Y.: Perception and sentiment analysis of palliative care in Chinese social media: qualitative studies based on machine learning. Soc. Sci. Med. **379**, 118178 (2025)
22. Banik, D., Chalil Madathil, S., Lopes, A.J., Luna Fong, S.A., Mukka, S.K.: An evaluation of the maternal patient experience through natural language processing techniques: the case of twitter data in the united states during COVID-19. Appl. Sci. **14**(19), 8762 (2024)
23. Umair, A., Masciari, E., Ullah, M.H.: Vaccine sentiment analysis using BERT+ NBSVM and geo-spatial approaches. J. Supercomput. **79**(15), 17355–17385 (2023)

24. Nagaraj, P., Deepalakshmi, P., Muneeswaran, V., Muthamil Sudar, K.: Sentiment analysis on diabetes diagnosis health care using machine learning technique. In: Congress on Intelligent Systems: Proceedings of CIS 2021, Volume 1, pp. 491–502. Springer (2022)
25. Gaurav, H., Yadav, D., Sharma, S.: Development and evaluation of an audio-based sentiment analysis system for stress monitoring in individuals with chronic health conditions. Int. Res. J. Mod. Eng. Technol. Sci. **05**(05) (2023). https://doi.org/10.56726/IRJMETS41171
26. Dai, J., Lyu, F., Yu, L., Zhou, Z., He, Y.: Medical service quality evaluation based on LDA and sentiment analysis: examples of seven chronic diseases. Digit. Health **10**, 20552076241233864 (2024)

A Systematic Evaluation of Preprocessing, Feature Selection and Machine Learning Models for Sentiment Analysis

Kamal Walji[1(✉)], Allae Erraissi[2], Abdelali Zakrani[1], and Mouad Banane[1]

[1] Laboratory of Artificial Intelligence & Complex Systems Engineering, Hassan II University, ENSAM, Casablanca, Morocco
kamal.walji-etu@etu.univh2c.ma
[2] Chouaib Doukkali University, El Jadida, Morocco

Abstract. Sentiment analysis is a core task in natural language processing (NLP), widely applied to product reviews, customer feedback, and social media monitoring. While transformer-based models such as BERT and RoBERTa now dominate the field, classical machine learning approaches remain valuable due to their efficiency, simplicity, and interpretability—especially in constrained environments. This study presents a systematic evaluation of classical sentiment analysis pipelines on the IMDb Large Movie Review Dataset (50,000 balanced reviews). We explore 33 combinations of preprocessing techniques (none, stopword removal, stemming), feature selection methods (none, Chi^2, PCA), and classifiers (Logistic Regression, SVM, Naive Bayes, Random Forest). Results show that linear models (SVM and Logistic Regression), particularly when combined with TF-IDF and Chi^2 feature selection, consistently outperform other configurations. The best-performing pipeline—stemming + Chi^2 + SVM—achieved a macro-F1 score of 0.897. PCA consistently degraded performance, and stopword removal offered minimal benefit. These findings reaffirm the value of classical pipelines as strong and interpretable baselines for sentiment classification, particularly when computational resources are limited or interpretability is a priority.

Keywords: Sentiment analysis · IMDb dataset · TF-IDF · feature selection · Logistic Regression · SVM · Naive Bayes · Random Forest

1 Introduction

Sentiment analysis—also known as opinion mining—has become a foundational task in natural language processing (NLP), with wide-ranging applications in product reviews, customer feedback, social media monitoring, politics, and healthcare. The ability to automatically extract sentiment from unstructured text provides valuable insights for decision-makers and researchers alike.

Historically, sentiment classification has relied on classical machine learning algorithms such as Naive Bayes, Logistic Regression, Support Vector Machines (SVM),

A. E. Hassanien et al. (Eds.): AISDS 2025, LNDECT 287, pp. 129–138, 2027.
https://doi.org/10.1007/978-3-032-18927-1_13

and Random Forest. Combined with sparse text representations like Bag-of-Words or TF-IDF, these methods remain popular due to their simplicity, interpretability, and computational efficiency. In practice, such models are still widely deployed in environments where resources are limited or explainability is essential.

In recent years, deep learning approaches—such as Recurrent Neural Networks (RNNs), Convolutional Neural Networks (CNNs), and especially transformer-based models like BERT and RoBERTa—have achieved state-of-the-art performance on sentiment analysis benchmarks. These models leverage contextual embeddings and attention mechanisms to outperform traditional models in terms of accuracy. However, they come with trade-offs: they require significant computational power, are harder to interpret, and may not be suitable for real-time or embedded systems.

Despite the rise of neural models, few studies have conducted a systematic evaluation of classical pipelines that includes variations in preprocessing, feature selection, and classification algorithms. Most prior work compares models in isolation or lacks a structured exploration of pipeline components. This limits our understanding of how individual steps impact overall performance.

This study fills that gap by systematically evaluating 33 classical sentiment analysis pipelines on the IMDb Large Movie Review Dataset. We analyze the interaction of three preprocessing strategies (none, stopword removal, stemming), three feature selection techniques (none, Chi^2, PCA), and four classifiers (Logistic Regression, SVM, Naive Bayes, Random Forest), all using TF-IDF representations. The objective is to clarify the trade-offs between accuracy, efficiency, and interpretability in low-resource settings and reaffirm the role of classical pipelines as strong baselines in modern sentiment analysis.

2 Related Works

Sentiment analysis has been widely investigated over the past decade, evolving from traditional statistical approaches to modern deep learning methods [1, 2]. Early research relied on Bag-of-Words and TF-IDF vectorization combined with classical machine learning algorithms such as Naive Bayes, Logistic Regression, Support Vector Machines (SVM), and Random Forest [3, 4]. These methods offered strong baselines due to their simplicity, reproducibility, and relatively low computational cost [5].

Several recent studies (2020–2025) have confirmed the competitiveness of these classical approaches. Salah et al. [6] demonstrated that Logistic Regression with TF-IDF outperforms Naive Bayes and Random Forest on Twitter-based climate change data. Szwoch [7] showed that SVM consistently surpasses Random Forest in propaganda detection tasks. Similarly, Chowdhury et al. [8] reported that Chi^2 feature selection combined with linear models significantly improves classification accuracy in e-commerce reviews. Gupta and Goel [12] reached the same conclusion in the context of emotion analysis, where Logistic Regression and SVM were superior to Random Forest.

Other comparative works emphasize the efficiency–accuracy trade-off. Naive Bayes is consistently reported as the fastest algorithm, but its F1-scores remain below those of linear models [9–11]. Logistic Regression and SVM, by contrast, achieve higher accuracy and macro-F1 values across multiple domains such as customer feedback [10], Arabic sentiment [9], and movie reviews [11]. Manth et al. [13] further showed that traditional ML approaches (especially SVM and LogReg) can rival deep learning models in fake news detection when trained with optimized features.

In parallel, deep learning methods such as CNNs and LSTMs have gained traction for sentiment analysis, due to their ability to model sequential dependencies and contextual patterns in text [14]. More recently, transformer-based architectures (e.g., BERT, RoBERTa, DistilBERT) have achieved state-of-the-art performance by leveraging self-attention and large-scale pretraining [14, 15]. Although these models outperform classical baselines in terms of accuracy, they require significant computational resources and are less interpretable.

In summary, the literature highlights a clear evolution in sentiment analysis techniques: from efficient but simple TF-IDF + classical ML pipelines, to more sophisticated neural architectures, and finally to transformer-based models. This progression is summarized in Fig. 1 which conceptually illustrates the shift from traditional approaches to deep learning and transformers.

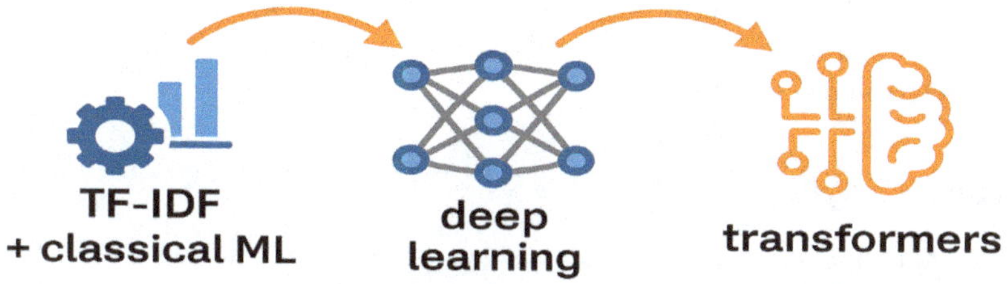

Fig. 1. Evolution of sentiment analysis approaches

Figure 1 illustrates the evolution of sentiment analysis methods, from simple Bag-of-Words and classical machine learning techniques to complex deep learning and transformer-based models. This shift highlights the increasing sophistication and computational demands of modern approaches.

This overview provides the necessary background to interpret our results in Section, where we evaluate the role of preprocessing, feature selection, and classifier choice on the IMDb dataset.

3 Methodology

This section describes the experimental setup used to evaluate 33 classical sentiment analysis pipelines on the IMDb Large Movie Review Dataset. It covers the dataset, preprocessing techniques, feature representation, feature selection methods, classifiers, evaluation metrics, and overall pipeline design.

3.1 Dataset

The experiments were conducted on the IMDb Large Movie Review Dataset, which contains 50,000 English-language reviews labeled as either positive or negative. The dataset is balanced, with 25,000 reviews for training and 25,000 for testing. Only the supervised portion of the dataset was used. This dataset provides a realistic binary sentiment classification task with diverse review lengths and linguistic styles.

3.2 Preprocessing

We evaluated three preprocessing strategies. In the first setting, only lowercasing and tokenization were applied. The second approach removed common English stopwords using the NLTK stopword list. The third strategy applied stemming using the Porter Stemmer to reduce tokens to their root forms. This range of strategies allows us to assess the impact of different preprocessing levels on classification performance.

3.3 Feature Representation

All text data was transformed into TF-IDF vectors using unigrams and bigrams. The vocabulary size was limited to 20,000 features. This limit was chosen to maintain a balance between informativeness and computational efficiency, following established practice in sparse text classification.

3.4 Feature Selection

We evaluated three approaches to feature selection. The first retained all TF-IDF features without additional filtering. The second used the Chi-Square (Chi^2) statistical test to select the 10,000 most discriminative features. The third applied Principal Component Analysis (PCA) using TruncatedSVD, reducing the feature space to 500 latent dimensions. The choice of 500 components was based on common thresholds in prior literature, aiming to reduce dimensionality while retaining sufficient variance.

It is important to note that pipelines combining PCA and Naive Bayes were excluded. This is because the Multinomial Naive Bayes algorithm assumes that all input features are non-negative, as it treats them as frequencies or counts [16]. However, PCA produces real-valued projections that include negative values, which are incompatible with the probabilistic assumptions of Naive Bayes.

3.5 Classification Algorithms

Four supervised learning models were evaluated. Logistic Regression with L2 regularization served as a linear baseline. Support Vector Machines (SVM) were implemented using a linear kernel. The Multinomial Naive Bayes classifier was chosen for its speed and simplicity [17, 18]. Lastly, a Random Forest classifier with 200 decision trees was included as an ensemble-based method. These models cover a range of classical learning paradigms commonly used in sentiment analysis.

3.6 Evaluation Metrics

Pipeline performance was assessed using accuracy, macro-averaged F1 score, and execution time. Macro-F1 was chosen to reflect balanced performance across both sentiment classes. Execution time was included to evaluate computational efficiency under realistic conditions. All experiments were conducted in a controlled environment to ensure consistency across runs.

3.7 Experimental Design

We constructed 33 unique pipelines by combining the three preprocessing strategies, three feature selection methods, and four classifiers. The entire experimental process is illustrated in Fig. 2, which presents the sequential stages of pipeline construction: preprocessing, vectorization, feature selection, model training, and performance evaluation.

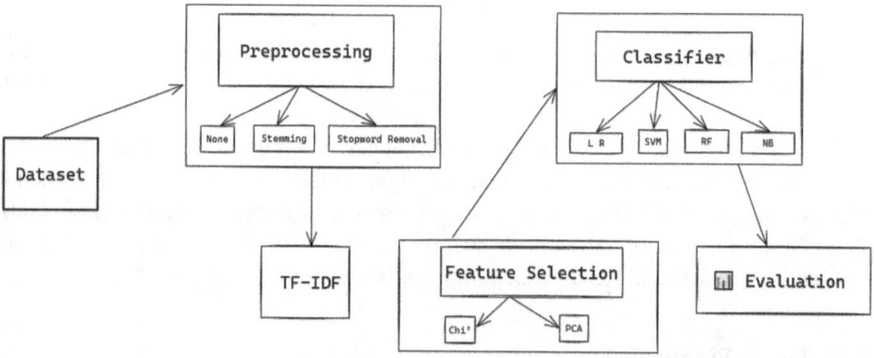

Fig. 2. Overview of the sentiment analysis pipeline used in this study. The diagram illustrates the modular design, allowing interchangeable components at each stage.

Figure 2 outlines the modular design of the sentiment analysis pipeline. It highlights how preprocessing, TF-IDF vectorization, feature selection, and classifier modules are combined to create 33 distinct configurations. This visual representation reinforces the systematic structure of the experimental framework.

4 Results and Discussion

4.1 Overall Performance

The results confirm that linear classifiers consistently outperform other classical models for sentiment classification on the IMDb dataset. As shown in Table 1, the best-performing configuration was the pipeline using stemming, Chi^2 feature selection, and SVM, which achieved both an accuracy and macro-F1 score of 0.897.

Table 1. The 10 best performing pipelines on the IMDb dataset

Preprocessing	Feature Selection	Model	Accuracy	F1
Stemming	Chi^2	SVM	0.897	0.897
None	None	Logistic Regression	0.896	0.896
Stemming	None	Logistic Regression	0.893	0.893
None	Chi^2	SVM	0.894	0.894
Stemming	Chi^2	Logistic Regression	0.892	0.892
None	None	SVM	0.890	0.890
Stopword removal	None	Logistic Regression	0.888	0.888
Stopword removal	PCA	SVM	0.883	0.883
None	PCA	SVM	0.888	0.888
Stemming	PCA	SVM	0.888	0.888

Logistic Regression with no preprocessing also performed nearly as well, reaching an F1 score of 0.896, highlighting the robustness of TF-IDF representations even with minimal text processing.

These results support the idea that TF-IDF, when combined with strong linear classifiers and Chi^2 feature selection, remains a highly effective and efficient approach to sentiment analysis. Differences among top pipelines were relatively small (within 0.005 points), suggesting marginal gains from preprocessing strategies and reinforcing the importance of model and feature selection over text cleaning.

4.2 Effect of Preprocessing

Preprocessing had a limited but measurable impact on performance. Stemming slightly improved performance for both SVM and Logistic Regression, especially when combined with Chi^2. In contrast, stopword removal showed no significant advantage, and in some configurations, slightly degraded results.

This suggests that TF-IDF already captures essential signal from the text, making heavy preprocessing optional. Practically, this implies that minimal text cleaning can suffice when using linear models with TF-IDF features.

4.3 Effect of Feature Selection

Feature selection played a more substantial role than preprocessing. Chi^2 selection improved F1 scores across linear models by filtering out non-discriminative terms, thus enhancing model focus and reducing noise.

Conversely, PCA reduced performance across all classifiers. This result reinforces that projection-based dimensionality reduction is less effective for high-dimensional, sparse text data like TF-IDF, especially when it introduces negative values (which are problematic for models like Naive Bayes).

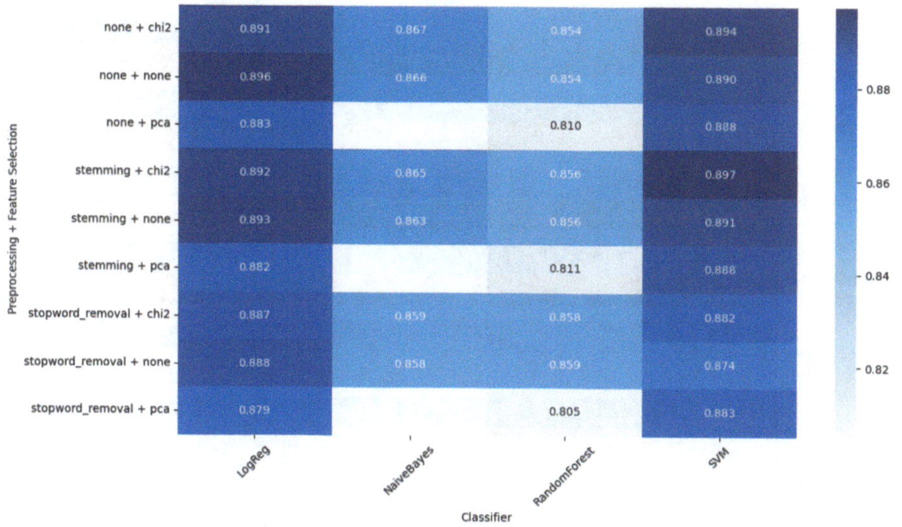

Fig. 3. Heatmap showing F1 score by pipeline and model. Pipelines using Chi^2 show higher scores than those using PCA, especially for SVM and Logistic Regression.

Figure 3 confirms that pipelines using Chi^2 feature selection consistently outperform those using PCA or no selection, especially when paired with linear classifiers. This pattern supports the conclusion that statistical feature filtering enhances discriminative power, particularly for sparse textual data.

4.4 Speed Versus Accuracy Trade-off

Figure 4 visualizes the trade-off between execution time and macro-F1. Naive Bayes was the fastest classifier (around 0.1 s per run) but had lower F1 scores (~0.86). SVM and Logistic Regression offered a better balance, reaching F1 scores above 0.89 with execution times of only a few seconds.

Random Forest was both slower (over 2 min) and less accurate (F1 between 0.81 and 0.86), making it a less practical choice in time-sensitive or large-scale applications.

Figure 4 visualizes the trade-off between accuracy and computational efficiency. Pipelines using SVM or Logistic Regression with Chi^2 occupy the top-right quadrant, demonstrating that high accuracy can be achieved with minimal runtime, making them suitable for real-time or resource-constrained environments.

4.5 Comparison with Deep Learning Approaches

While transformer-based models such as BERT and RoBERTa have surpassed classical models in absolute accuracy—often exceeding 0.92 macro-F1 on similar sentiment tasks—they require substantial resources and are difficult to interpret.

For example, training BERT from scratch or fine-tuning on IMDb often takes several hours on a GPU, while inference speed can be 10 to 50 times slower than linear models.

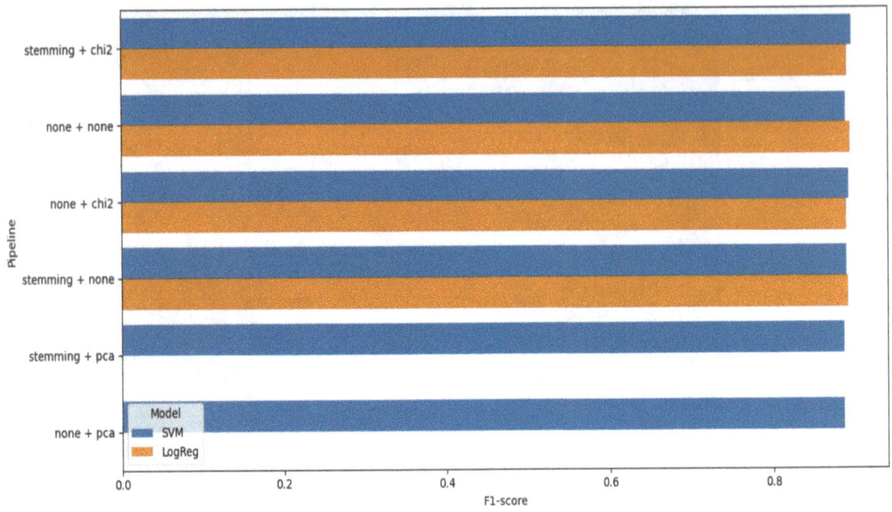

Fig. 4. Top 10 pipelines plotted by accuracy and execution time Linear models dominate the top-right quadrant (high performance, low time).

In contrast, our best pipeline (SVM + Chi^2) reaches 0.897 F1 in just a few seconds on CPU, with full transparency and interpretability.

These results suggest that classical pipelines remain highly competitive baselines, particularly for reproducibility, interpretability, and low-resource applications such as edge devices or real-time monitoring systems.

4.6 Summary of Finding

SVM and Logistic Regression with TF-IDF + Chi^2 deliver the best trade-offs in performance and speed.

- Stemming offers slight gains, while stopword removal is largely unnecessary.
- Chi^2 improves performance, but PCA degrades it, especially for sparse data.
- Classical models are still valuable in contexts where interpretability, speed, and resource efficiency matter more than marginal accuracy gains.

5 Conclusion and Future Work

This study presented a systematic evaluation of 33 classical sentiment analysis pipelines on the IMDb Large Movie Review Dataset. By combining three preprocessing methods, three feature selection techniques, and four supervised learning models, we assessed their interactions in terms of accuracy, macro-F1, and execution time. The results clearly show that linear classifiers, particularly SVM and Logistic Regression, combined with TF-IDF vectorization and Chi^2 feature selection, offer the best trade-off between accuracy and efficiency. Stemming produced slight improvements, while stopword removal offered no consistent benefits. PCA consistently reduced performance due to its distortion of sparse text representations.

These findings reaffirm the value of classical machine learning models as interpretable and resource-efficient baselines for sentiment analysis, especially in real-world applications constrained by time, memory, or explainability requirements. Rather than being obsolete, these pipelines provide a pedagogically valuable and practically deployable alternative to large transformer-based models.

Future work will expand this benchmark by including deep learning architectures such as CNNs, LSTMs, and transformer-based models like BERT and RoBERTa. We also aim to develop hybrid pipelines that combine classical feature engineering (e.g., Chi^2) with contextual embeddings from pre-trained models to strike a better balance between performance and interpretability. Finally, we plan to test pipeline generalization on domain-specific and multilingual sentiment datasets to assess robustness across contexts beyond English movie reviews.

References

1. Daneshfar, F., Rahman, P., Parvin, H.: Multi-objective manifold representation for opinion mining. In: Expert Systems. Wiley (2025)
2. Chandan, M.K., Mandal, S.: A comprehensive survey on sentiment analysis: framework, techniques, and applications. Comput. Sci. Rev. (2025)
3. Nakamura, Y., Romano, I.: Deep learning for text classification: integrating BiGRU with CNNs. Int. J. Intell. Data Mach. Learn. (2025)
4. Soy, A., Balkrishna, S.M.: Deep learning applications in natural language processing for sentiment analysis: unlocking consumer behavior trends. In: Proc. IEEE Conf. on Security and Artificial Intelligence (2025)
5. Mehta, S., Kumar, R.: Federated deep learning for multiclass sentiment analysis: privacy-preserving CNNs with optimized aggregation. In: Proc. IEEE Int. Conf. on Computing and Communication Systems (2024)
6. Salah, A., Mahdi, M., Omar, M.: Analyzing public opinion on climate change via twitter: a machine learning approach using historical data. J. Comput. Inf. Sci. (2025)
7. Szwoch, J.E.: Study on Propaganda Detection in Polish Online News. Hokkaido University Repository (2025)
8. Chowdhury, M.S., et al.: Optimized approaches for real-life sentiment analysis in E-commerce. In: Optimizing Solutions for Real-Life Problems, pp. 93–117. Singapore, Springer Nature Singapore (2025)
9. Alharbi, A., Sharma, N., Hussain, F.: Arabic sentiment analysis with social network data: a comparative study. In: Proc. IEEE ICMLA (2024)
10. Bhuiyan, R.J., et al.: Sentiment analysis of customer feedback in the banking sector: a comparative study of ML models. Am. J. Technol. Appl. Sci. (2024)
11. AbdElminaam, D.S., et al.: Improving movie review sentiment classification through advanced ML models. In: Proc. IEEE ICCES (2024)
12. Gupta, S.C., Goel, N.: Emotion analysis of textual contents using NLP and text mining. J. Theor. Appl. Inf. Technol. (2024)
13. Manth, A., Mitra, A., Bachani, A., Jha, A.K.: TruthScan: Fake news detection using machine learning and NLP techniques, SSRN preprint (2025)
14. Kulkarni, S.S., et al.: Explainable depression detection in social media using transformer-based models: a comparative analysis of ML. In: Proc. IEEE ICMI (2024)

15. Maria, M.N., et al.: Sentiment analysis using NLP libraries and machine learning. In: Proc. IEEE SMACS (2024)
16. Elbayed, Z., Qadi El Idrissi, A.: Deep learning in financial modeling: predicting European put option prices with neural networks. Algorithms **18**(3), 161 (2025). https://doi.org/10.3390/a18030161
17. Khaldi, M.I., Erraissi, A., Hain, M., Banane, M.: In-Depth Comparison of supervised classification models - performance and adaptability to practical requirements. Int. J. Adv. Comput. Sci. Appl. **16**(8) (2025). https://doi.org/10.14569/IJACSA.2025.0160862
18. Walji, K., Erraissi, A., Zakrani, A., Banane, M.: Sentiment analysis revisited: a multi-metric comparative study. Int. J. Adv. Comput. Sci. Appl. **16**(9) (2025). https://doi.org/10.14569/IJACSA.2025.0160965

Data Analysis and Collection Methods for Fake News Detection

Hind Chelh[1(✉)], Mohamed Nabil[1], Abdelkaher Ait Abdelouahad[1], Abdelmajid Hajami[2], and Abdelkrim Haqiq[3]

[1] LAROSERI, Computer Science Department, Faculty of Sciences, University of Chouaib Doukkali, El Jadida, Morocco
Chelhhind1996@gmail.com

[2] Faculty of Sciences and Techniques, LAVETE Laboratory, Hassan First University of Settat, Settat, Morocco
abdelmajid.hajami@uhp.ac.ma

[3] Faculty of Sciences and Techniques, Computer, Networks, Mobility and Modeling Laboratory (IR2M), Hassan First University of Settat, Settat, Morocco
abdelkrim.haqiq@uhp.ac.ma

Abstract. Corpora play a key role in developing a good fake news detection model. To this end, this article presents a comprehensive review of the different techniques used to collect new data in the form of text (articles, comments, tweets) for fake news detection applications based on machine learning and deep learning approaches, covering the period 2017–2024. We explore several data acquisition methods, including web scraping, the use of public Application Programming Interfaces (API), manual collection, fact-checking platforms, and manual annotation. Our analysis focuses on the structure and characteristics of the datasets used in these studies, paying particular attention to the diversity of languages, thematic domains, as well as the length and format of the main textual content. We also examine the evolution of data collection strategies over time, highlighting a notable shift toward more automated and scalable methods. At the same time, we highlight the increasing inclusion of low-resource languages, such as Arabic, in datasets previously dominated by English. These developments demonstrate the growing adaptation potential of fake news detection systems to multilingual and multicultural contexts, thereby improving their reliability and applicability globally.

Keywords: Collection method · Corpus · Data collection · Dataset · Fake news · Fake news detection

1 Introduction

The spread of misinformation represents a crucial issue for modern societies, endangering social life, health and democracy [1].

Fake news is fabricated or distorted information disseminated with the intention of misleading, manipulating public opinion, or serving economic, political, or ideological interests [2]. This content can take various forms: text articles, doctored images,

A. E. Hassanien et al. (Eds.): AISDS 2025, LNDECT 287, pp. 139–148, 2027.
https://doi.org/10.1007/978-3-032-18927-1_14

manipulated videos (e.g., deepfakes), or even synthetic social messages. Its dissemination often relies on the sharing dynamics of social networks and the absence of strict editorial control, making it a powerful vector of large-scale disinformation through the disinformation propagation channels, illustrated in Fig. 1.

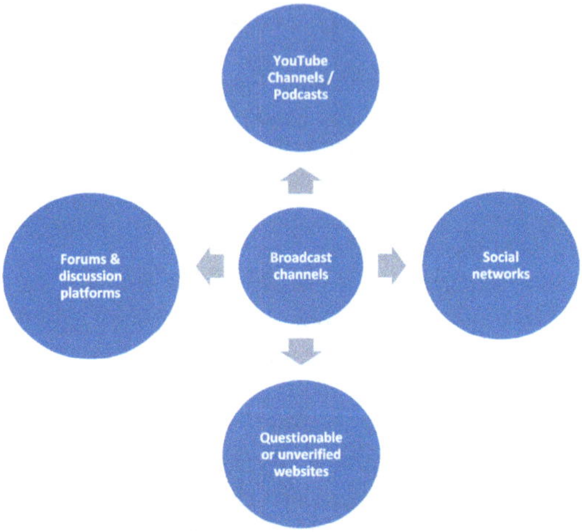

Fig. 1. Channels for spreading fake news

The rise of artificial intelligence technologies, particularly machine learning (ML) and deep learning (DL), has led to the creation of increasingly effective fake news detection systems. However, the quality and representativeness of the datasets used to train these models remain a key element of their performance [1].

Datasets are the most important part of any artificial intelligence-based approach to fake news detection. They determine not only the models' ability to learn relevant patterns, but also their ability to generalize across varied and realistic contexts. A rich, diverse, and well-balanced database can better capture the nuances between truthful and false content, whether it involves text, images, or multimodal content [3]. Conversely, datasets lacking linguistic diversity, thematic bias, or insufficient size can lead to illusory performance that is not transferable to real-life situations [4]. The importance of rigorous annotation, reliable sources, and representative sampling is therefore crucial to ensure the reliability of automated verification systems. In this sense, the quality of the dataset is often as decisive as the complexity of the model used.

Given the importance of the datasets on which the system or automated model will be trained, our paper attempts to cover the data collection techniques used in recent research (2017–2024), highlighting their characteristics, strengths, and constraints. We aim to guide researchers in selecting the most suitable collection methods according to their objectives.

The retained datasets in this survey were selected to ensure representativeness through diversity, which is a crucial element in reviewing data collection for fake news detection. The selection reflects multiple dimensions of diversity, including modality (text, images, videos, and metadata), domain coverage (politics, health, and social issues), and annotation scheme. Datasets were also chosen from various sources such as social media, news outlets, and fact-checking platforms, introducing contextual and linguistic variety. In addition, priority was given to multilingual and large-scale corpora to capture cultural richness and enhance global applicability. By considering datasets along these dimensions, this survey provides a representative overview of existing resources and establishes a solid basis for analyzing their strengths, limitations, and emerging trends.

The remainder of the paper is structured as follows: Sect. 2 presents a comparative analysis of the datasets used in existing papers between 2017 and 2024, mainly in text form (article, tweets, comments), Sect. 3 discusses the results obtained, and Sect. 4 concludes with the main contributions and perspectives.

2 Comparative Analysis of Datasets

To provide a comparative and systematic analysis of the data collection methods used in fake news detection articles using machine or deep learning, we identify the main data collection approaches used, classify the collection techniques, as well as the linguistic diversity and size of the datasets.

Table 1 summarizes the datasets used for fake news detection in our study. These datasets are chronologically ordered according to the corpus size (number of news or text tweets), collection method (API, scraping, manual selection), main language (English, Arabic), and domain.

2.1 Corpus Size

The datasets sizes (Table 1) vary considerably; MuMiN has the largest size, which is commensurate with large-scale analyses, while Clickbait and AraFacts are defined as smaller datasets.

These size differences reflect fake news detection methods' heterogeneity: big corpora allow better generalization over models, while smaller, often manually labeled, datasets allow closer contextual scrutiny. This depth of analysis and quantity of data complementarity is a central issue in recent research [9, 29].

2.2 Language

The English language remains dominant in the majority of datasets (Fig. 2), particularly those from Twitter platforms and international media. However, the emergence of several Arabic-language games (AraNews, AFND, Satirical, Clickbait, ANS, LebFake, AraFacts) indicates a growing desire to extend disinformation research to the Arab world. This multilingual openness is essential to improve the robustness of models in varied linguistic contexts.

Table 1. Fake news detection datasets.

Year	Reference	Data base	Language	Size	Domain	Method used
2017	[5]	LIAR	English	12,836	Politics	Manual collection
2018	[6]	FEVER	English	185,000	Multi-Domain	Manual annotation
2019	[7]	FakeNewsAMT	English	9,400	Politics	Manual collection
2020	[8]	AraNews	Arabic	5,187,957	News	Web scraping
2020	[9]	FakeNewsNet (Politifact + GossipCop)	English	23,921	Politics & Entertainment	Web scraping + annotation
2020	[10]	Satirical-Fake-News-Dataset	Arabic	6,895	Politics	Manual collection
2020	[11]	FNC-1	English	75,385	Multi-Domain	Manual reuse of news
2020	[12]	ISOT Fake News Dataset	English	44,898	News	Manual collection
2021	[13]	AraFacts	Arabic	6,222	Multi-Domain	Manual collection
2021	[14]	AFND	Arabic	606,912	News	Hybrid collection
2021	[15]	COVID-19 Infodemic Twitter dataset	English/Arabic	500	COVID-19	Manual annotation
2021	[16]	COVID-19-disinformation	English	4,542	COVID-19	Manual annotation
			Arabic	4,966		
			Bulgarian	3,697		
			Dutch	2,665		
2021	[17]	NewsBag++	English	589,000	Multi-Domain	Annotation + fact -checking
2021	[18]	FakeNewsNet	English	20,800	Politics	Manual collection
2021	[19]	COVID-19 Fake News Dataset	English	4,072	COVID-19	Manual collection

(continued)

Table 1. *(continued)*

Year	Reference	Data base	Language	Size	Domain	Method used
2021	[20]	Covid19-fake-news-detection	English	10,700	Health	Fact-checking + manual annotation
2022	[21]	Clickbait	Arabic	3,000	Multi-Domain	Web scraping
2022	[22]	Unnamed	English	30,278	Multi-Domain	News Trends, Kaggle and Reuters websites
2022	[23]	MuMiN	English	21,000,000	Social media data	Twitter API
2022	[24]	ANS	Arabic	4,627	News	Manual collection
2023	[25]	Unnamed	English	125,709	COVID-19	Web scraping
2024	[26]	Multi-Fake-DetectiVE	English	920,054	Politics	Twitter API
2024	[27]	3 combined datasets (WELFake + FakeNewsNet + FakeNewsPrediction	English	101,665	Multi-Domain	Web scraping + annotation
2024	[28]	ConFake Dataset	English	72,413	Multi-Domain	Hybrid collection

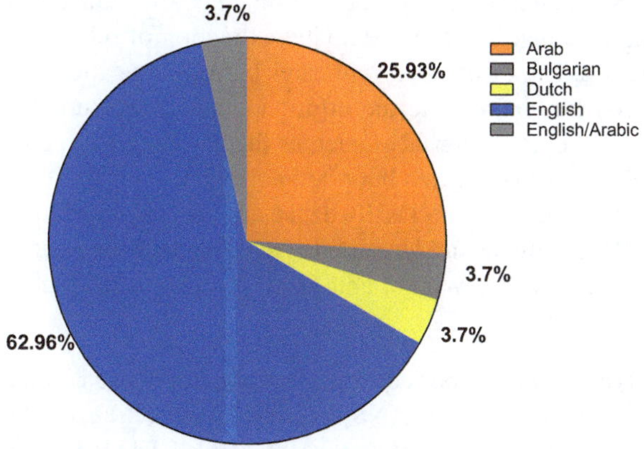

Fig. 2. Share of languages in the Fake News detection corpora.

2.3 Areas Covered

Several datasets focus on specific domains: FakeNewsNet and MULTI Fake Detective on politics, the COVID-19 corpus on health, and Clickbait on media sensationalism. Others like FNC-1 or ConFake are multidomain, which favours more general applications, but makes annotation more complex.

Fake news is particularly prevalent in certain areas with a strong emotional or political charge (FakeNewsNet and MULTI Fake Detective). The political sector is one of the most affected, as false information is often used to influence public opinion, destabilize opponents or manipulate electoral results [30]. Similarly, the health sector (COVID-19 corpus), particularly during the COVID-19 pandemic, has seen an explosion of misinformation around vaccines, unvalidated treatments, and conspiracy theories [31]. This proliferation is explained by the high collective anxiety and the urgent need for information, which opens the door to manipulation. International conflict constitutes another fertile ground for fake news, as it can be used as weapons of psychological warfare, aimed at influencing public opinion or demoralizing the enemy [29], and Clickbait on media sensationalism. Others like FNC-1 or ConFake are multidomain, which favors more general applications, but makes annotation more complex.

There is also a significant presence of false information in areas such as economics, technology or environmental issues, often linked to a poor understanding of the subject or a desire to polarize debates [30, 31].

2.4 Collection Methods

Before comparing the corpora, it is important to define the main collection methods used in constructing datasets:

1. Automated verification: Fact-checking models assess the veracity of a claim based on the available evidence. High performance of these models reflects their ability to reason about the evidence [32].
2. Manual collection: Data is extracted directly from specific sources by humans. This method ensures high quality control but is limited by the human resources required.
3. Web scraping: This involves the systematic extraction of relevant content from the web. A software agent, or crawler, replicates human navigation to access websites, analyze their content, extract relevant information, and structure them [33].
4. Social media API: Use of interfaces such as the Twitter API to collect posts in real time. This approach provides rich but often noisy data.
5. Manual annotation: Annotators classify data using specialized platforms and methods such as RUN-AS (Reliable and Unreliable News Annotation Scheme) or the 5W1H (What, Why, Who, When, Where, and How) method, allowing for linguistic and contextual adaptation [34].

We can observe that the corpora differ according to the used method of collection. Some datasets, such as Satirical and ANS, use manual collection, has low numbers with high-quality annotations. Others, such as MuMiN and MULTI Fake Detective, use APIs to generate large corpora, and AraFacts and AFND employ automatic scraping of news sites.

3 Discussion

The evolution of data collection strategies between 2017 and 2024 reflects a constant tension between reliability and completeness. The manual approaches prioritized precision; their limited scope limited the representativeness of the observed phenomena. Conversely, recent approaches based on scraping and APIs have made it possible to capture the real dynamics of the social web, but at the cost of a deterioration in the quality of annotations. This shift reveals a paradigm shift: research is no longer focused solely on the authenticity of each instance, but on the ability to model discursive regularities and propagation mechanisms from large volumes.

However, quantitative growth does not guarantee proportional scientific progress. As Jindal et al. (2020) states that expanding corpora size can lead to stagnant performance because of the inconsistency of internal data. Therefore, the real challenge is not the size, but intelligent data curation: identifying rigorously, diverse, and representative labeled subsets. However, hybrid strategies (combining human validation and automatic collection) illustrate this shift toward a more reflexive data science, where methodological quality prevails over simple accumulation [17].

Linguistically, English dominance refers to a structural imbalance in the disinformation studies: the majority of NLP tools, training corpora, and fact-checking platforms are based on Western sociolinguistic contexts [35]. This imbalance limits the transferability of models to low-resource languages, such as Arabic, where diglossia and dialectal diversity complicate automatic analysis [36]. Recent projects such as AraFacts, AFND, and Clickbait [13, 14, 21] thus represent important attempts at epistemic rebalancing, but remain fragmented.

Finally, the linguistic specificities of fake news, having strong emotional charge, the use of hyperbole, and the modal formulations [37, 38], confirm that automatic detection cannot be purely statistical: it needs socio-discursive and contextual understanding. In this context, ethical considerations (consent, anonymization, privacy protection) are not only a legal imperative, but a methodological condition assuring the scientific reliability of the corpora [39].

4 Conclusion and Perspectives

This review confirms that the construction of corpora plays a decisive role in advancing fake news detection models. Between 2017 and 2024, data collection methodologies shifted from small, manually curated datasets to more automated, and large-scale with multilingual approaches.

Looking forward, the creation of high-quality Arabic corpora remains a key challenge and opportunity. Such resources will ideally combine multiple sources (news outlets, social media platforms, and regional fact-checking initiatives) while reflecting the variety of dialects. Meanwhile, it is also vital that future efforts should be dedicated to analyse and model the sources of fake news, since understanding the origin, credibility, and dissemination patterns of information is as important as detecting false content itself.

References

1. Kuntur, S., et al.: Fake News Detection: It's All in the Data! arXiv (2024). http://arxiv.org/abs/2407.02122, last accessed 2025/08/28.
2. Haq, B.: Fake news: understanding its impact and implications. Glob. Media J. **22**(70), 1–3 (2024)
3. Hamed, S.K., Aziz, M.J.A., Yaakub, M.R.: A review of fake news detection approaches: a critical analysis of relevant studies and highlighting key challenges associated with the dataset, feature representation, and data fusion. Heliyon. **9**(10) (2023)
4. Baym, G.: Real news/fake news: beyond the news/entertainment divide. In: The Routledge Companion to News and Journalism. Routledge (2009)
5. Wang, W.Y.: "Liar, Liar pants on fire": a new benchmark dataset for fake news detection. In: Barzilay, R., Kan, M.-Y. (eds.) Proceedings of the 55th Annual Meeting of the Association for Computational Linguistics (Volume 2: Short Papers), pp. 422–426. Association for Computational Linguistics, Vancouver (2017)
6. Hanselowski, A., et al.: UKP-Athene: multi-sentence textual entailment for claim verification. In: Thorne, J., Vlachos, A., Cocarascu, O., Christodoulopoulos, C., Mittal, A. (eds.) Proceedings of the First Workshop on Fact Extraction and VERification (FEVER), pp. 103–108. Association for Computational Linguistics, Brussels (2018)
7. Pérez-Rosas, V., Kleinberg, B., Lefevre, A., Mihalcea, R.: Automatic detection of fake news. In: Bender, E.M., Derczynski, L., Isabelle, P. (eds.) Proceedings of the 27th International Conference on Computational Linguistics, pp. 3391–3401. Association for Computational Linguistics, Santa Fe (2018)
8. Nagoudi, E.M.B., et al.: Machine generation and detection of Arabic manipulated and fake news. arXiv (2020). http://arxiv.org/abs/2011.03092, last accessed 2025/08/25.
9. Shu, K., Mahudeswaran, D., Wang, S., Lee, D., Liu, H.: FakeNewsNet: a data repository with news content, social context, and spatiotemporal information for studying fake news on social media. Big Data. **8**(3), 171–188 (2020)
10. Saadany, H., Mohamed, E., Orasan, C.: Fake or real? A study of arabic satirical fake news. arXiv (2020). http://arxiv.org/abs/2011.00452, last accessed 2025/08/25.
11. Umer, M., Imtiaz, Z., Ullah, S., Mehmood, A., Choi, G.S., B.-W.: On: fake news stance detection using deep learning architecture (CNN-LSTM). IEEE Access. **8**, 156695–156706 (2020)
12. Ahmad, I., Yousaf, M., Yousaf, S., Ahmad, M.O.: Fake news detection using machine learning ensemble methods. Complexity. **2020**(1), 8885861 (2020)
13. Sheikh Ali, Z., Mansour, W., Elsayed, T., Al-Ali, A.: AraFacts: the first large Arabic dataset of naturally occurring claims. In: Habash, N., et al. (eds.) Proceedings of the Sixth Arabic Natural Language Processing Workshop, pp. 231–236. Association for Computational Linguistics, Kyiv (2021)
14. Khalil, A., Jarrah, M., Aldwairi, M., Jaradat, M.: AFND: Arabic fake news dataset for the detection and classification of articles credibility. Data Brief. **42**, 108141 (2022)
15. Alam, F., et al.: Fighting the COVID-19 Infodemic in social media: a holistic perspective and a call to arms. In: Proceedings of the International AAAI Conference on Web and Social Media, vol. 15, pp. 913–922 (2021)
16. Alam, F., et al.: Fighting the COVID-19 Infodemic: Modeling the perspective of journalists, fact-checkers, social media platforms, policy makers, and the society. In: Moens, M.-F., Huang, X., Specia, L., Yih, S.W. (eds.) Findings of the Association for Computational Linguistics: EMNLP 2021, pp. 611–649. Punta Cana, Association for Computational Linguistics (2021)

17. Jindal, S., Sood, R., Singh, R., Vatsa, M., Chakraborty, T.: NewsBag: a multimodal benchmark dataset for fake news detection. In: CEUR Workshop Proc, pp. 138–145 (2020)
18. Kaliyar, R.K., Goswami, A., Narang, P.: FakeBERT: fake news detection in social media with a BERT-based deep learning approach. Multimed. Tools Appl. **80**(8), 11765–11788 (2021)
19. Koirala, A.: COVID-19 fake news dataset. Mendeley (2021). https://data.mendeley.com/datasets/zwfdmp5syg/1, last accessed 2025/08/26.
20. Patwa, P., et al.: Fighting an Infodemic: COVID-19 fake news dataset. In: Chakraborty, T., Shu, K., Bernard, H.R., Liu, H., Akhtar, M.S. (eds.) Combating Online Hostile Posts in Regional Languages during Emergency Situation, pp. 21–29. Springer International Publishing, Cham (2021)
21. Bsoul, M.A., Qusef, A., Abu-Soud, S.: Building an optimal dataset for Arabic fake news detection. Procedia Comput. Sci. **201**, 665–672 (2022)
22. Kaur, S., Kumar, P., Kumaraguru, P.: Automating fake news detection system using multi-level voting model. Soft. Comput. **24**(12), 9049–9069 (2020)
23. Nielsen, D.S., McConville, R.: MuMiN: a large-scale multilingual multimodal fact-checked misinformation social network dataset. In: Proceedings of the 45th International ACM SIGIR Conference on Research and Development in Information Retrieval, pp. 3141–3153. Association for Computing Machinery, New York (2022)
24. Sorour, S.E., Abdelkader, H.E.: AFND: Arabic fake news detection with an ensemble deep cnn-lstm model. J. Theor. Appl. Inf. Technol. **100**(14), 5072–5086 (2022)
25. Akhtar, M.M., Karunanayake, I., Sharma, B., Masood, R., Ikram, M., Kanhere, S.S.: Towards automatic annotation and detection of fake news. In: 2023 IEEE 48th Conference on Local Computer Networks (LCN), pp. 1–9. IEEE, Daytona Beach (2023)
26. Bondielli, A., Dell'Oglio, P., Lenci, A., Marcelloni, F., Passaro, L.: Dataset for multimodal fake news detection and verification tasks. Data Brief. **54**, 110440 (2024)
27. Hashmi, E., Yayilgan, S.Y., Yamin, M.M., Ali, S., Abomhara, M.: Advancing fake news detection: hybrid deep learning with FastText and explainable AI. IEEE Access. **12**, 44462–44480 (2024)
28. Jain, M.K., Gopalani, D., Meena, Y.K.: ConFake: fake news identification using content based features. Multimed. Tools Appl. **83**(3), 8729–8755 (2024)
29. Zhou, X., Zafarani, R.: A survey of fake news: fundamental theories, detection methods, and opportunities. ACM Comput. Surv. **53**(5), 1–40 (2021)
30. Shu, K., et al.: Fake news detection on social media: a data mining perspective. arXiv (2017). http://arxiv.org/abs/1708.01967, last accessed 2025/08/25.
31. Ahmed, W., Vidal-Alaball, J., Downing, J., Seguí, F.L.: COVID-19 and the 5G conspiracy theory: social network analysis of twitter data. J. Med. Internet Res. **22**(5), e19458 (2020)
32. Hansen, C., Hansen, C., Chaves Lima, L.: Automatic fake news detection: are models learning to reason? In: Zong, C., Xia, F., Li, W., Navigli, R. (eds.) Proceedings of the 59th Annual Meeting of the Association for Computational Linguistics and the 11th International Joint Conference on Natural Language Processing (Volume 2: Short Papers), pp. 80–86. Association for Computational Linguistics (2021)
33. Glez-Peña, D., Lourenço, A., López-Fernández, H., Reboiro-Jato, M., Fdez-Riverola, F.: Web scraping technologies in an API world. Brief. Bioinform. **15**(5), 788–797 (2014)
34. Wang, R.: Data annotation methodologies for fake news. Trans. Comput. Sci. Intell. Syst. Res. **9**, 185–190 (2025)
35. Lazer, D.M.J., et al.: The science of fake news. Science. **359**(6380), 1094–1096 (2018)
36. Albtoush, E.S., Gan, K.H., Alrababa, S.A.A.: Fake news detection: state-of-the-art review and advances with attention to Arabic language aspects. PeerJ Comput. Sci. **11**, e2693 (2025)

37. Rashkin, H., Choi, E., Jang, J.Y., Volkova, S., Choi, Y.: Truth of varying shades: Analyzing language in fake news and political fact-checking. In: Palmer, M., Hwa, R., Riedel, S. (eds.) Proceedings of the 2017 Conference on Empirical Methods in Natural Language Processing, pp. 2931–2937. Association for Computational Linguistics, Copenhagen (2017)
38. Shu, K., Wang, S., Liu, H.: Beyond news contents: the role of social context for fake news detection. In: Proceedings of the Twelfth ACM International Conference on Web Search and Data Mining, pp. 312–320. Association for Computing Machinery, New York (2019)
39. Dalieva, M.: Methods, challenges, and ethical considerations in data collection of corpus compilation. Innov. Technol. Methodical Res. J. **3**(3), 1–6 (2024)

Audio Deepfake: Literature Review and Detection Limitations

Bouzid Mahmoud(✉) and Bouzidi Driss

Smart Systems Laboratory (SSL), ENSIAS, Mohammed V University, Rabat, Morocco
mahmoud_bouzid@um5.ac.ma, driss.bouzidi@ensias.um5.ac.ma

Abstract. The rapid development of artificial intelligence technologies has led to significant growth in the production of models for detecting false voices, voice imitations, or so-called audio deepfakes, created through text-to-speech (TTS) and voice conversion (VC) technologies, offer valuable applications in accessibility and media but also pose serious threats to security, privacy, and society, but the models generating audio deepfakes also have a great capacity to bypass detection models. This paper is an easy entry to this area of research, it offers a global overview on generative AI, a review of literature of the various recent surveys which have gathered a lot of proposed solutions in terms of generation and detection, as well as the various limitations and problems that detection solutions face.

Keywords: Audio deepfake · generation · detection · generative AI · text-to-speech · voice conversion

1 Introduction

Over the past decades, technological progress has enabled researchers to design increasingly advanced methods for generating texts, images, music, and other types of content. Among these breakthroughs, autoregressive language models (such as GPT), deep neural networks, particularly Generative Adversarial Networks (GANs), known for their power in generating realistic images for example, marked a major turning point, making the creation of high-quality and realistic content possible [1]. Deepfakes, which represent a major category of generative artificial intelligence, are generated or altered in a realistic way by artificial intelligence, especially through deep neural networks such as autoencoders and Generative Adversarial Networks (GANs). Becoming increasingly sophisticated, they pose a major risk of disinformation on a global scale. Ultimately, their use as informational weapons could seriously undermine trust in public institutions, the media, and other actors, as it is becoming more and more difficult for the public to distinguish between authentic content and manipulated forgeries [2]. Deepfake technology covers various categories, including visual deepfakes, which encompass video deepfakes and image deepfakes, text deepfakes, multimodal deepfakes that combine different types, and audio deepfakes, which are the focus of this paper. Considering the numerous implications of deepfake technology, it is important to highlight that research

A. E. Hassanien et al. (Eds.): AISDS 2025, LNDECT 287, pp. 149–157, 2027.
https://doi.org/10.1007/978-3-032-18927-1_15

and development efforts have so far focused primarily on video deepfakes. This is due to their strong visual impact and their high potential for misuse. Many studies have therefore been conducted, addressing detection techniques as well as ethical and legal dimensions, in order to mitigate the harmful effects of these manipulated contents. In contrast, audio deepfakes have received much less attention, both in the scientific community and in public debate, despite the risks they also pose [3]. Nevertheless, most excellent scientific research on audio deepfakes particularly on audio deepfake detection has made significant progress, with the aim of minimizing the damage and risks caused by the power and complexity of audio deepfake generation techniques. In this paper, we will present audio deepfakes, generation & detection in Sect. 2, and we will provide a state-of-the-art overview of the most well-known and recent scientific surveys that have deployed major efforts in collecting, gathering, analyzing, and categorizing scientific articles on audio deepfake generation and detection in Sect. 3. In Sect. 4, we will presenting and identifying the various gaps in the literature in order to improve people's security and privacy regarding protection against all forms of manipulated or deepfake audio for malicious purposes.

2 Fundamentals of Audio Deepfakes

An audio deepfake is a voice recording created or manipulated using AI techniques, in order to reproduce an almost real voice or one that closely resembles the voice of an existing or well-known person. According to [3], publications on video deepfakes consistently outnumber those on audio deepfakes, which suggests that research is more focused on visual deception within deepfake technology. Audio deepfakes play an important role in assistive technologies by enabling the creation of personalized synthetic voices for people with speech disorders and fostering the development of tools that enhance accessibility in various domains. However, alongside these advantages, audio deepfakes also present significant drawbacks that raise serious ethical and security concerns. They can be exploited to bypass voice authentication systems, thereby threatening the reliability of biometric security measures. Their misuse poses risks to public security and can facilitate social engineering attacks, where individuals are manipulated into divulging sensitive information. Additionally, audio deepfakes contribute to the spread of fake news and disinformation, with the potential to control or mislead public opinion for political purposes or vested interests. Beyond these societal risks, they also create challenges in cybersecurity and can be weaponized for cyber harassment, amplifying their negative impact on individuals and communities.

2.1 Audio Deepfake Generation

Audio deepfakes are divided into several categories, defined by their methods of creation and the objectives they pursue. Among them, two main and most widespread approaches stand out: Voice conversion (VC) and Text-to-Speech (TTS).

Voice Conversion (VC): Voice conversion consists of transforming speech produced by a source speaker so that it reproduces the vocal characteristics of a target speaker, while preserving the linguistic content of the source. In practice, this method modifies elements such as timbre, prosody, or intonation of speaker X's speech to replace it with that of speaker Y, without altering the spoken words [4]. This is essentially digital voice cloning. To achieve this result, VC systems rely on deep learning models, particularly Generative Adversarial Networks (GANs), which are capable of accurately capturing and reproducing the prosodic and stylistic traits of the target voice, thereby generating high-quality imitations.

Text-to-Speech (TTS) Synthesis: Text-to-Speech synthesis, or TTS, enables the production of an artificial voice directly from written text. Unlike voice conversion, it does not rely on an existing audio signal but instead generates entirely new speech. TTS models are trained on large corpora of recordings of the target speaker's voice in order to learn their distinctive features (accent, timbre, rhythm, intonation). Once trained, these models can produce new sentences never spoken by the original speaker, with natural prosody and convincing expressiveness, creating the illusion of a genuine vocal performance. However, TTS is computationally expensive [5].

It is worth noting that the first well-known case of deepfake audio fraud was reported in mid-2019, when hackers used AI-based software to imitate the voice of a CEO and request a massive transfer of 243,000 US dollars [6].

2.2 Audio Deepfake Detection

Threats related to voice identity theft are rapidly increasing, creating risks for systems relying on voice biometrics and undermining societal stability. Consequently, the detection of audio deepfakes has become a critical and urgent task to preserve the integrity and security of audio content in technology and to foster social trust among individuals. Several studies have focused on developing new models and algorithms for detecting deepfake audio or improving existing ones in order to keep pace with the advances in generation techniques. To address these challenges, various detection approaches have been proposed. According to the literature, audio deepfake detection can be divided into two main categories: detection based on acoustic features and detection based on non-acoustic features, the latter being less explored and requiring greater attention from the research community to address this gap. After research, it can be concluded that current audio deepfake detection methods rely on several implemented approaches: Detection methods based on spectral and acoustic analysis, which aim to identify anomalies in the signal such as pitch shifts, inconsistencies in vocal characteristics, as well as irregularities in mel-spectrograms, prosody, or background noise. Detection methods based on traditional machine learning, such as the use of SVM or KNN, and deep learning approaches using CNN, RNN, and Transformers, which are capable of automatically classifying manipulated audio segments. These models learn to distinguish real audio from synthetic audio by leveraging large amounts of data. Such detection models can also be referred to as classifier-based models. The detection of audio deepfakes relies on the use of several reference datasets for training and evaluating models. Among them:

- The ASVSpoof series (2015, 2017, 2019, 2021) are the main benchmarks for studying spoofing attacks [7].
- WaveFake [8], offer recordings generated by modern vocoders and TTS systems such as WaveGlow, MelGAN, and ParallelWaveGAN, in order to analyze the ability of detectors to generalize across different generators.

- The Fake or Real (FoR) corpus [9], meanwhile, was designed to evaluate the robustness of approaches when faced with synthetic speech produced by various voice systems.
- In the Wild [10] is a dataset designed to evaluate machine learning models dedicated to detecting deepfakes and preventing voice spoofing. In particular, it measures the ability of models to generalize when faced with realistic audio samples that are close to natural conditions.

3 A Comprehensive Review of Audio Deepfake Surveys

Most of the well-known scientific research on audio deepfakes, both for generation and detection, has made significant progress with the aim of presenting the power and potential damages caused by the complexity and capabilities of audio deepfake generation techniques, as well as the valuable efforts in developing detection methods. Nevertheless, this field continues to face limitations and challenges due to the rapid evolution of artificial intelligence. In this section, we will present scientific papers, specifically surveys and literature reviews from the past five years, that provide a wealth of very useful information covering the widest possible range of research articles, whether for understanding the domain of audio deepfakes from generation to detection, or for highlighting the major detection techniques.

In [5], the authors presented the various AI methods that focus on the generation and detection of deepfakes, and concluded that CNNs (Convolutional Neural Networks), DNNs (Deep Neural Networks), and GANs (Generative Adversarial Networks) are common approaches for creating and detecting deepfakes. The authors presented different types of audio deepfake attacks, including replay attacks, speech synthesis like (Char2Wav, WaveNet, WaveGlow, Tacotron…), and voice conversion like (CycleGAN-VC, StarGAN-VC, SINGAN…) or impersonation. The authors presented directions for future research, including blockchain, more robust models, emotion analysis, and, above all, further research into audio.

In [11], which according to the authors is the first study specifically targeting detection methods for spoofed and synthetically generated audio, three types of audio deepfake attacks are presented, along with different machine learning- and deep learning-based methods for detecting spoofed and synthetically generated audio deepfakes, as well as a comparative description of available audio deepfake datasets. The authors provided a quantitative comparison of recent state-of-the-art audio deepfake detection methods and their weaknesses.

In [12], this paper is an improvement of the previously cited article [5]. The authors presented a section on essential preliminaries to properly begin research in the field of audio deepfakes, providing explanations of Machine Learning, Deep Learning, encoder-decoder architectures, GANs, CNNs, and RNNs (Recurrent Neural Networks). They categorized audio deepfakes into two classes: Audio deepfakes not generated by artificial intelligence, which include audio spoofing, replay attacks, and mimicking & Audio deepfakes generated by artificial intelligence, which include speech synthesis and voice conversion (VC) or impersonation. The authors also presented the different types of detectors like CNN/ResNet, DenseNet-BiLSTM, Temporal Convolutional Networks

(TCN), and feature-engineering–enhanced models (spectral, perceptual, or vocal tract–based)., which are based on Machine Learning and Deep Learning. The authors also presented the different datasets most using in this domain like ASVspoof Challenge Datasets, Fake or real dataset, WaveFake dataset.

In [4], the authors present the main differences between the types of audio deepfakes, identifying five categories: [speech synthesis / voice conversion / fake emotion / fake scene / partially fake]. They provide a systematic overview focused on learning common discriminative audio features for detecting audio deepfakes, a detailed summary of the most recent audio deepfake detection datasets, and benchmarks on these datasets, particularly the ASVSpoof and ADD challenges. The authors carried out a unified comparison of representative detection methods. They concluded the paper with two major findings: the lack of large-scale datasets, and the poor generalization of existing detection methods to previously unseen attacks, as well as challenges in the interpretability of detection results.

In [13], the authors present the technologies involved in creating and detecting audio deepfakes, as well as detection methods based on machine learning and deep learning, including [SVM / Decision Trees / CNN / DNN / RNN]. They also describe the main audio deepfake generation methods, namely Text-to-Speech (TTS) and Voice Conversion (VC). This article provides an overview of the latest audio synthesis methods and detection approaches, both handcrafted feature-based and deep learning-based. The authors conducted a comparison of the effectiveness of traditional machine learning models and deep learning models in detecting audio deepfakes, based on model accuracy and success rates. They also compared the performance of different audio deepfake detection methods and datasets according to [size / format / sample types / languages / speaker accents].

The paper [14], the authors outlined the types of audio deepfake attacks, namely Text-to-Speech (TTS) and Voice Conversion (VC). They discussed differences among existing surveys in the literature and categorized datasets into "fully deepfake speech datasets" and "partially deepfake speech datasets." They also reviewed evaluation metrics used in audio deepfake detection literature. The authors provided an in-depth explanation of complete audio deepfake detection, covering feature extraction categories, classifier architectures, classifier types, as well as detection models not based on machine learning. They further discussed advances in training and robustness in complete detection, the integration of ASV (Automatic Speaker Verification) into countermeasures against deepfakes, and partial deepfake detection.

In [15], the authors introduce recent techniques for audio deepfake generation aimed at non-experts, with a focus on Text-to-Speech (TTS) and Voice Conversion (VC). They also present current trends in TTS, highlighting autoregressive and non-autoregressive models, as well as trends in VC, including deep learning-based models, text-based models, VAEs (Variational Autoencoders), and GANs. The authors also discuss evaluation metrics for TTS like (), the datasets used like (CSTR VCTK dataset), and present different open-source for TTS and VC like (ESPNet, Nvidia NeMo, TortoiseTTS...) and closed-source tools for TTS and VC like (ElevenLabs, ResembleAI, VoiceAI...).

In [16], the authors provides an introduction to non-specialists on the field of audio deepfakes, particularly in detection. The authors argue that a good detection model should be characterized by robustness, generalizability, explainability, and interpretability. This paper discusses audio deepfake attacks such as TTS and VC, like most other surveys. The authors also outline countermeasures against audio deepfakes, including (digital watermarking, adversarial attacks on audio manipulation systems), detection methods, datasets used for training detection models, and evaluation metrics for detectors. They classify detection methods based on (handcrafted features / deep learning-based representations / end-to-end models / self-supervised learning), and discuss additional aspects of audio deepfake detection such as [multilingual capability / cross-dataset performance / explainability / use of vocal emotions for detection, etc.].

In [3], the authors present a global overview of the audio deepfake field. They begin by tracing the evolution of deepfakes and note that research on video deepfakes far outweighs that on audio deepfakes, stressing the need for the scientific community to focus more on audio deepfakes in order to keep pace with advances in generative AI. In this paper, the authors explain audio deepfakes and their generation process, presenting TTS (with GAN-based approaches) and VC (also with GAN-based approaches). They also describe the detection process, distinguishing between feature extraction–based detection and classifier-based detection, and compare various detection strategies such as GAN-based, SSL (Self-Supervised Learning), Deep Learning, and Transfer Learning. The paper reviews the datasets used for audio deepfake detection and the evaluation metrics applied.

In [17], the authors present a comparative study of audio deepfake detection surveys, highlighting their contributions and limitations. This article, considered the most recent survey, provides a quantitative comparison of audio deepfake detection models and analyzes privacy, fairness, adaptability, and explainability in detection. It offers an overview of audio deepfake methodologies, datasets and their characteristics, as well as performance evaluation metrics such as Equal Error Rate (EER) and Tandem Detection Cost Function (t-DCF). The authors give a detailed explanation of detection approaches, categorized into two groups: front-end features (handcrafted and learning-based) and back-end features (machine learning and deep learning-based).

The following Table 1 gives a summary of surveys papers, that contain limitations and challenges, presented in this section.

Table 1. Summary of papers, that contain limitations and challenges, presented in related work

Papers and year	Key Findings	Limitations and challenges

(continued)

Table 1. (*continued*)

Papers and year	Key Findings	Limitations and challenges
[11] - 2022	A comparative description of the available faked audio datasets. A review of existing Audio deepfake detection methods was conducted. Limited Audio deepfake detection methods with Respect to Non-English languages. Excessive Preprocessing to Build deepfake detection Models. Limited Assessment of Noisy Audio in Existing Audio Deepfake Detection Methods Limited Audio Deepfake Imitation-Based Detection Methods	Limited Audio Deepfake Detection Methods with Respect to Non-English Languages like Arabic and Chinese / A Lack of Accent Assessment in Existing Audio Deepfake Detection Methods / Excessive Preprocessing to Build Deepfake Detection Models / Limited Assessment of Noisy Audio in Existing Audio Deepfake Detection Methods / Limited Audio Deepfake Imitation-Based Detection Methods.
[4] - 2023	Identify key differences between audio deepfake types and review competitions, datasets, features, classifiers, and evaluations. / Offer a unified comparison of features and classifiers on ASVSpoof 2021, ADD 2023, and In-the-Wild datasets.	Collecting audio datasets in the wild / Designing large-scale multilingual datasets / Improving generalization ability and robustness of detection models / Dealing with rapid development of deepfake technologies / Improving the interpretability of detection results / Exploring more reasonable evaluation metrics
[14] - 2024	Comprehensive review of the detection pipeline components: Model architectures / Optimization techniques / Generalizability / Evaluation metrics / Performance comparisons / Available datasets / Open-source resources	Reproducibility of detection models / Cross-dataset generalization / Interpretability of detection results / Robustness to multilingual with multiple accents / Robustness to adversarial attacks / Robustness to neural codec-based Deepfake speech / Streaming, Real-time detection / Privacy preservation
[3] - 2024	Present the evolution of audio deepfake/ Describe audio deepfake generation, pipeline & models without GAN and with GAN for TTS and VC / Describe audio deepfake detection, pipeline & models of detection for TTS & VC / Describe datasets for audio deepfake detection / Describe the evaluation metrics applied	Replay-based attacks out of scope / Lack of Comprehensive Datasets / Dataset Bias / Realism and Plausibility / Environmental Noise Interference / Real-Time Detection Challenges / Interpretability and Explainability

(*continued*)

Table 1. (*continued*)

Papers and year	Key Findings	Limitations and challenges
[17] - 2025	Provides a quantitative comparison of detection models / Analyze privacy, fairness, adaptability, and explainability in audio deepfake detection / Comparison of speech synthesis and voice conversion techniques. / Overview of additional datasets and their characteristics / Baseline performance of ASVSpoof and ADD challenges	Robustness Against Compression / Real-World Implications / Enhancing generalization and adaptability / Education and training / Developing lightweight models / Improving dataset diversity and fairness / Ethical frameworks and policies / Improving Model explainability

4 Discussion and Future Works

In this research on deepfake audio detection, several limitations remain and continue to hinder its effectiveness and widespread use. Most current methods are developed and tested primarily on the English language, which reduces their performance when applied to other languages, particularly Arabic and low-resource languages. In addition, speakers' accents, which are crucial to the perceived authenticity of a voice, are largely neglected in existing approaches. Detection pipelines also suffer from often excessive preprocessing, which can bias results and limit their relevance in real-world conditions, where audio is generally disturbed by background noise that is not sufficiently taken into account. Furthermore, little work has been done on specific scenarios such as voice imitation or proofreading, even though these constitute credible threats. In light of these limitations, several avenues for improvement are emerging: the creation of richer and more representative databases, including real-world environments, as well as the creation of large-scale multilingual corpora. In addition, strengthening the robustness and generalization capacity of models, improving the interpretation of results, and developing more appropriate evaluation metrics appear to be essential for increasing the reliability of detection systems. Finally, a major challenge remains the issue of "false positives" during voice authentication or identification. To remedy this, it is necessary to go beyond the acoustic characteristics of the voice alone and also incorporate non-acoustic features, such as analysis of the vocabulary and linguistic style specific to the person being authenticated.

5 Conclusion

This paper presents an entry in the domain of audio deepfake. It discusses the basics of audio deepfake, we are presenting the two categories of audio deepfake generation, Text-To-Speech, and Voice Conversion, and presenting a generic pipeline to illustrate the steps of creation of an audio deepfake. The paper also presents a state-of-the-art overview of the most well-known and recent scientific surveys. We have collected the most important surveys papers, which in turn have collected and analyzed a huge number of academic articles interested in audio deepfake generation and detection propositions. We have also revealed a point of similarity between these articles to open another, equally important door, which is the detection of audio deepfake with using the non-acoustic, with using the linguistic features, vocabulary of speech and the semantic of voice contents.

References

1. Qu'est-ce que l'IA generative: définition, histoire et cas d'usage. https://data-ai.theodo.com/fr/parlons-data/quest-ce-que-l-ia-generative-definition-histoire-et-cas-d-usage. Accessed 21 Sept 2025
2. Masood, M., Nawaz, M., Malik, K.M., Javed, A., Irtaza, A., Malik, H.: Deepfakes generation and detection: state-of-the-art, open challenges, countermeasures, and way forward. Appl. Intell. **53**, 3974–4026 (2023)
3. Wani, T.M., Qadri, S.A.A., Wani, F.A., Amerini, I.: Navigating the soundscape of deception: a comprehensive survey on audio Deepfake generation, detection, and future horizons. Found Trends® Priv Secur. **6**, 153–345 (2024)
4. Yi, J., et al.: Audio Deepfake Detection: A Survey (2023). https://doi.org/10.48550/ARXIV.2308.14970
5. Khanjani, Z., Watson, G., Janeja, V.P.: How Deep Are the Fakes? Focusing on Audio Deepfake: A Survey. ArXiv (2021)
6. Gupta, G., Raja, K., Gupta, M., Jan, T., Whiteside, S.T., Prasad, M.: A comprehensive review of DeepFake detection using advanced machine learning and fusion methods. Electronics. **13**, 95 (2023)
7. Wu, Z., et al.: ASVspoof: the automatic speaker verification spoofing and countermeasures challenge. IEEE J. Sel. Top. Signal Process. **11**, 588–604 (2017)
8. Frank, J., Schönherr, L.: WaveFake: A data set to facilitate audio DeepFake detection (2021). https://doi.org/10.5281/ZENODO.4904579
9. Reimao, R., Tzerpos, V.: FoR: a dataset for synthetic speech detection. Int. Conf. Speech Technol. Hum. Comput. Dialogue SpeD. **2019**, 1–10 (2019)
10. Müller, N.M., et al.: Does Audio Deepfake Detection Generalize? (2022). https://doi.org/10.48550/ARXIV.2203.16263
11. Almutairi, Z., Elgibreen, H.: A review of modern audio Deepfake detection methods: challenges and future directions. Algorithms. **15**, 155 (2022)
12. Khanjani, Z., Watson, G., Janeja, V.P.: Audio deepfakes: A survey. Front. Big Data. **5**, 1001063 (2023)
13. Shaaban, O.A., Yildirim, R., Alguttar, A.A.: Audio Deepfake approaches. IEEE Access. **11**, 132652–132682 (2023)
14. Li, M., Ahmadiadli, Y., Zhang, X.-P.: A survey on speech Deepfake detection. ACM Comput. Surv. **57**, 1–38 (2025)
15. Choi, J.-E., Schäfer, K., Zmudzinski, S.: Introduction to audio Deepfake generation: academic insights for non-experts. In: 3rd ACM Int Workshop Multimed AI Disinformation, pp. 3–12 (2024)
16. Schäfer, K., Choi, J.-E., Zmudzinski, S.: Explore the world of audio Deepfakes: a guide to detection techniques for non-experts. In: 3rd ACM Int Workshop Multimed AI Disinformation, pp. 13–22 (2024)
17. Zhang, B., Cui, H., Nguyen, V., Whitty, M.: Audio Deepfake detection: what has been achieved and what lies ahead. Sensors. **25**, 1989 (2025)

Phonetic-Based Arabic Text Articulation System for Children Using Machine Learning

Ahmed Mohamed[1], Ziad Ihab[1], Alaa Gamal[1], Eslam Salah[1], Mohamed Firas[1], Yehia Abo Bakr[1], Ali Eldeen Tarek[1], and Gehad Ismail Sayed[2(✉)]

[1] School of Computer Science, Canadian International College (CIC), Cairo, Egypt
[2] Faculty of Informatics and Computer Science, The British University in Egypt (BUE), Cairo, Egypt
Gehad.Ismail@bue.edu.eg

Abstract. Learning to read Arabic correctly provides a particular difficulty for young learners due to its rich phonetic structure, complex script, and diacritical marks that substantially alter pronunciation. However, many students struggle to evaluate their development without direct teacher evaluation. This paper presents a machine learning-based system for automatically assessing phonetic articulation of Arabic text, primarily for educational purposes with children. The system has several stages: first, the child chooses a reading and a verse to practice. Following repeated listening and memory, the child records their reading using a microphone. The recorded audio is then subjected to two major preprocessing steps: noise reduction and segmentation, in which the audio is cleaned and separated into smaller sections to increase analysis accuracy. Next, machine learning classification model is used to compare the child's reading to the reference audio. Finally, a feedback module provides the evaluation results, indicating whether the reading is correct and highlighting mistakes. The K-Nearest Neighbors (KNN) algorithm outperformed the other classifiers, with a final accuracy of 90%, precision and recall of 66.67%, and an F1-score of 80%, making it the best-balanced model for the task. Other models, such as MLP and Random Forest, achieved 73% and 70% accuracy, respectively, whilst SVM and Decision Tree performed at 60% and 50%. These findings indicate the suggested system's efficiency in giving intelligent and reliable feedback to help children improve their Arabic text articulation on their own.

Keywords: Audio Classification · Arabic Text Articulation · Machine Learning · Children Education · Speech Processing

1 Introduction

Learning to read Arabic accurately presents a particular difficulty for young learners due to its rich phonetic structure, complex script, and diacritical marks (tashkeel) that influence pronunciation. Traditional teaching approaches frequently rely significantly on human supervision and may not successfully adjust to individual learning paces [1]. Conventional teaching methods often rely on in-person instruction, which is not always

A. E. Hassanien et al. (Eds.): AISDS 2025, LNDECT 287, pp. 158–166, 2027.
https://doi.org/10.1007/978-3-032-18927-1_16

accessible, especially for learners in remote areas or countries. The existing digital tools tend to focus on theoretical explanations rather than helping learners create a natural and realistic flow of reading. Furthermore, these technologies usually lack real-time feedback, making it difficult for students to self-correct and track their progress [2]. Many learners confront interrelated obstacles such as unfamiliarity with Arabic phonemes, a lack of tailored assistance, and a lack of motivation owing to ineffective content. These difficulties together impede the development of fluent and confident phonetic-based Arabic text articulation, which necessitates more than just theoretical rule memorization; it necessitates a thorough phonetic and rhythmic knowledge of phonetic-based Arabic text articulation, as revealed [3].

The most current phonetic articulation of Arabic text recognition systems are employed a variety of deep learning and speech processing algorithms. Al-Harere and Al-Jallad (2023) proposed an end-to-end deep learning technique that uses CNN-BiGRU and a CTC loss function to achieve recognition with a character error rate of 2.42% [4]. In another paper, they reported a mispronunciation detection system for phonetic-based Arabic text articulation rules that used Mel-Frequency Cepstral Coefficients (MFCC) features and Long Short-Term Memory (LSTM) networks to achieve up to 96% accuracy [5]. In one of the identification papers, audio was turned into MFCC-based visuals, and multiple deep learning architectures were tested, with NASNetLarge achieving an accuracy of 98.5% [6]. The system used a PocketSphinx toolkit to identify phonetic-based Arabic text articulation verse recitals and convert them to text with 100% accuracy and 0% word error [7]. Another research team created a transformer-based Automatic Speech Recognition (ASR) system with a 6.16% word error rate and a 1.98% character error rate, which was exceptionally effective and well-suited to phonetic-based Arabic text articulation recognition [8]. Alagrami and Eljazzar developed Smartajweed, which used Support Vector Machine (SVM) to automatically detect violations of rules in reading Arabic text articulation [9]. Another study developed a rule-based and Hidden Markov Model (HMM) hybrid phoneme classification method to improve the recognition of compliant reading of Arabic text articulation [10]. Moustafa and Aly utilized Wav2Vec2.0 and HuBERT on a group of Arabic text articulation readers, achieving 98% and 97.1% speaker identification rates, respectively [11]. Another study employed MFCC and Gaussian Mixture Models in Arabic text articulation reading measuring learning software [12].

This paper presents a machine learning-based model to automatically assess phonetic articulation of Arabic text. It is made up of two primary parts: the front-end and the backend. In the front-end, the user selects a reader, followed by a specific Arabic script to practice. To help with remembering, the system constantly plays the selected verse. When the child is ready, he or she uses a microphone to record his/her reading. After recording, the system examines the input and offers quick feedback, indicating if the reading is valid or contains errors, as well as a complete report to guide future improvements. The proposed model at the back-end contains three major phases: preprocessing, classification, and assessment. During the preprocessing phase, the child's recorded audio is segmented and cleaned using noise reduction. Data augmentation and oversampling techniques are then used to improve the dataset. The processed data is utilized to train machine learning classifiers like KNN and Random Forest that compare the child's reading to the original audio. Finally, the evaluation step gives performance indicators like as accuracy, precision, recall, and F1-score to generate useful feedback and aid in the learning process. This paper's key contribution is summarized below:

(a) Presenting an end-to-end system that uses machine learning algorithms to analyze phonetic articulation of Arabic text in real-time.
(b) Creating a feedback system based on phonetics that emphasizes the accuracy of reading in general.
(c) Using an accessible and adaptable learning platform to address the difficulties that children and non-native Arabic speakers encounter during studying.
(d) Assessing how well the system works to increase child engagement and the correctness of Arabic text articulation in various learning scenarios.

The rest of the paper is structured as follows: A description of the adopted dataset is given in Sect. 2. The proposed system is covered in depth in Sect. 3. The results of the analysis and discussion are shown in Sect. 4. A summary of the key conclusions and suggestions for additional research are provided at the end of Sect. 5.

2 Dataset Description

In order to train and assess the proposed phonetic articulation of Arabic text system, the dataset utilized in this paper was privately collected. One hundred child audio recordings, featuring both male and female voices, make up the collected dataset. Fifty of the recordings are identified as correct readings, and the other fifty are labeled as incorrect. A child reading of a particular Arabic script can be found in each audio recording. Mel Frequency Cepstral Coefficients (MFCC), a popular method for collecting the essential audio characteristics and characterizing the short-term power spectrum of sound, were extracted from all recordings. From every recording, 32 MFCC features were taken out. In order to prepare the audio data for classification, this preprocessing step is crucial. Overfitting and generalization were carefully taken into account during model training because of the dataset's small size and balanced nature (equal numbers of correct and incorrect readings). The dataset is confidential and not publically accessible, hence all tests and analyses were carried out locally on a desktop computer.

3 The Proposed System

This paper proposes a machine learning-based system for evaluating the phonetic articulation of Arabic text, specifically designed for children. The system is composed of two primary components: the front-end and the back-end. Audio selection and learning, capturing the child's reading, audio preprocessing (cleaning and segmentation), machine learning model classification, and final feedback generation are all included in the back-end portion. After that, a thorough explanation of each component.

3.1 Front-End Part

The user selects an Arabic text to study after choosing a reader in the front-end section. The chosen script is played over and over again to aid in the child's learning and memorization. The child uses the device's microphone to record their own reading when they're ready. The input is processed by the system after it has been recorded, and it provides an instructive report to encourage further learning as well as comments indicating if the reading is accurate or contains errors.

3.2 Back-End Part

In this section, the proposed system first records the child reading before proceeding with audio preprocessing, which entails noise reduction and audio segmentation. Data augmentation is then used to expand the amount of training data after the data has been prepared by separating it into training, testing, and validation sets. The model is then trained and tested using machine learning classifiers such as KNN and random forest classifier (RFC) to determine whether the reading is accurate by comparing it to the original, accurate audio. In order to enhance the learning process, the system provides feedback and displays evaluation results based on accuracy, precision, recall, and F1-score. The block diagram for the back-end portion is displayed in Fig. 1.

3.2.1 Data Preprocessing Phase

Audio preprocessing is the initial stage of the system, which begins after collecting children's recorded audio. The data is manually gathered using the system's recording capability because there isn't a publicly accessible dataset for this task. Following data collecting, the audio undergoes crucial preprocessing procedures to get it ready for the following phases.

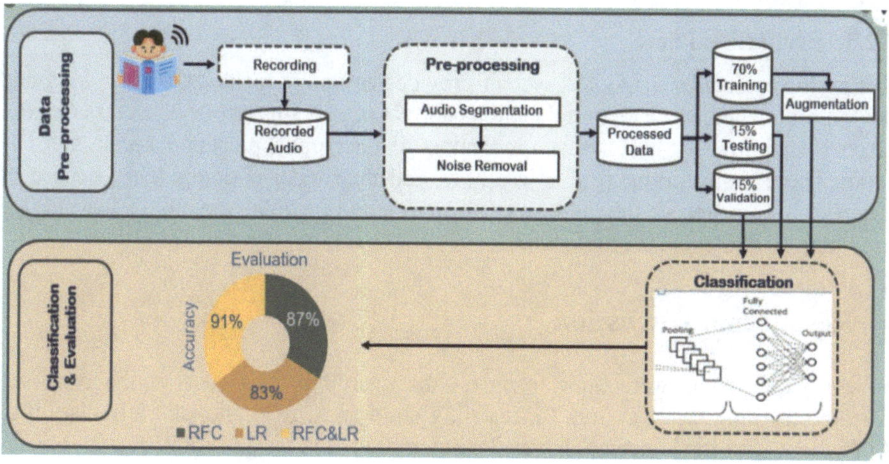

Fig. 1. The block diagram of the back-end part

Audio segmentation is the first step, in which each recorded audio clip is broken up into smaller segments. This enhances the processing quality and makes handling the audio easier. Following segmentation, noise removal is the following stage, which helps to clean the audio by eliminating unwanted noise and background sounds. This improves the accuracy and clarity of the audio for the learning model. Following these steps, the audio dataset is processed and divided into three parts: 70% for training, 15% for validation, and 15% for testing. Each part is essential to the model's development. While validation data aids in fine-tuning the model's parameters, training data is utilized to teach the model. Evaluation of the final performance is done using the testing data. By producing additional variations of the current recordings, data augmentation is also used to expand the quantity of training data. This increases the system's overall efficacy and helps the model better identify the various ways children might read phonetic articulation of Arabic text.

3.2.2 Classification Phase

The preprocessed and enhanced audio data is fed into machine learning classifiers during the classification stage. In this stage, the Random Forest Classifier and K-Nearest Neighbors (KNN) are the two primary algorithms used. After experimenting with several algorithms, these models were chosen because they demonstrated strong performance in determining whether or not the child's reading was accurate. To increase the models' accuracy and performance, they are trained using the training dataset and validated using the validation dataset. The classification's purpose is to compare the child's recorded audio to the original correct audio and determine if they match or have errors. The trained classifiers learn to recognize auditory patterns and attributes that allow them to categorize recordings as correct or incorrect reading. This classification stage is critical to the system because it allows for accurate and helpful feedback to the child based on their ability to interpret phonetic articulation of Arabic text.

3.2.3 Evaluation Phase

During the evaluation stage, the system provides complete feedback by checking and determining whether the recorded audio is correct or contains errors. The classification result is supported by performance indicators like accuracy, precision, recall, and F1-score. These indicators help to assess how well the model is doing. The purpose of this phase is to improve the model's performance and facilitate the child's learning by providing accurate and useful feedback.

4 Results and Discussion

Several experiments were done to assess the overall performance of the proposed machine learning-based system for teaching children to read phonetic articulation of Arabic text. The implementation was done in Python and tested against five classification algorithms. Table 1 compares KNN, Multilayer Perceptron (MLP), Random Forest, SVM, and Decision Tree based on a variety of performance criteria, including accuracy (before and after preprocessing), precision, recall, F1 score, and training parameters. As indicated in the table, KNN is the best technique for this problem, with a final accuracy of 90%, precision and recall of 66.67%, and F1-score of 80%. The high performance was attained with 5 neighbors and Euclidean distance. The MLP model, on the other hand, achieved an accuracy of 73% and an F1-score of 50%, indicating that it improved after preprocessing but had significantly lower consistency. Random Forest did relatively well, with an accuracy of 70% and an F1-score of 57%, while SVM and Decision Tree achieved lower results, with accuracy levels of 60% and 50%, respectively. These findings, as presented in Table 1, confirm that KNN is the best algorithm for providing accurate and reliable feedback in Arabic text reading evaluation tasks.

Table. 1. A comparison between different machine leaning algrithms with different paramater setting

Model	Accuracy	Precision	Recall	F1-Score	Parameters
Random Forest	0.7000	0.5000	0.6667	0.5714	n_estimators = 100
SVM	0.6000	0.4000	0.6667	0.5000	kernel = 'linear', probability = True
KNN	0.9000	0.6667	0.6667	0.8000	k = 5, Distance Metric = 'Euclidean'
Decision Tree	0.5000	0.4444	0.4444	0.4444	random_state = 42
Neural Network	0.7300	0.5800	0.5800	0.5000	hidden_layer_sizes = (100,), max_iter = 200

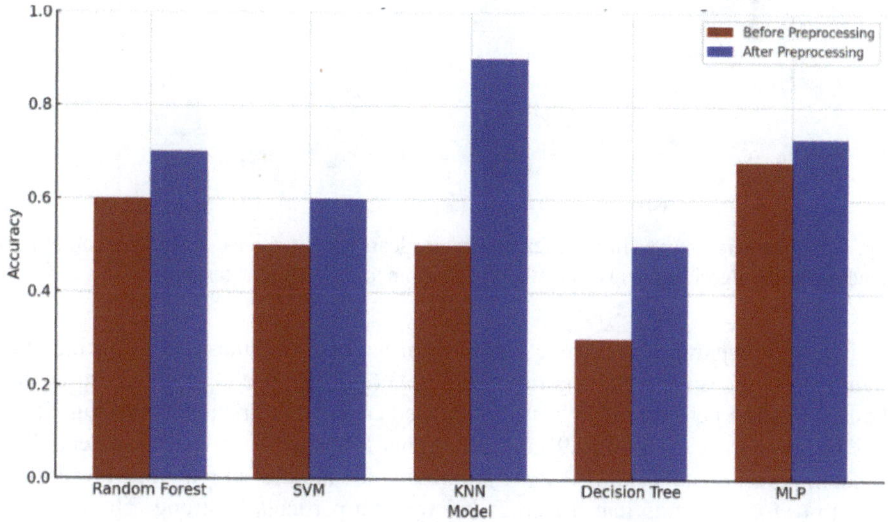

Fig. 2. Comparison between accuracy before and after preprocessing

Figure 2 shows accuracy before and after preprocessing for all five machine learning algorithms, with KNN showing the most substantial improvement, increasing from 50% to 90%, making it the most successful model after preprocessing. Random Forest increased from 60% to 70%, whilst SVM improved from 50% to 60%. Finally, the Decision Tree model had the lowest overall performance but improved slightly, going from 30% to 50%. Based on these findings, it is clear that preprocessing greatly improves accuracy, with KNN benefiting the most, confirming its selection as the best-performing machine learning algorithm for classification task.

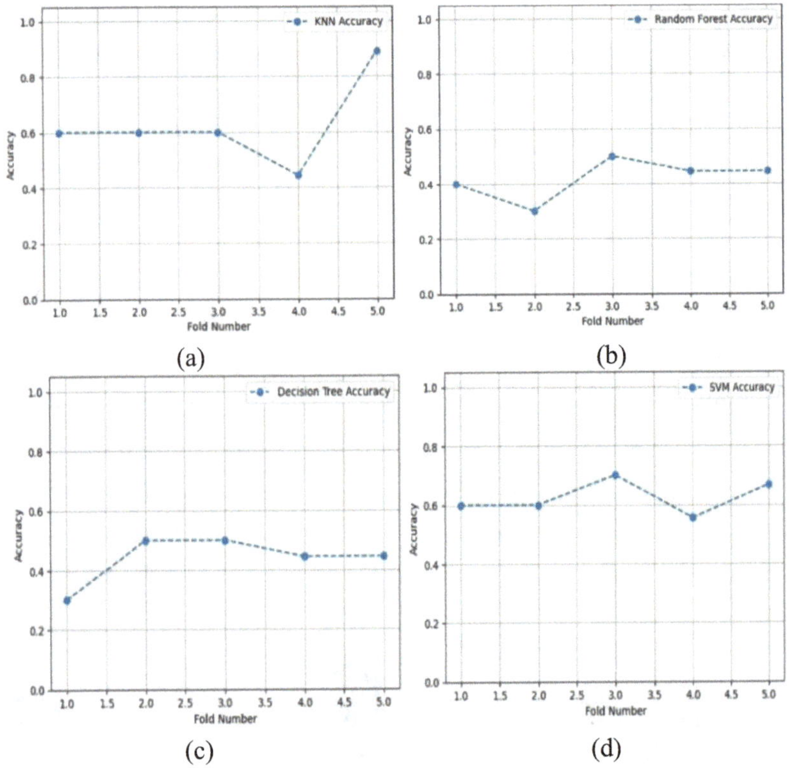

Fig. 3. Convergence curves from different machine learning algorithms, (a) KNN algorithm (b) Random Forest algorithm, (c) Decision Tree algorithm and (d) SVM algorithm.

Figure 3 displays convergence curves from various machine learning techniques. Figure 3(a) depicts the KNN algorithm. Figure 3 (a), (b), (c), and (d) demonstrate that the KNN model performed the best. It remained constant at first until finishing with a very high accuracy close to (0.9), indicating that it learned very well in the end. The Random Forest model exhibited some progress in the middle but remained low around (0.4) to (0.5), indicating that it learned but was not particularly strong. The Decision Tree model started low, improved slightly at fold 2, but sank again and remained weak, resulting in poor performance. The SVM algorithm was more stable, starting at (0.6), improving slightly to (0.7), but not going very high, indicating small progress but not very strong learning. Overall, the KNN model performed the best of the four, with the highest final accuracy and convergence.

Figure 4 displays Receiver Operating Characteristic (ROC) curves from several machine learning algorithms. Figure 4 (a), (b), (c), and (d) demonstrate how accurately the models classify classes. Starting with Fig. 4 (a), the SVM model produced the best curve with a high area under curve (AUC) value of (0.86), indicating that it was very good at proper classification. The Random Forest model, shown in Fig. 4 (b), also had a good curve with an AUC of 0.74, indicating that it worked well but was not as powerful as SVM. The Fig. 4 (c) Decision Tree model had the weakest curve, with an AUC of 0.55, which was extremely near to random guessing, indicating that it was not good at class separation. The KNN model in Fig. 4 (d) had a good curve with an AUC of 0.71, indicating that it outperformed Decision Tree but fell short of SVM and Random Forest.

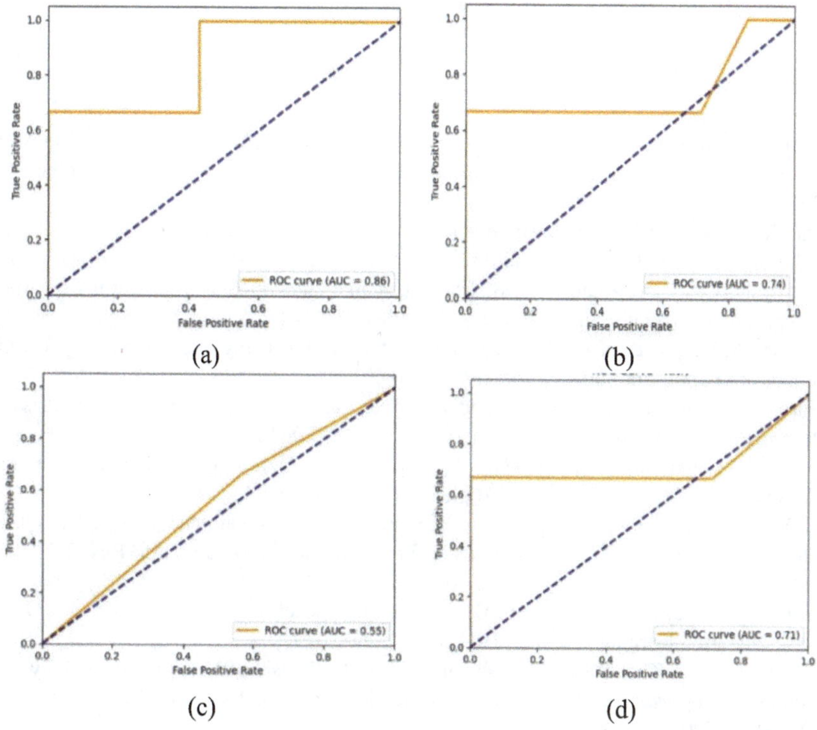

Fig. 4. ROC curves from different algorithms (a) SVM algorithm, (b) Random Forest algorithm, (c) Decision Tree algorithm and (d) KNN algorithm

5 Conclusion

This paper presents a machine learning-based system to help children learn and improve their phonetic articulation of Arabic text. The proposed system tackles the difficulty of giving accurate and timely feedback without requiring ongoing teacher supervision. Audio selection, recording, preprocessing (noise reduction and segmentation), feature extraction, and classification are all steps in the system's systematic process. Extensive experiments with several machine learning classifiers were carried out. The KNN algorithm had the best performance, with an accuracy of 90%, an F1-score of 80%, precision of 66.67%, and recall of 66.67%. In comparison, other models such as MLP and Random Forest achieved accuracy scores of 73% and 70%, respectively. SVM and Decision Tree classifiers have lower accuracies of 60% and 50%, respectively. These results demonstrate the proposed KNN-based system's ability to provide intelligent and reliable feedback. The system empowers children by allowing for self-directed practice, constant evaluation, and engagement. Future enhancements may include the use of deep learning architectures, real-time microphone integration, and dataset expansion to increase feedback quality and classification performance. Future work could concentrate on increasing the dataset, incorporating real-time recording via microphone, and investigating deep learning architectures to increase classification accuracy and feedback quality.

References

1. Asadi, I.A., Kawar, K., Tarabeh, G.: Development of verb and noun word patterns in Arabic: a comparison between typically developing children and those with reading difficulties. J. Speech Lang. Hear. Res., 1–13 (2025)
2. Sari, R., Sidik, A.: Digital media innovation for Istima'Teaching in Arabic language education: literature study. Al-Himam: Jurnal Ilmu Ilmu Pendidikan Bahasa Arab. **4**(2), 40–72 (2025)
3. Pasaribu, A.R.I., Al Khalili, M.M.: Phonetics in digital media: implications for speech recognition technology. Jurnal Recoms. **1**(2), 68–79 (2024)
4. Harere, A.A., Jallad, K.A.: Quran recitation recognition using end-to-end deep learning. (2023) arXiv preprint https://arxiv.org/abs/2305.07034.
5. Harere, A.A., Jallad, K.A.: Mispronunciation detection of basic quranic recitation rules using deep learning. arXiv preprint https://arxiv.org/abs/2305.06429 (2023)
6. Saber, H.A., Younes, A., Osman, M., Elkabani, I.: Quran reciter identification using NASNetLarge. Neural Comput. & Applic. **36**(12), 6559–6573 (2024)
7. Al-Kaf, H.A.G., Sulong, M.S., Joret, A., Aminuddin, N.F., Mohammad, C.A.: QVR: Quranic verses recitation recognition system using PocketSphinx. J. Quranic Sci. Res. **2**(2), 35–41 (2021)
8. Putra, B., Atmaja, B.T., Prananto, D.: Developing speech recognition system for Quranic verse recitation learning software. IJID. **1**(2), 1–8 (2012)
9. Nahar, K.M., Al-Khatib, R.M., Al-Shannaq, M.A., Barhoush, M.M.: An efficient holy Quran recitation recognizer based on SVM learning model. Jordanian J. Comput. Inf. Technol. **6**(04), 394–414 (2020)
10. Nahar, K.M., Abu Shquier, M., Al-Khatib, W.G., Al-Muhtaseb, H., Elshafei, M.: Arabic phonemes recognition using hybrid LVQ/HMM model for continuous speech recognition. International Journal of Speech Technology. **19**(3), 495–508 (2016)
11. Moustafa, A., Aly, S.A.: Towards an efficient voice identification using Wav2Vec2. 0 and HuBERT Based on the Quran Reciters Dataset. (2021). arXiv preprint https://arxiv.org/abs/2111.06331.
12. Ait Mait, H., Aboutabit, N.: HMM-GMM acoustic Modeling for Arabic speech recognition system. In: International Conference on Mathematics Data Science, pp. 1–13. Springer Nature Switzerland, Cham (2024)

Healthcare and Smart Diagnostics

Explainable Machine Learning Models for Diabetes Mellitus Prediction: A Comparison of SMOTE-Enhanced Neural Networks and a Genetic Algorithm Optimized XGBoost Models

Solomon Nwaneri[1(✉)] and Oluwakemi Odukoya[2]

[1] Department of Biomedical Engineering, Faculty of Engineering, University of Lagos, Akoka, Lagos 101017, Nigeria
snwaneri@unilag.edu.ng

[2] Department of Community Health and Primary Care, College of Medicine, University of Lagos, Idi-Araba, 101014 Lagos, Nigeria

Abstract. Diabetes mellitus (DM) is a complex illness that can be effectively managed if detected early. Conventional machine learning models for DM prediction are often simplistic and lack interpretability. This study aimed to develop explainable machine learning models for DM prediction, including SMOTE-enhanced Neural Networks and a Genetic Algorithm (GA) optimized XGBoost models. The models were implemented in Python using Spyder Integrated Development Environment and trained on the PIMA Indian diabetes dataset. Hyperparameter optimization using GA improved model performance, achieving a best ROC-AUC of 0.8438 for the XGBoost model, compared to the initial population. The GA-optimized XGBoost model achieved an accuracy of 0.75, outperforming the Synthetic Minority Over-Sampling Technique (SMOTE) enhanced Neural Networks, which achieved an accuracy of 0.721. To enhance transparency and interpretability, SHAP (Shapley Additive Explanations) was applied. Overall, GA-based hyperparameter tuning led to a more reliable and interpretable XGBoost model, with ROC-AUC improvement from 0.8402 to 0.8438. Therefore, GA-optimized XGBoost model showed better performance compared to SMOTE-enhanced model for DM prediction.

Keywords: Artificial Neural Networks · Diabetes Mellitus · Explainable Machine Learning Model · Genetic Algorithm · XGBoost Model

1 Introduction

Diabetes Mellitus (DM) is a disease that either prevents a person's ability to produce or utilize the insulin already produced by the pancreas, making glycemic control difficult [1]. Long-term effects of uncontrolled DM could lead to various complications such as damage to blood vessels and organ damage [2–4]. Insulin resistance is common in persons suffering from T2DM. This condition is often associated with impaired glucose

A. E. Hassanien et al. (Eds.): AISDS 2025, LNDECT 287, pp. 169–178, 2027.
https://doi.org/10.1007/978-3-032-18927-1_17

regulation [5, 6]. Over 537 million adults (20–79 years) currently suffer from DM, with a prevalence of 10.5% [7]. Common risk factors for DM include sedentary lifestyle, family history, unhealthy dietary habits, tobacco smoking, older age, and obesity [8]. The conventional methods of diagnosis involve the use of painful, invasive diagnostic methods, of fasting blood glucose, which require piercing the skin with a needle. To address this problem, there is a need to apply machine learning (ML) in non-invasive point-of-care DM detection in affected people to enable early commencement of medications and prevention of long-term complications, as well as increase their life expectancy. Various studies in machine learning have led to improvements in disease diagnosis. They serve as assistive tools to medical practitioners in the early prediction of diseases, including DM. Common examples of machine learning approaches used in disease prediction are found in [1, 9]. Of the numerous ML approaches used in disease prediction, ANNs are easy to implement and consist of a system of interconnected units called nodes, and are implemented by a network structure that is constituted by the interaction of layers and a local function [10]. Of the various types of ANNs, the Feed-forward ANN topology is the most widely [11].

Some studies have previously investigated ML models for DM prediction. Jayalakshmi and Santhakumaran [12] developed an ANN classification model for DM prediction and investigated different missing values imputation and preprocessing techniques. However, the study lacked explainability and insight into how preprocessing techniques led to improved accuracy. In a study [13], a classification approach was proposed using an ANN and logistic regression T2DM prediction in rural adults without relying on biochemical parameters. However, the positive predictive value of the model was very low (31.86%) despite having strong sensitivity and AUC. The model also lacked interpretability. Also, in [14], three ML models were developed and assessed: logistic regression, ANN, and a decision tree in predicting DM based on various features in a Nigerian population, but the model was not explainable. In [15], a classification model was developed for the detection of DM and identification of its type. The study did not address the problem of class imbalance associated with datasets. Class imbalance occurs when the majority of the output of a dataset belongs to a particular class, while the minority belongs to another class. This problem is common in ML classification problems as the models are likely to predict the majority class [16]. Pima Indian datasets, a well-known DM dataset, usually suffer from class imbalance. Similarly, the majority of previous machine learning models were not explainable. This paper aimed to develop explainable ML models for DM prediction, comprising SMOTE-enhanced Neural Networks and Hybrid XGBoost with Genetic Algorithm (GA) Optimisation. The objectives of the study are to: (1) develop an explainable neural network model for early diabetes mellitus detection to address class imbalance. (2) Develop an explainable Hybrid XGBoost model for Diabetes Mellitus Prediction. (3) Compare the performance of the SMOTE-enhanced Neural Networks and the Hybrid GA optimized XGBoost models.

2 Methodology

The models were developed using the Python programming language due to its readability and versatility in scientific computing. The programmes were implemented using multiple essential libraries, including TensorFlow, Keras, scikit-learn, NumPy, Pandas,

Matplotlib, Seaborn, and joblib. Model coding and debugging were performed in Spyder, an open-source integrated development environment (IDE). All experiments were conducted on a Windows 11 (64-bit) computer. The flowchart in Fig. 1 describes the methodology for the study.

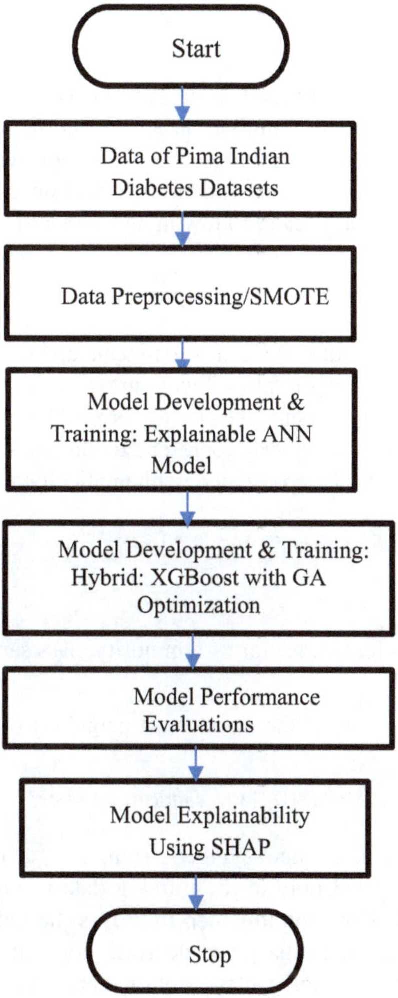

Fig. 1. Flowchart showing the Methodology

2.1 Data Description

The study used the PIMA Indian Diabetes Dataset, a secondary dataset obtained from the Kaggle Repository. This dataset contains 768 samples with eight input features and one output feature. The input features include: Plasma Glucose Concentration, Age, Diabetes

Pedigree Function, Skin Thickness, Diastolic Blood Pressure, 2-h Serum Insulin, Body Mass Index (BMI), and Number of Times Pregnant [17]. The single output feature represents the patient's diabetic status, with 1 denoting diabetic and 0 denoting non-diabetic. Of the 768 patients, 268 (34.9%) were diabetic and 500 (65.1%) were non-diabetic, demonstrating class imbalance.

2.2 Data Preprocessing

Initial data exploration was performed using Pandas functions, which confirmed the dataset size as 768 × 9. Missing numbers in any of the features were replaced with the mean of the feature. To mitigate differences in feature scales, the StandardScaler function from scikit-learn was applied. This transformation ensured the normalization of the input values to a standard scale, following the standardization equation:

$$z = \frac{x - \mu}{\sigma} \tag{1}$$

where x represents the input feature, μ is the mean, and σ is the standard deviation.

The Synthetic Minority Oversampling Technique (SMOTE) was applied to the training set of the PIMA Indian Diabetes dataset to correct the class imbalance. SMOTE generates synthetic examples for poorly represented outcomes (diabetic patients) using an interpolation method. This is represented mathematically as:

$$x_{new} = x_i + \delta \bullet (x_{zi} - x_i) \tag{2}$$

where:

$$x_i = \text{a feature vector of a minority class sample,}$$

$$x_{zi} = \text{one of the k} - \text{nearest neighbors of } \mathrm{x_i}$$

$$\delta \sim U(0, 1) = \text{random number}$$

This equation implies that a new synthetic sample x_{new}, causes variability in the datasets. SMOTE was applied only to the training data to minimize the risk of data leakage in the test set. This resampling step improves the model's sensitivity for the minority class while maintaining balanced decision boundaries. Thus, it reduces the bias of the model toward the majority class (non-diabetic patients).

2.3 Explainable Neural Network Model

A feedforward ANN was developed with the Keras framework in TensorFlow to perform binary classification after partitioning the dataset into training (80%) and testing (20%) subsets. The model was built with eight neurons in the input layer, sixteen neurons in the first hidden layer with a ReLU activation function, eight neurons in the second hidden layer with a ReLU activation function, and one neuron in the output layer (8-16-8-1). The model was implemented with the Adam optimizer (learning rate = 0.001) with a loss function based on binary classification. While accuracy was the primary evaluation metric, training was conducted for 35 epochs with a batch size of 16, and 20% of the training data was used for validation. To ensure transparency and interpretability, SHAP (Shapley Additive exPlanations) was applied to the model. SHAP values revealed how each input feature affected the model's predictions. SHAP summary plots were generated to visualize the most important features affecting diabetes risk predictions.

2.4 Explainable Hybrid XGBoost with Genetic Algorithm Optimization

A Genetic Algorithm (GA) was implemented to optimize the hyperparameters of the Extreme Gradient Boosting (XGBoost) classifier. The GA mimics the process of natural evolution through selection, crossover, and mutation to explore the hyperparameter search space as shown in Table 1.

Table 1. Hyperparameter Search Space

Hyperparameter	Datatype	Upper and Lower Limits
Number of estimators	Integer	[50, 500]
Maximum Depth	Integer	[2, 10]
Learning Rate	Float	[0.01, 0.3]
Subsample	Float	[0.5, 1.0]
colsample_bytree	Float	[0.5, 1]
Minimum Child Weight	Integer	[1, 10]
Gamma	Float	[0.0, 5.0]

Each individual in the population represented a candidate solution with an initial population size of 25 individuals generated randomly, because it gave reasonable results after a few trials with other population sizes within that range. The fitness of each individual was evaluated by training an XGBoost classifier on the training-validation set and computing the mean Area Under the Receiver Operating Characteristic Curve (ROC-AUC) across three-fold stratified cross-validation. ROC-AUC was chosen as it provides a robust measure of discrimination performance in binary classification problems with class imbalance. The selection method adopted was Tournament selection with a tournament size of three. A uniform crossover operator exchanged hyperparameter values between two parents with a crossover probability of 0.9, producing two offspring. The mutation rate was 0.25, and the hyperparameter values were perturbed within $\pm 20\%$. Elitism ensured that the two best-performing individuals from each generation were preserved directly into the subsequent generation. The GA was executed for 25 generations. At each iteration, the population was evaluated, ranked, and evolved according to the operators above. The optimal hyperparameter set was selected at the end of the evolutionary process. SHapley Additive exPlanations (SHAP) library in Python was employed to implement explainability.

3 Results and Discussion

3.1 Model Performance for Artificial Neural Networks

Table 2 shows the model performance of the Explainable ANN model. The results demonstrate that the model can effectively discriminate between diabetic and non-diabetic cases. The training and validation accuracies of the model versus the number of epochs are shown in Fig. 2. The average accuracy of the model is 0.721.

Table 2. Results of Evaluation Metrics of Smote-Enhanced ANN

	Precision	Recall	F1-Score	Support
0	0.80	0.760	0.779	100
1	0.593	0.648	0.619	54

Figure 2 presents the accuracy of the ANN optimized with the Adam algorithm over 35 epochs. The graph indicates that both the training and validation accuracy curves exhibit a strong upward trend in the initial epochs (approximately 0–10). This implies that the model is effectively learning relevant features from the training data.

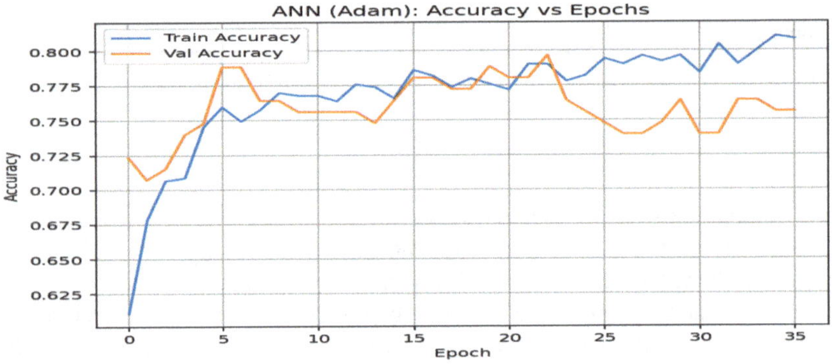

Fig. 2. Training and Validation Accuracy of the ANN Model versus Epochs

The final accuracies were stable around ~0.775–0.80 for training and ~ 0.75–0.775 for validation. The results establish a successful model training process with good generalization and minimal model overfitting. Figure 3 describes the ROC curve for the proposed explainable neural network model. The area under the ROC curve (AUC) was 0.88, reflecting a strong predictive performance and robustness of the classifier. The high AUC suggests that the model achieves an excellent balance between correctly identifying positive cases while minimizing false alarms. The ROC analysis, therefore, confirms that the proposed SMOTE-enhanced, explainable neural network is a reliable tool for early diabetes screening, complementing its interpretability through SHAP explanations.

3.2 Model Performance for GA Optimized XGBoost

Table 2 presents the progressive improvement of model discrimination (ROC-AUC) across generations in GA-Optimized XGBoost. The initial population achieved a best ROC-AUC of 0.8402, indicating a reasonably strong baseline model. For each successive generation, a gradual improvement of the model was observed up to the 25th generation. The initial population achieved a best ROC-AUC value of 0.8402, which indicates a relatively strong baseline model.

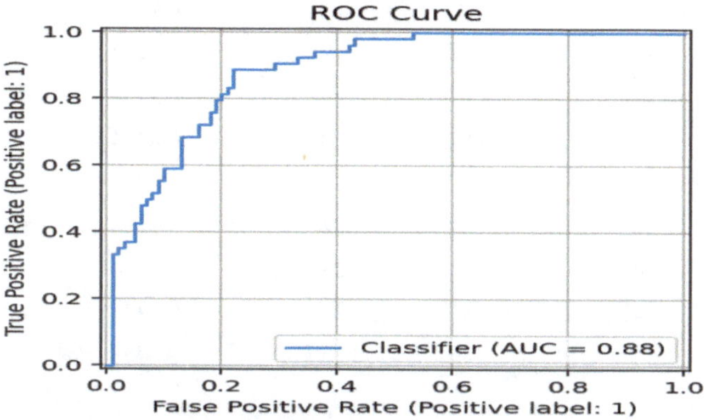

Fig. 3. ROC curve for ANN model

The GA achieved a best ROC-AUC of 0.8438, which is a significant improvement over the initial population. The average ROC-AUC was 0.8386, confirming that the population as a whole converged to high-performing hyperparameter sets. The use of GA for hyperparameter optimization can lead to a reliable XGBoost model with ROC-AUC improvement from 0.8402 to 0.8438. This demonstrates the effectiveness of evolutionary optimization for enhancing diabetes prediction (Table 3).

The GA-optimized XGBoost model results in Table 4 revealed an overall accuracy of 0.75. The accuracy is comparable to previously published studies that used PIMA Indian datasets for their evaluation. In a study by Ahmed et al. [18], the accuracies obtained for various models were: Logistic Regression (0.72), Naive Bayes (0.76), Decision Tree (0.71), and Random Forest (0.75).

Table 4. Results of Evaluation Metrics of GA optimized XGBoost model

	Precision	Recall	F1-Score	Support
0	0.79	0.84	0.81	100
1	0.66	0.57	0.61	54

Table 3. ROC-AUC) Across generations in GA-optimized XGBoost

Generation	Best ROC-AUC	Average ROC-AUC	Generation	Best ROC-AUC	Average ROC-AUC
1	0.8402	0.8273	13	0.8434	0.8369
2	0.8402	0.8321	14	0.8434	0.8372
3	0.8402	0.8346	15	0.8434	0.8376
4	0.8402	0.8343	16	0.8434	0.8389
5	0.8412	0.8356	17	0.8434	0.8395
6	0.8432	0.8354	18	0.8434	0.8385
7	0.8432	0.8363	19	0.8434	0.8389
8	0.8432	0.8362	20	0.8434	0.8394
9	0.8432	0.8377	21	0.8434	0.8392
10	0.8432	0.8375	22	0.8434	0.8385
11	0.8432	0.8371	23	0.8434	0.8391
12	0.8434	0.8378	24	0.8434	0.8390
			25	0.8438	0.8386

3.3 Model Explainability

To enhance transparency and interpretability, SHAP (Shapley Additive Explanations) was applied to the SMOTE-enhanced Neural Networks model and the GA optimized XGBoost model in Figs. 3 and 4, respectively. The SHAP plots reveal the most important features that influenced the model prediction as glucose level, BMI, and age. Glucose level is the strongest predictor of DM. In contrast, BMI is the second most important feature after glucose that contributes to DM. High BMI is common for overweight or obese persons. This finding is in agreement with previous studies indicating that obesity is a significant contributor to DM [19]. A comparison of Figs. 4 and 5 confirms the clinical relevance of the features of DM based on their importance with Glucose, BMI, and Diabetes Pedigree Function as the 3 strongest contributors to the ANN model, while Glucose level, BMI, and age were the strongest contributors to the GA optimized XGBoost model, which is similar to the results obtained in [20]. This explainability framework provides clinicians with clear insights into why the model predicted diabetes or not, so that they can determine if the predictions are reasonable. Thus, this makes it more trustworthy for real-world healthcare settings.

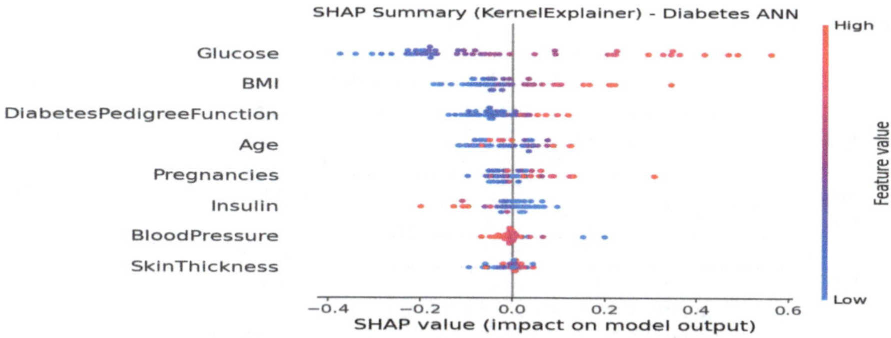

Fig. 4. SHAP Plot for the Explainable Neural Network Model

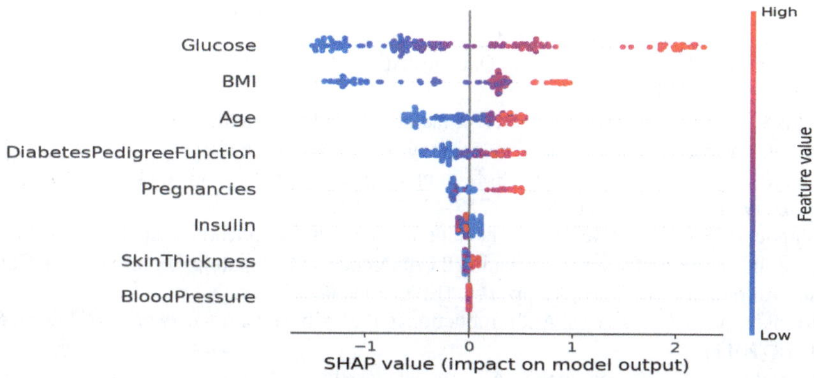

Fig. 5. SHAP Plot for the Explainable GA Optimized XGBoost

This study is limited by the quality and size of the PIMA Indian Diabetes Dataset used to train the models, which may affect the generalizability of the models. Also, the models were not validated by other external sources of data.

4 Conclusion

This study developed and evaluated explainable ML models for early diabetes mellitus prediction by training SMOTE-enhanced Neural Network and GA-optimized XGBoost model. The GA-optimized XGBoost model showed better performance compared to SMOTE-enhanced Neural Network. Model explainability was achieved using SHAP, which revealed glucose level, BMI, and age as the most influential factors. Overall, the integration of SMOTE, GA-based XGBoost optimization, and SHAP explainability resulted in reliable, interpretable, and clinically meaningful models that can support early diabetes detection in real-world healthcare settings.

References

1. Sengamuthu, R., Abirami, R., Karthik, D.: Various data mining techniques analysis to predict diabetes mellitus. Int. Res. J. Eng. Technol. **5**(5), 676–679 (2018)
2. Iyer, A., Jeyaldiatha, S., Sumbaly, R.: Diagnosis of diabetes using classification mining techniques. Int. J. Data Mining Knowl. Manag. Process. **5**(1), 1–14 (2015)
3. Sethi, H., Goraya, A.: Artificial intelligence-based ensemble model for diagnosis of diabetes. Int. J. Adv. Res. Comput. Sci. **8**(4), 1540–1548 (2017)
4. Borgharkar, S.S., Das, S.S.: Real-world evidence of glycemic control among patients with type 2 diabetes mellitus in India: the TIGHT study. BMJ Open Diabetes Res. Care. **7**, e000654 (2019). https://doi.org/10.1136/bmjdrc-2019-000654
5. Holzinger, A., Langs, G., Denk, H., Zatloukal, K., Müller, H.: Causability and explainability of artificial intelligence in medicine. Wiley Interdiscipl. Rev. Data Mining Knowl. Discovery. **9**(4), e1312 (2019)
6. Nwaneri, S., Yinka-Banjo, C., Uregbulam, U., Odukoya, O., Imoize, A.: Explainable neural networks in diabetes mellitus prediction. In: Imoize, A., Hemanth, J., Nath Sur, S. (eds.) Explainable Artificial Intelligence in Medical Decision Support Systems, pp. 313–334. The IET Digital Library, London (2023)
7. International Diabetes Federation, Diabetes Atlas. 10th ed. Brussels, Belgium: International Diabetes Federation; (2021).
8. Uloko, A.E., et al.: Prevalence and risk factors for diabetes mellitus in Nigeria: a systematic review and meta-analysis. Diabetes Therapy. **9**(3), 1307–1316 (2018)
9. Dilip, K.C., Sanchita, P., Joy, B.: Soft computing approaches for diabetes disease diagnosis: a survey. Int. J. Appl. Eng. Res., 11715–11726 (2014)
10. Shinde, P., Shah, S.: A review of machine learning and deep learning applications. In: Proceedings of the 2018 Fourth International Conference on Computing Communication Control and Automation (ICCUBEA), pp. 1–6. IEEE, Pune (2018)
11. Krenker, A., Bešter, J., Kos, A.: Introduction to the artificial neural networks. INTECH. **6**(2), 1–18 (2011)
12. Jayalakshmi, T., Santhakumaran, A.: A novel classification method for diagnosis of diabetes mellitus using artificial neural networks. In: Proceedings of the 2010 International Conference on Data Storage and Data Engineering (DSDE), pp. 159–163. IEEE, Bangalore (2010)
13. Wang, C., et al.: Evaluating the risk of type 2 diabetes mellitus using an artificial neural network: an effective classification approach. Diabetes Res. Clin. Pract. **100**(1), 111–118 (2013)
14. Odukoya, O., et al.: Development and comparison of three data models for predicting diabetes mellitus using risk factors in a Nigerian population. Healthcare Inf. Res. **28**(1), 58–67 (2022)
15. Pradhan, N., Rani, G., Dhaka, V.S., Poonia, R.C.: Diabetes prediction using artificial neural network. In: Deep Learning Techniques for Biomedical and Health Informatics, pp. 49–62. Springer, Singapore (2020)
16. Buda, M., Maki, A., Mazurowski, M.A.: A systematic study of the class imbalance problem in convolutional neural networks. Neural Netw. **106**, 249–259 (2018)
17. Blake, C.L., Merz, C.J.: UCI Repository of Machine Learning Databases. University of California, Department of Information and Computer Science, Irvine (1998)
18. Ahmed, A., et al.: Machine learning algorithm-based prediction of diabetes among female population using PIMA dataset. Healthcare (Basel). **13**(1), 37 (2024). https://doi.org/10.3390/healthcare13010037
19. Klein, S., Gastaldelli, A., Yki-Järvinen, H., Scherer, P.E.: Why does obesity cause diabetes? Cell Metab. **34**(1), 11–20 (2022)
20. Chang, V., Bailey, J., Xu, Q.A., Sun, Z.: Pima Indians diabetes mellitus classification based on machine learning (ML) algorithms. Neural Comput. Appl., 1–17 (2022)

A Systematic Literature Review of Fetal Health Prediction Models: Development, Challenges, and Future Directions

Samaa Ahmed(✉), Doaa Elzanfaly, and Ahmed Yakoub

Capital University (Formerly Helwan University), Cairo 11795, Egypt
{samaaahmed252,doaa.saad,ahmedYakoup}@fci.capu.edu.eg

Abstract. Predicting fetal health during pregnancy is critical to ensuring the fetus remains healthy throughout its growth. Cardiotocography (CTG) shows the pregnancy's quiet patterns, providing a non-invasive insight into fetal health. Machine learning and deep learning models understand these complicated signals, identifying indicators of danger before they become severe. This combination of medical knowledge and artificial intelligence turns prenatal care into a more creative, accurate, and lifesaving procedure. This research outlines a Systematic Literature Review (SLR) of fetal health prediction using machine learning and deep learning models. It intends to investigate the datasets, techniques, and frameworks utilized, alongside the evaluation criteria utilized to evaluate model performance. Applicable major articles were picked from digital repositories such as IEEE Xplore, Scopus, SpringerLink, and the ACM Digital Library. A collection of forty-seven (47) studies published up to June 2025 was conducted. The research results emphasize the need to increase the dataset size and utilize data augmentation techniques to increase model applicability. The examination of effective deep learning frameworks improves prediction accuracy significantly. These approaches collaborate to improve the reliability and proactiveness of fetal health examinations.

Keywords: Fetal Health Prediction · Cardiotocography · Machine Learning · Deep Learning · Systematic Literature Review.

1 Introduction

According to the World Health Organization (WHO), Fetal health refers to the fetus's general well-being and healthy growth throughout pregnancy [1]. Tracking fetal health is an important part of early pregnancy care because it helps detect potential problems that might lead to negative outcomes. Cardiotocography (CTG) represents a popular, simple technique for tracking uterine contractions and the fetal heart rate throughout pregnancy. Early detection of prenatal health concerns remains a considerable challenge. Predicting fetal health accurately and quickly is critical to improving outcomes for both mothers and new babies. Frameworks powered by AI and deep learning are becoming increasingly known for their contribution to improving prediction abilities. Recent research indicates that these prediction models are effective in identifying fetuses that are at risk, which

A. E. Hassanien et al. (Eds.): AISDS 2025, LNDECT 287, pp. 179–189, 2027.
https://doi.org/10.1007/978-3-032-18927-1_18

facilitates more informed treatment decisions and eventually leads to improved prognosis for both the mother and the fetus [2]. Due to small, unbalanced datasets and poor data augmentation, fetal health prediction still has significant gaps despite breakthroughs in ML and DL. Accurate and timely monitoring requires more resilient, flexible, and comprehensible DL models trained on more extensive, high-quality data.

2 Research Methodology

Attempting to identify techniques, evaluate performance measures, and identify research gaps in prenatal health prediction, this SLR examines recent publications. Planning and conducting are their two main phases.

2.1 Planning Phase

An SLR's planning step includes creating a structured review process by defining the scope of the study, creating questions, choosing databases, and developing criteria for inclusion/exclusion and quality assessment.

2.1.1 Research Questions

For finding, evaluating, and incorporating papers concerning prenatal health prediction, this SLR employs research questions as a framework. The PICOC model guarantees that these inquiries are precise, targeted, and in line with the review's goals. The study questions were created to precisely address each framework component based on the PICOC factors.

- **Population:** Perinatal Care, Fetal Health Prediction, Cardiotocography (CTG) Data.
- **Intervention:** Enhanced Deep Learning Model, Data Augmentation Techniques, Improved Prediction Model, Hybrid Model, and Evaluation Measures.
- **Comparison:**Deep Learning Models without Data Augmentation, Traditional Machine Learning Prediction Models.
- **Outcome:**Improved the fetal health prediction Result, Better Handling of Imbalanced Data, and Enhanced Medical Decision Support.
- **Context:** Maternal-Fetal Healthcare Facilities, Fetal Health Risk Assessment Conditions, Cardiotocography (CTG) Data Analysis.

RQ1:What datasets are utilized to predict fetal health in studies that apply machine learning and deep learning techniques?
RQ2:Which data augmentation strategies are best suited to handling data imbalance and variability in fetal health datasets?
RQ3:Which machine learning and deep learning models are employed for fetal health prediction?
RQ4:Which evaluation measures are utilized to assess the efficacy of fetal health prediction models?
RQ5:What limitations and directions have been identified in current studies on fetal health prediction using machine learning and deep learning?
RQ6:Was the Explainability technique used in the fetal health prediction model?

2.1.2 Search Methodology

The methodical procedure for finding relevant research for the SLR is described in the digital library approach. Because of their extensive coverage of AI, engineering, and medical research, reputable databases like IEEE Xplore, Scopus, SpringerLink, and ACM were selected. The pertinent studies were then found using keyword searches based on the PICOC framework.: **("Deep Learning" OR "Recurrent Neural Network" OR "Convolutional Neural Network" OR "Deep Belief Network" OR "Long Short Term Memory" OR "Generative Adversarial Network") AND ("Cardiotocography" OR "CTG") AND ("Fetal" AND "Health" AND "Prediction")**. Articles are retrieved based on title, abstract, and keywords. All digital library studies are integrated, and duplicates are deleted. Also, selected studies were published from January 2019 to June 2025. A total of 141 research papers were retrieved from the four digital sources. Articles are retrieved based on title, abstract, and keywords. All digital library studies are integrated, and duplicates are deleted. Also, selected studies were published from January 2019 to June 2025. A total of 141 research papers were retrieved from the four digital sources.

2.1.3 Inclusion and Exclusion Criteria

The inclusion and exclusion criteria provide explicit guidelines for choosing papers for the SLR. The criteria for selecting the main relevant studies are presented in Table 1.

Table 1. Inclusion and Exclusion Criteria

Inclusion Criteria☑
Articles conducted from January 2019 to June 2025 in the domain of fetal health prediction.
Articles from journals or conferences.
Articles illustrate experimental results with different performance metrics.
Articles utilize machine learning or deep learning techniques for predicting fetal health.
Exclusion Criteria☒
Articles not written in English.
Articles concentrate on unrelated fields rather than machine learning or deep learning techniques.
Articles without a comprehensive prediction model.
Theoretical articles, reviews, or conceptual frameworks without practical confirmation.
Articles with inadequate analytical details or improved performance.

2.2 Conducting Phase

Using the PICOC framework as a guide, it starts by crafting exact search terms and executing them in particular digital libraries. Following the retrieval of the research, duplicates are eliminated, and the remaining papers are evaluated using predetermined inclusion and exclusion criteria.

2.2.1 Conduct Primary Research

According to the PRISMA flow diagram, the full database search from the four conducting sources yielded 141 articles. After eliminating eight duplicates and eighty-six irrelevant publications, forty-seven distinct studies remained, indicating the most significant and relevant for the current review.

2.2.2 Quality Assessment

The research clarifies the dataset, preprocessing, model construction, and evaluation methodologies, alongside data augmentation and performance indicators. It also compares results with other methodologies, examines limitations and future advances, and tackles model explainability in healthcare settings.

2.2.3 Data Extraction

This stage often includes documenting the study's dataset source (e.g., CTG or ultrasound), the usage of machine learning and deep learning algorithms, and any data augmentation approaches used to overcome insufficient data. It also involves addressing the limits, future enhancement recommendations, and evaluation criteria employed in the studies, as presented in Table 2.

Table 2. Data extraction features aligned with research questions.

Feature	Research Question
Fetal Health datasets	RQ1
Data Augmentation Techniques	RQ2
Machine and deep learning techniques for fetal health prediction	RQ3
Evaluation Metrics	RQ4
Limitations and Future Recommendations	RQ5
Explainability Approaches	RQ6

2.2.4 Reporting Results Stage

This step covers the datasets, data augmentation methods, machine learning and deep learning models, assessment measures, and performance outcomes employed in fetal health prediction, while also addressing limitations such as tiny datasets and limited usage of explainable AI approaches. Table 3 presents a complete summary of the findings, summarizing the essential aspects and conclusions of the examined studies to lay the basis for answering research questions and guiding future advancement in prenatal health prediction.

A. Datasets

The Datasets phase lists the public repositories, medical records, and monitoring instruments like CTG, fECG, and ultrasound that were used in the reviewed research. Of the forty-seven investigations, 23 used the Kaggle/UCI Fetal Health Classification Dataset (CTG), while three studies used the CTU-UHB CTG dataset. Private clinical data or other datasets, like ultrasonography, only showed up once. A heavy dependence on the Kaggle/UCI dataset restricts diversity and may have an impact on the generalizability of the model, underscoring the need for additional diverse data sources in subsequent studies.

B. Data Augmentation Techniques

Data augmentation approaches generate modified samples to expand and balance datasets, solving problems with incompleteness and class imbalance in prenatal health prediction. Most research used SMOTE, which generates synthetic minority samples to increase training and accuracy. Several studies used GAN-based algorithms (CGANs, GANs) to generate realistic CTG signals or prenatal data while maintaining medical relevance. One study simulated missing CTG segments to help with real-world interruptions. Overall, SMOTE and GANs, when paired with vision-based augmentations, significantly improve model performance and generalization.

C. Machine and Deep Learning Techniques for Fetal Health Prediction

Throughout systematic literature reviews (SLR) research on prediction modeling, an extensive number of machine learning and deep learning techniques have been utilized, with deep learning frameworks dominating in recent years.

D. Evaluation Metrics for Fetal Health Prediction Models

Evaluation metrics are critical for measuring model performance and guaranteeing consistent outcomes. Most research used traditional classification metrics like accuracy, precision, recall, and F1-score, while AUC-ROC was used for imbalanced datasets. Regression-focused works used MAE, MSE, and RMSE.

E. Limitations and Future Directions

Many of the analyzed studies used machine learning methods, while deep learning was rarely used, despite its high promise in healthcare. One significant issue was the reliance on tiny datasets, which hampered model dependability and generalization. Few studies clearly justified their data augmentation approaches, revealing gaps in clinical relevance. Most people focused solely on correctness, ignoring difficulties like interpretability, consistency, and class imbalance. To fully benefit from deep learning, future work should use larger and more diverse datasets, well-defined augmentation methodologies, and integrate explainable AI (XAI) to improve transparency, reliability, and clinical acceptance of prediction models.

Table 3. Results from analyzing the conducted studies.

Ref NO.	Best Result
[3]	93.1% (Voting Ensemble)
[4]	97.81%
[5]	99.98% (GB)
[6]	–
[7]	97.2%.
[8]	95.6%
[9]	97.5% (Hybrid LSTM-DNN)
[10]	90.73% (LGBM)
[11]	93% (RF)
[12]	94% (CNN)
[13]	97.3% (GB)
[14]	97.8%
[15]	100% (NB, DT), 99.93% (RF)
[16]	96.00% (Stacked Model)
[17]	99.8% (AWD)
[18]	93.6% (two-class)
[19]	92.59%
[20]	94.02%
[21]	99%
[22]	97.3% (XGBoost)
[23]	93% (RF)
[24]	98.9% (Stacking)
[25]	97.35% (RF)
[26]	94% (SVM)
[27]	98.2% (RF)
[28]	91.94%
[29]	~ >90%
[30]	98.59%
[31]	99.78% (voting ensemble)
[32]	96%
[33]	≈ >0.88
[34]	>90%
[35]	95.39%

(continued)

Table 3. (*continued*)

[36]	90.77%
[37]	≈ 81.6 %
[38]	99.66%
[39]	91.32%
[40]	97.3%
[41]	96%
[42]	98.34%
[43]	>90%
[44]	F1 score: 0.67 ± 0.03, AUC: 0.73 ± 0.04
[45]	99.97%
[46]	>90%
[47]	99%
[48]	90%
[49]	99.02%

3 Conclusion

This paper examines significant issues and developments in fetal health prediction models, a critical area for early detection and better pregnancy outcomes. Despite increased attention, it still confronts challenges such as restricted datasets, small sample sizes, and flaws in model construction and validity. While machine learning is widely employed, its efficiency is frequently limited by manual feature dependency and ambiguous data augmentation. Deep learning has the potential to improve complex data management and automated feature extraction; however, it is currently underutilized. This review identifies these shortcomings and recommends focusing on deep learning, explainable AI, better augmentation approaches, and larger datasets to increase the accuracy and therapeutic use of fetal health prediction.

References

1. WHO: WHO recommendations on antenatal care for a positive pregnancy experience. Who.int (2016). [Online]. Available: https://www.who.int/publications/i/item/9789241549912
2. Research Google: Predicting fetal well-being from cardiotocography signals using AI. Research. Google. [Online]. Available: https://research.google/blog/predicting-fetal-well-being-from-cardiotocography-signals-using-ai/
3. Abinaya, P.S., Geo, A.V.A.: Enhanced fetal health prediction using deep learning and ensemble methods. In: 2024 5th International Conference on Data Intelligence and Cognitive Informatics (ICDICI), pp. 1303–1307 (2024). https://doi.org/10.1109/ICDICI62993.2024.10810940

4. Ramkumar, G., Swarna, M., Madhini, M., Devi, P.K., Giri, P., Al Sherideh, A.: GDVM: A deep learning assisted fetal health prediction system using gradient distance vector model. In: 2024 2nd International Conference on Emerging Trends in Engineering and Medical Sciences (ICETEMS), pp. 592–596 (2024). https://doi.org/10.1109/ICETEMS64039.2024.10965032
5. Mohanty, R., et al.: Comparitive analysis of machine learning models for prediction of fetal health. In: 2024 International Conference on Advancements in Smart, Secure and Intelligent Computing (ASSIC) (2024). https://doi.org/10.1109/ASSIC60049.2024.10508030
6. Nitin Moses, S., Priyadarsine, S., Lalitha Ambigai, V.D., Logavarshini, S., Madhanlal, U., Kanchana, D.: Automated classification of CTG signals using deep learning based scalogram analysis. In: 2022 IEEE Global Conference on Computing, Power and Communication Technologies (GlobConPT) (2022). https://doi.org/10.1109/GlobConPT57482.2022.9938304
7. Puri, C., Reddy, K.T.V.: Smart healthcare: early diagnosis of fetal health conditions using machine learning and cardiotocographic data. In: 2024 2nd DMIHER International Conference on Artificial Intelligence in Healthcare, Education and Industry (IDICAIEI), pp. 1–5 (2024). https://doi.org/10.1109/IDICAIEI61867.2024.10842704
8. Moutaib, M., Fattah, M., Farhaoui, Y., Aghoutane, B.: Machine learning in fetal health: improving ECG analysis with random forest. In: 2024 International Conference on Circuit, Systems and Communication (ICCSC), vol. 2, pp. 1–5 (2024). https://doi.org/10.1109/ICCSC62074.2024.10617320
9. Kaushik, T., Sri Vani, T.A., Vithyatharshana, N., Khare, S.: A comparative analysis of ML & DL models for fetal health classification using cardiotocographic data. In: 2024 Asia Pacific Conference on Innovation in Technology (APCIT), pp. 1–6 (2024). https://doi.org/10.1109/APCIT62007.2024.10673488
10. Bansal, R., Sonowal, I.B., Bhukya, R.K.: Predicting fetal health using cardiotocography and machine learning: a study on prenatal care. In: 2025 International Conference on Innovation in Computing and Engineering (ICE), pp. 1–6 (2025). https://doi.org/10.1109/ICE63309.2025.10984296
11. Pradhan, A.K., Rout, J.K., Maharana, A.B., Balabantaray, B.K., Ray, N.K.: A machine learning approach for the prediction of fetal health using CTG. In: 2021 19th OITS International Conference on Information Technology (OCIT), Theor. Decis. (2021). https://doi.org/10.1109/OCIT53463.2021.00056
12. Marandi, A.K., Shah, H.: A convolution neural network-based prediction of fetal health. In: 2022 11th International Conference on System Modeling & Advancement in Research Trends (SMART), pp. 1369–1374 (2022). https://doi.org/10.1109/SMART55829.2022.10047304
13. Ukrit, M.F., Aluru, L.R., Chandana, V.: Fetal health classification and visceral fat level prediction using gradient boosting and deep learning techniques. In: 2024 2nd International Conference on Networking and Communications (ICNWC) (2024). https://doi.org/10.1109/ICNWC60771.2024.10537531
14. Neelakandan, S., Brinda, A., Dheekshitha, S., Divya Priya, M.S.: Extreme gradient boosting based fetal health classification. In: 2023 International Conference on Applied Intelligence and Sustainable Computing (ICAISC), pp. 1–6 (2023). https://doi.org/10.1109/ICAISC58445.2023.10200526
15. Kadarina, T.M., Basari, Gunawan, D.: ML-based interpretation of cardiotocography data: Current state and future research. In: 2023 International Conference of Computer Science and Information Technology (ICOSNIKOM), pp. 1–6 (2023). https://doi.org/10.1109/ICoSNIKOM60230.2023.10364517
16. Kori, A.: Enhancing fetal health prediction using machine learning and ensemble methods. In: 2025 International Conference on Pervasive Computational Technologies (ICPCT), pp. 116–120 (2025). https://doi.org/10.1109/ICPCT64145.2025.10940887

17. Shastry, K.A., Gounder, M.S., Kumar, T.G.M., Karthik, D.U., Sushma, V., Subashree, D.: Adaptive weighted diversity ensemble learning approach for fetal health classification on cardiotocography data. IEEE Access. **12**, 190690–190710 (2024). https://doi.org/10.1109/ACCESS.2024.3516880
18. Wong, W.K., Juwono, F.H., Apriono, C., Fitri, I.R.: Fetal health prediction from cardiotocography recordings using kolmogorov–Arnold networks. IEEE Open J. Eng. Med. Biol. **6**, 345–351 (2025). https://doi.org/10.1109/OJEMB.2025.3549594
19. Marques, J.A.L., et al.: IoT-based smart health system for ambulatory maternal and fetal monitoring. IEEE Internet Things J. **8**(23), 16814–16824 (2021). https://doi.org/10.1109/JIOT.2020.3037759
20. Swain, S., Patra, K.J., Jayasingh, S.K., Mohanty, M.N., Pattanayak, B.K.: The role of ensemble learning in advancing child fetal health: a focus on voting and bagging techniques. In: 2024 1st International Conference on Cognitive, Green and Ubiquitous Computing (IC-CGU) (2024). https://doi.org/10.1109/IC-CGU58078.2024.10530780
21. Suresh, K., Jayanth, K.L., Rajaa, S.L., Jebin, S., Yaashuwanth, C., Prathibanandhi, K.: Fetal health management using ensemble learning techniques and providing data analysis. In: 2025 International Conference on Computing and Communication Technologies (ICCCT), pp. 1–9 (2025). https://doi.org/10.1109/ICCCT63501.2025.11020115
22. Sudharson, D., Sushmita, V., Jananiksha, P., Vaishali, V., Mukilarasu, R., Kanagaraj, S.: AI powered monitoring and risk prediction for maternal health to ensure fetal well-being. In: 2025 3rd International Conference on Advancements in Electrical, Electronics, Communication, Computing and Automation (ICAECA), pp. 1–5 (2025). https://doi.org/10.1109/ICAECA63854.2025.11012457
23. Ramesh, A., Pandian, D.P., Nagarajan, D., Murugesan, B., Andrew, A.M.: Fetal health prediction using machine learning. In: AIP Conference Proceedings, vol. 3291, p. 030016 (2025). https://doi.org/10.1063/5.0269960
24. Ahmed, S.S., Mahmoud, N.M.: Early detection of fetal health status based on cardiotocography using artificial intelligence. Neural Comput. Appl. (2025). https://doi.org/10.1007/s00521-025-11343-x
25. Ramachandran, R., Abhijith, K., Karthik, J.J.: Fetal health prediction from cardiotocographic data. In: Smart Innovation, Systems and Technologies, pp. 179–190. Springer Nature Singapore, Singapore (2025). https://doi.org/10.1007/978-981-96-1210-9_16
26. Alharbi, N., et al.: Fetal hypoxia detection using machine learning: a narrative review. AI (Basel). **5**(2), 516–532 (2024). https://doi.org/10.3390/ai5020026
27. Abishek, A., Mathialagan, H., Swarup, O., Goyal, M., Mannar Mannan, J.: An intelligent fetal prognostic system using machine learning techniques. In: 2024 IEEE International Conference on Interdisciplinary Approaches in Technology and Management for Social Innovation (IATMSI) (2024). https://doi.org/10.1109/IATMSI60426.2024.10503455
28. Aswathi Mohan, P.P., Uma, V.: Enhanced prediction of fetal hypoxia through GAN-synthesized data and advanced ensemble feature fusion. In: 2024 15th International Conference on Computing Communication and Networking Technologies (ICCCNT), pp. 1–7 (2024). https://doi.org/10.1109/ICCCNT61001.2024.10724351
29. Kim, H.Y., Cho, G.J., Kwon, H.S.: Applications of artificial intelligence in obstetrics. Ultrasonography. **42**(1), 2–9 (2023). https://doi.org/10.14366/usg.22063
30. Takale, D.G., Pangaonkar, S., Jadhav, T., Deshmukh, G.B., Shinde, M.: Enhancing fetal health monitoring: utilizing WWPA based BiGRU with dropout layer regulation to classify fetal health conditions on cardiotocography data. Multiscale Multidiscip. Model. Exp. Des. **8**(7) (2024). https://doi.org/10.1007/s41939-025-00885-1
31. Mondal, S., Maity, R., Nag, A., Ghosh, S.: Fetal health risk prediction using ensemble-based machine learning approaches. Knowl. Inf. Syst. (2025). https://doi.org/10.1007/s10115-025-02442-1

32. Saxena, S., Saxena, S., Tahilramani, N., Charanarur, P.: Blockchain enhanced smart healthcare management for chronic diseases. Discov. Comput. **28**(1) (2025). https://doi.org/10.1007/s10791-025-09574-6
33. Chiou, N., et al.: Development and evaluation of deep learning models for cardiotocography interpretation. npj Womens Health. **3**(1) (2025). https://doi.org/10.1038/s44294-025-00068-w
34. Mushtaq, G., Veningston, K.: A comprehensive review on fetal health surveillance using ultrasound imagery. Discov. Appl. Sci. **7**(5) (2025). https://doi.org/10.1007/s42452-025-06644-8
35. Feng, J., et al.: A hybrid stacked ensemble and Kernel SHAP-based model for intelligent cardiotocography classification and interpretability. BMC Med. Inform. Decis. Mak. **23**(1), 273 (2023). https://doi.org/10.1186/s12911-023-02378-y
36. Cao, Z., et al.: Intelligent antepartum fetal monitoring via deep learning and fusion of cardiotocographic signals and clinical data. Health Inf. Sci. Syst. **11**(1), 16 (2023). https://doi.org/10.1007/s13755-023-00219-w
37. Lin, Z., et al.: Deep learning with information fusion and model interpretation for long-term prenatal fetal heart rate data. npj Womens Health. **2**(1) (2024). https://doi.org/10.1038/s44294-024-00033-z
38. Talukder, M.S.H., Akter, S.: An improved ensemble model of hyper parameter tuned ML algorithms for fetal health prediction. Int. J. Inf. Technol. **16**(3), 1831–1840 (2024). https://doi.org/10.1007/s41870-023-01447-9
39. Barnova, K., Martinek, R., Vilimkova Kahankova, R., Jaros, R., Snasel, V., Mirjalili, S.: Artificial intelligence and machine learning in electronic fetal monitoring. Arch. Comput. Methods Eng. **31**(5), 2557–2588 (2024). https://doi.org/10.1007/s11831-023-10055-6
40. Ranaei-Zamani, N., et al.: Saving babies and families from preventable harm: a review of the current state of fetoplacental monitoring and emerging opportunities. npj Womens Health. **2**(1) (2024). https://doi.org/10.1038/s44294-024-00015-1
41. Fahad Alhasson, H., Elhag, N., Saleem Alharbi, S., Adam, I.: Application of machine learning in identifying risk factors for low APGAR scores. BMC Pregnancy Childbirth. **25**(1), 548 (2025). https://doi.org/10.1186/s12884-025-07677-y
42. Zhao, Z., Deng, Y., Zhang, Y., Zhang, Y., Zhang, X., Shao, L.: DeepFHR: intelligent prediction of fetal Acidemia using fetal heart rate signals based on convolutional neural network. BMC Med. Inform. Decis. Mak. **19**(1), 286 (2019). https://doi.org/10.1186/s12911-019-1007-5
43. Abel, J.D.K., Dhanalakshmi, S., Kumar, R.: A comprehensive survey on signal processing and machine learning techniques for non-invasive fetal ECG extraction. Multimed. Tools Appl. **82**(1), 1373–1400 (2023). https://doi.org/10.1007/s11042-022-13391-0
44. Ogasawara, J., et al.: Deep neural network-based classification of cardiotocograms outperformed conventional algorithms. Sci. Rep. **11**(1), 13367 (2021). https://doi.org/10.1038/s41598-021-92805-9
45. Das, S., Mukherjee, H., Obaidullah, S.M., Roy, K., Saha, C.K.: Ensemble based technique for the assessment of fetal health using cardiotocograph – a case study with standard feature reduction techniques. Multimed. Tools Appl. **79**(47–48), 35147–35168 (2020). https://doi.org/10.1007/s11042-020-08853-2
46. Kaur, K., Kaur, N.: Machine learning applications in healthcare systems: a comprehensive analysis of benefits and challenges. J. Inst. Eng. (India) Ser. B. (2025). https://doi.org/10.1007/s40031-025-01252-x
47. Sheron, S., Sreejith, N., Chandran, A.B., Murugaveni, S.: Fetal health prediction: comparative analysis of machine learning models. In: 2025 International Conference on Recent Advances in Electrical, Electronics, Ubiquitous Communication, and Computational Intelligence (RAEEUCCI), pp. 1–6 (2025). https://doi.org/10.1109/RAEEUCCI63961.2025.11048286

48. Jain, E., Kaushik, P., Kukreja, V., Sakshi, A.D., Goyal, B.: Fetal diagnostics using vision transformer for enhanced health and severity prediction in ultrasound imaging. Curr. Med. Imaging Rev. (2025). https://doi.org/10.2174/0115734056360199250227053012
49. Bertieaux, J., Shateri, M., Labeau, F., Dutoit, T.: Cardiotocography signal abnormality detection based on deep semi-unsupervised learning. In: Proceedings of the IEEE/ACM 10th International Conference on Big Data Computing, Applications and Technologies, pp. 1–10 (2023). https://doi.org/10.1145/3632366.3632391

EEG-Based Epilepsy Detection Using DWT-PCA and Machine Learning

Omar Bamaarouf[1,2(✉)], Said Agounad[1], Said Charfi[3], Anas Mazid[1], and Bouazza Faiz[1]

[1] Physics Department, Laboratory of Metrology and Information Processing, Faculty of Sciences, Ibn Zohr University, 80000 Agadir, Morocco
s.agounad@uiz.ac.ma, anas.mazid.82@edu.uiz.ac.ma, b.faiz@uiz.ac.ma

[2] Electrical Engineering, Interdisciplinary Applied Research Laboratory, International University of Agadir—Universiapolis, 80000 Agadir, Morocco
omar.bamaarouf@e-polytechnique.ma

[3] Department of Computer Science, Laboratory of Computing Systems and Vision, Faculty of Sciences, Ibn Zohr University, 80000 Agadir, Morocco
s.charfi@uiz.ac.ma

Abstract. Epilepsy is a common chronic neurological disorder characterized by recurrent and abnormally synchronized neuronal activity. The detection and classification of seizures are critical for effective clinical care. The complexity and variability of seizure patterns have driven the development of automated detection systems to assist neurophysiologists. Electroencephalogram (EEG) signals are most commonly used, as they contain valuable brain information, though revealing subtle but critical changes in EEG remains challenging. This study proposes a combination of the Discrete Wavelet Transform (DWT), Principal Component Analysis (PCA), and machine learning classifiers. EEG segments are first decomposed into sub-bands using DWT, and relevant statistical features are extracted from each sub-band. PCA is then applied to reduce feature redundancy while preserving the most informative components, the resulting low-dimensional feature extraction are fed into machine learning classifiers to distinguish between epileptic and non-epileptic signals using cross-validation scheme. The publicly available EEG database from the University of Bonn is employed to validate the proposed method. Experimental results demonstrate that the classification accuracy reaches 99.00%, 98.50%, 96.00%, 97.33% and 97.69% for the scenarios A vs. E, B vs. E, C vs. E, D vs. E, CD vs. E and ABCD vs. E, respectively.

Keywords: Epilepsy seizure detection · EEG · DWT · PCA · Machine learning

S. Agounad, S. Charfi, A. Mazid and B. Faiz—These authors contributed equally to this work.

A. E. Hassanien et al. (Eds.): AISDS 2025, LNDECT 287, pp. 190–200, 2027.
https://doi.org/10.1007/978-3-032-18927-1_19

1 Introduction

Epilepsy is a chronic neurological disorder marked by repeated episodes of transient brain dysfunction[1]. According to the World Health Organization (WHO), epilepsy affects more than 50 million people worldwide, with nearly 80% of patients living in low- and middle-income countries [2]. Epilepsy significantly affects the quality of life of patients, including their daily activities and mental health, while also representing a considerable threat to their physical safety.

Electroencephalography is a useful technique for monitoring the nonlinear electrical function of brain nerve cells, and it continues to be an important clinical tool for the evaluation and treatment of epilepsy. EEG signals can be obtained using either invasive electrodes implanted at selected regions of the cerebral cortex or non-invasive electrodes attached to the scalp [3]. Currently, the identification and classification of epileptic seizures are mainly performed by trained neurophysiologists through visual inspection of these EEG recordings [4]. However, the traditional method requires considerable time and effort, and the diagnostic process can be influenced by human subjectivity, which may lead to potential errors during evaluation [5]. Consequently, there is a growing need for the development of accurate and resilient approaches to automatically identify epileptic activity in EEG recordings. As EEG signals represent typical nonlinear and non-stationary weak biological signals, their feature extraction and classification are currently a research an important research topic for researchers worldwide [6].

Various signal analysis and processing techniques have been proposed in the literature for the detection and identification of epileptic seizures [7,8]. Generally speaking, the process of EEG-based methods in three steps, namely preprocessing, feature extraction, and classification. In preprocessing, artifact detection and rejection are performed to remove artifactual signals, such as eye blinks and muscle activity, which may compromise EEG analysis and classification [9–12]. Feature extraction captures useful information from EEG signals [13]. Once the feature sets are extracted, classification constitutes a crucial step in EEG signal analysis, as it determines the class labels of the input features, such as ictal or inter-ictal states.

Recently, many automated methods for detecting epileptic activity have been proposed [14]. Most of them employ Fourier spectral analysis to transform EEG signals from the time to the frequency domain. However, since EEG signals are non-stationary, analyses limited to either domain provide only partial information. Timefrequency methods address this issue by capturing both temporal and spectral features. The Short-Time Fourier Transform (STFT) is widely used [15], while the Discrete Wavelet Transform (DWT) offers a flexible multiresolution decomposition. DWT extracts features at different frequency bands, is less sensitive to noise, and mitigates mode mixing, making it suitable for EEG-based seizure detection.

Mingyang et al. [16] proposed a wavelet-based envelope analysis with neural network ensemble to classify normal, inter-ictal, and epileptic EEG signals. Atangana et al. [17] employed the WT to process EEG signals, applying LDA for feature selection and dimensionality reduction, then used the resulting features as input for a Multi-Layer Perceptron (MLP). Swami et al. [4] proposed an automated classification system with limited supervised training, where EEG signals were decomposed into timefrequency sub-bands up to the sixth level using DTCWT. Aliyu et al. [7] proposed an approach for automatic seizure detection using DWT for noise removal and feature extraction, establishing 20 eigenvalue features, and refining selection through correlation and P-value analysis to reduce dimensionality and LSTM parameters.

Based on previous studies, DWT has proven to be highly effective for signal decomposition and feature extraction in EEG analysis. However, it often produces a large number of features, leading to high dimensionality and increased computational cost, which highlights the need for efficient feature reduction. This study presents an automatic method for detecting epileptic activity using DWT to decompose EEG signals for feature extraction. To overcome the drawback of high feature dimensionality, PCA is employed to reduce redundancy and enhance computational efficiency. This integration exploits the strengths of DWT for robust signal decomposition while ensuring conditions more suitable for real-time implementation. Once the low-dimensional feature sets are obtained, classification methods are applied to determine the class of input signals.

The rest of this paper is structured as follows. Section 2 provides a description of the EEG datasets, the proposed method, the feature extraction and the classification techniques employed. The results and discussion appear in Sect. 3. Finally, Sect. 4 summarizes the paper.

2 Materials and Methods

2.1 Bonn Dataset

This study employs the public EEG dataset from the University of Bonn, Germany [18]. It includes five subsets (AE), each with 100 single-channel EEG segments of 23.6 s. Subsets A and B were obtained extracranially from healthy subjects with eyes open (A) and closed (B), whereas C, D, and E were collected intracranially from epilepsy patients: C and D correspond to inter-ictal recordings, and E to ictal recordings. Signals were acquired with a 128-channel amplifier at 173.6 Hz and 12-bit resolution. Figure 1 shows sample EEG data.

2.2 Proposed Method

The flowchart of the proposed DWTPCA algorithm for EEG-based seizure detection is presented in Fig. 2. The approach comprises three main blocks:

- **Block 1:** EEG signal decomposition using DWT
- **Block 2:** feature extraction and feature dimensionality reduction through PCA
- **Block 3:** classification for distinguishing seizure from non-seizure EEG segments.

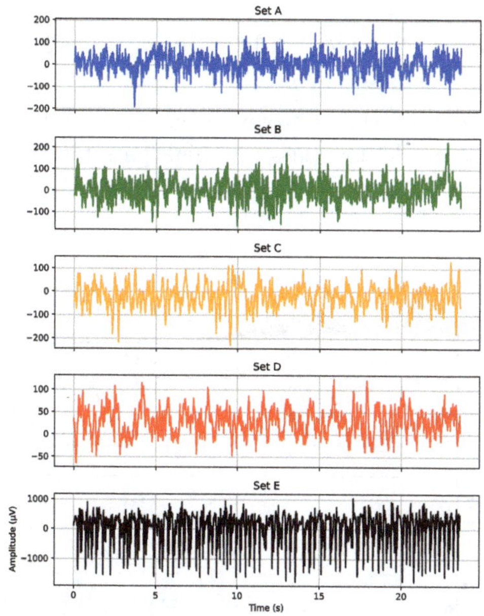

Fig. 1. Example of EEG recordings of the five Bonn dataset types.

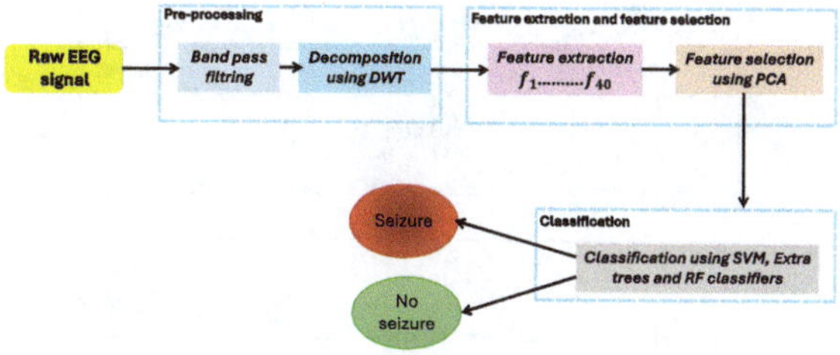

Fig. 2. Proposed methodology for epileptic seizure detection from EEG recordings.

2.2.1 Bandpass Filtering and EEG Signal Decomposition

Before decomposition, a bandpass filter (0.5 $\sim$ 40 Hz) is applied to remove artifacts and enhance EEG signal quality. Accurately identifying sharp and spike waves, indicative of seizures, is then critical. Not all features across frequency bands show significant differences between ictal, pre-ictal, and inter-ictal states. DWT is effective in seizure detection for its high-resolution timefrequency analysis [19]. DWT decomposes an EEG signal $x[n]$ into approximation and detail coefficients at different levels [9,20]. The decomposition at level j can be expressed as:

$$A_j[n] = \sum_{k=-\infty}^{+\infty} g[k-2n]A_{j-1}[k], \quad j = 1, 2, \ldots, J \tag{1}$$

$$D_j[n] = \sum_{k=-\infty}^{+\infty} h[k-2n]A_{j-1}[k], \quad j = 1, 2, \ldots, J \tag{2}$$

where $A_j(n)$ and $D_j(n)$ are the approximation and detail coefficients at level j, respectively, g and h are the low-pass and high-pass filters, and $2n$ denotes the decimation by 2.

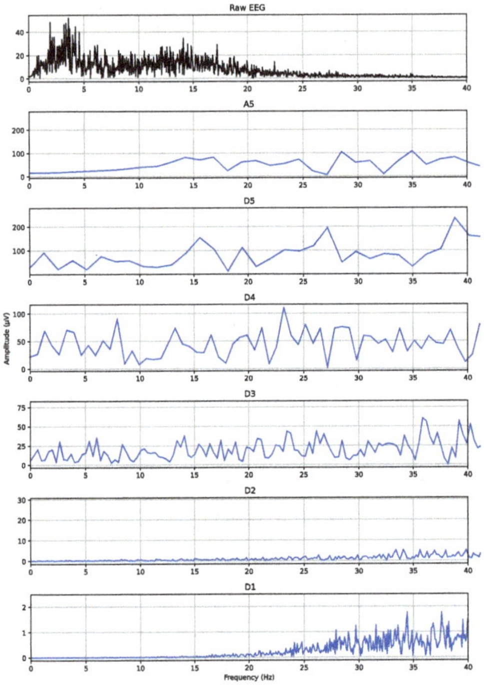

Fig. 3. 5th level DWT Decomposition of a Signal from Bonn dataset.

The DWT allows for multi-resolution analysis of EEG signals, enabling the extraction of informative features from different frequency bands while maintaining temporal localization. Figure 3 represents the evolution of the detail and approximation coefficients obtained by applying 5th level DWT on a Signal from Bonn dataset.

2.2.2 Feature Extraction and Dimensionality Reduction

Feature extraction is crucial for predicting epileptic seizures. After DWT decomposition, eight features are extracted from the last four detail coefficients ($d2, ..., d5$) and approximation ($A5$), yielding 40 features per EEG segment. The first detail coefficient ($d1$) is excluded to reduce high-frequency artifacts. PCA is then applied to reduce dimensionality and remove redundancy. This DWTPCA combination improves computational efficiency and focuses on the most relevant EEG patterns, enhancing seizure prediction accuracy. Table 1 summarizes the extracted features.

2.2.3 Classification

Following feature extraction and reduction, the classification stage is performed to distinguish between seizure and non-seizure states. In this study, three supervised machine learning algorithms are employed: Support Vector Machine (SVM), Extra Trees Classifier (ETC) and Random Forest (RF). These classifiers utilize the selected features to categorize EEG signals into their respective predefined classes.

Table 1. Extracted features and their mathematical formulations

Feature	Mathematical Formula	Description
Mean	$\mu = \frac{1}{N}\sum_{i=1}^{N} x_i$	Represents the EEG signal's average amplitude.
Standard Deviation (σ)	$\sigma = \sqrt{\frac{1}{N}\sum_{i=1}^{N}(x_i-\mu)^2}, \quad \mu = \frac{1}{N}\sum_{i=1}^{N} x_i$	Quantifies signal variability.
Variance	$\sigma^2 = \frac{1}{N}\sum_{i=1}^{N}(x_i-\mu)^2$	Measures EEG signal dispersion.
Min	$\min(x_i)$	Represents the EEG signal's minimum amplitude.
Max	$\max(x_i)$	Represents the EEG signal's maximum amplitude.
Peak-to-Peak (PTP)	$\text{PTP} = \max(x_i) - \min(x_i)$	Quantifies EEG signal range.
Skewness	$\text{Skew} = \frac{1}{N}\sum_{i=1}^{N}\left(\frac{x_i-\mu}{\sigma}\right)^3$	Measures EEG signal asymmetry.
Kurtosis	$\text{Kurt} = \frac{1}{N}\sum_{i=1}^{N}\left(\frac{x_i-\mu}{\sigma}\right)^4$	Describes the peakedness of the EEG signal.

3 Results and Discussion

3.1 Performance Evaluation

For measuring classification effectiveness, accuracy (ACC), sensitivity (SE), and specificity (SP) are employed as the main performance metrics. ACC indicates the overall detection accuracy, SE evaluates the capability to recognize seizures, and SP represents the ability to identify non-seizure instances. The classifier's performance is further assessed using true positives (TP), false positives (FP), true negatives (TN), and false negatives (FN). The equations for ACC, SE, and SP are given as follows:

$$\text{ACC} = \frac{TP + TN}{TP + TN + FP + FN}, \quad \text{SE} = \frac{TP}{TP + FN}, \quad \text{SP} = \frac{TN}{TN + FP} \tag{3}$$

3.2 Performance of Our Approach on Bonn Dataset

To evaluate the effectiveness of the proposed seizure detection approach, experiments were conducted using the Bonn EEG dataset. The classification was performed under the following scenarios:

(i) Combination of categories A and B vs. category E.
(ii) Category C vs. category E.
(iii) Category D vs. category E.
(iv) Combination of categories C and D vs. category E.
(v) Categories A, B, C, and D vs. category E.

Table 2. Classification results on the Bonn dataset without PCA

Scenarios	Classifiers	SE (%)	SP (%)	ACC (%)
AB vs. E	SVM	97.00	99.50	98.66
	RF	97.00	99.00	98.67
	ETC	97.00	100.00	98.67
C vs. E	SVM	98.00	96.00	96.00
	RF	99.00	99.00	97.50
	ETC	98.00	98.00	97.50
D vs. E	SVM	94.00	96.00	95.00
	RF	97.00	90.00	92.50
	ETC	96.00	96.00	95.49
CD vs. E	SVM	95.00	97.50	96.33
	RF	96.00	94.50	95.33
	ETC	93.00	98.05	96.67
ABCD vs. E	SVM	91.00	98.00	96.80
	RF	92.00	97.50	95.79
	ETC	93.00	99.00	97.50

Table 2 presents experimental results on the Bonn dataset without PCA, across five classification scenarios using three classifiers. Performance was evaluated in terms of SE, SP, and ACC. Sensitivity peaked at 99% for C vs. E and was

Table 3. Classification results on the Bonn dataset with extracted features (with PCA).

Scenarios	Classifiers	SE (%)	SP (%)	ACC (%)
AB vs. E	SVM	97.00	99.50	98.34
	RF	99.00	99.50	98.34
	ETC	97.00	100.00	99.00
C vs. E	SVM	98.00	96.00	97.00
	RF	99.00	99.00	97.50
	ETC	99.00	98.00	98.50
D vs. E	SVM	92.00	97.00	95.50
	RF	98.00	89.00	93.99
	ETC	96.00	97.00	96.00
CD vs. E	SVM	95.00	98.00	95.67
	RF	95.00	96.00	95.33
	ETC	94.00	95.50	97.33
ABCD vs. E	SVM	93.00	98.00	96.66
	RF	92.00	97.75	96.80
	ETC	99.00	93.00	97.69

lowest at 91% for ABCD vs. E. Specificity reached 100% for AB vs. E and 90% for D vs. E, as set D contains interictal signals from epileptogenic zones, increasing false positives. Overall accuracy was 98.67%, 97.50%, 95.49%, 96.67%, and 97.50% for AB vs. E, C vs. E, D vs. E, CD vs. E, and ABCD vs. E, respectively. The ExtraTrees Classifier consistently achieved the highest accuracy.

Table 3, presents the performance of our approach in the case where the feature selection is performed using PCA. Reducing redundancy among the extracted features was found to enhance classification performance. Applying PCA consistently improved accuracy across all scenarios, achieving 99%, 98.50%, 96.00%, 97.33%, and 97.69% for AB vs. E, C vs. E, D vs. E, CD vs. E, and ABCD vs. E, respectively. This improvement is attributed to the integration of PCA with DWT-based feature extraction, which reduces feature correlation and emphasizes the most informative components, thereby facilitating more robust and reliable seizure detection.

To further illustrate the effectiveness of the extracted features, we present in Fig. 4 a comparative visualization using Uniform Manifold Approximation and Projection (UMAP), a non-linear dimensionality reduction technique widely used to project high-dimensional data into two dimensions. The subfigure 4.a displays the UMAP projection of the raw EEG signals from set AB and set E (AB vs. E), whereas the subfigure 4.b shows the UMAP projection of the extracted feature vectors.

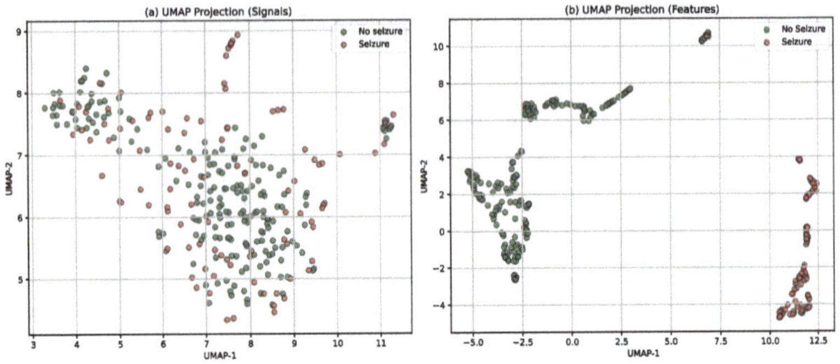

Fig. 4. UMAP projection of Bonn EEG signals: (a) AB vs. E before classification and (b) after classification. Non-seizure signals are shown in red, and seizure signals in green. (Color figure online)

3.3 Comparison With Related Works

EEG signals are non-linear and non-stationary, with epileptic events appearing as brief fluctuations. Signals were decomposed using DWT, and 40 features were extracted. PCA reduced feature dimensionality, retaining the most informative components. These features trained several supervised classifiers for binary classification with 5-fold cross-validation. Table 4 compares the proposed method with previous studies on the Bonn dataset for ABCD vs. E. The proposed method achieved 97.69% accuracy, outperforming others, thanks to DWT-based feature extraction combined with PCA, which reduces dimensionality and redundancy, enhancing discriminative power and minimizing noise.

Table 4. Comparison of the proposed method and previous works using Dataset Bonn

Authors	Methods	Combinasions	ACC (%)
P. Swami et al. [4]	DT-CWT+General regression neural network	ABCD vs. E	96.87
Y. Kumar et al. [21]	DWT + SVM	ABCD vs. E	97.38
M. Sameer et al. [22]	Short-time Fourier transform + Adaboost	ABCD vs. E	94.4
A. Zarei et al. [23]	OMP + SVM	ABCD vs. E	97
Proposed method	DWT-PCA +ETC	ABCD vs. E	97.69

4 Conclusion

Epileptic seizure detection is of significant clinical importance but remains challenging. This study introduces a feature extraction framework that employs

DWT to detect seizures in EEG recordings. Statistical and spectral features were extracted from each sub-band, and PCA was applied to reduce feature dimensionality and eliminate redundancy. The resulting feature set was classified using ETC, which achieves better performance due to its ensemble of randomized decision trees that effectively capture complex patterns and reduce overfitting. The integration of PCA not only improved accuracy but also reduced computational complexity and latency, making the proposed method suitable for real-time seizure detection. These results demonstrate that combining DWT with PCA and ETC offers a reliable and effective method for automated epileptic seizure identification.

References

1. Catarina Franco, A., Parreira, S., Bentes, C., Pimentel, J.: Management of a first unprovoked epileptic seizure in adolescence and adulthood. Epileptic Disord. **23**(4), 537–551 (2021)
2. World Health Organization: Epilepsy: Fact Sheets (2019). https://www.who.int/news-room/fact-sheets/detail/epilepsy. Accessed 29 July 2025
3. Moshé, S.L., Perucca, E., Ryvlin, P., Tomson, T.: Epilepsy: new advances. The Lancet **385**(9971), 884–898 (2015)
4. Swami, P., Gandhi, T.K., Panigrahi, B.K., Tripathi, M., Anand, S.: A novel robust diagnostic model to detect seizures in electroencephalography. Expert Syst. Appl. **56**, 116–130 (2016)
5. Yao, X., Li, X., Ye, Q., Huang, Y., Cheng, Q., Zhang, G.-Q.: A robust deep learning approach for automatic classification of seizures against non-seizures. Biomed. Signal Process. Control **64**, 102215 (2021)
6. Mazid, A., Moufassih, M., Agounad, S., Hamou, S., Tarahi, O., Bamaarouf, O.: Investigation on the effect of feature extraction block on the performance of eeg-based emotion recognition system. In: 2025 5th International Conference on Innovative Research in Applied Science, Engineering and Technology (IRASET), pp. 1–7. IEEE (2025)
7. Aliyu, I., Lim, C.G.: Selection of optimal wavelet features for epileptic eeg signal classification with lstm. Neural Comput. Appl. **35**(2), 1077–1097 (2023)
8. Zhang, S., et al.: A combination of statistical parameters for epileptic seizure detection and classification using vmd and nltwsvm. Biocybern. Biomed. Eng. **42**(1), 258–272 (2022)
9. Ait Yous, M., Agounad, S., Elbaz, S.: Automated detection and removal of artifacts from semg signals based on fuzzy inference system and signal decomposition methods. Biomed. Signal Process. Control **94**, 106307 (2024)
10. Agounad, S., Azami, H.I., Moufassih, M., Tarahi, O., Hamou, S.: Detection and removal of eog artifact from eeg signal using fuzzy logic and wavelet transform. In: 2022 27th International Conference on Automation and Computing (ICAC), pp. 1–6. IEEE (2022)
11. Agounad, S., Hamou, S., Tarahi, O., Moufassih, M., Islam, M.K.: Intelligent fuzzy system for automatic artifact detection and removal from eeg signals. J. King Saud Univ.-Comput. Inf. Sci. **34**(10), 9428–9441 (2022)

12. Agounad, S., Tarahi, O., Moufassih, M., Hamou, S., Mazid, A.: Advanced signal processing and machine/deep learning approaches on a preprocessing block for eeg artifact removal: A comprehensive review. Circuits Syst. Signal Process. **44**(5), 3112–3160 (2025)
13. Moufassih, M., Tarahi, O., Hamou, S., Agounad, S., Idrissi Azami, H.: Boosting motor imagery brain-computer interface classification using multiband and hybrid feature extraction. Multimed. Tools Appl. **83**(16), 49441–49472 (2024)
14. Polat, K., Günes, S.: Classification of epileptiform eeg using a hybrid system based on decision tree classifier and fast fourier transform. Appl. Math. Comput. **187**(2), 1017–1026 (2007)
15. Amiri, M., Aghaeinia, H., Amindavar, H.R.: Automatic epileptic seizure detection in eeg signals using sparse common spatial pattern and adaptive short-time fourier transform-based synchrosqueezing transform. Biomed. Signal Process. Control **79**, 104022 (2023)
16. Li, M., Chen, W., Zhang, T.: Classification of epilepsy eeg signals using dwt-based envelope analysis and neural network ensemble. Biomed. Signal Process. Control **31**, 357–365 (2017)
17. Atangana, R., Tchiotsop, D., Kenne, G., Chanel, L.: Eeg signal classification using lda and mlp classifier. Health Informat. Int. J **9**(1), 14–32 (2020)
18. Andrzejak, R.G., Lehnertz, K., Mormann, F., Rieke, C., David, P., Elger, C.E.: Indications of nonlinear deterministic and finite-dimensional structures in time series of brain electrical activity: Dependence on recording region and brain state. Phys. Rev. E **64**(6), 061907 (2001)
19. Shen, M., Wen, P., Song, B., Li, Y.: An eeg based real-time epilepsy seizure detection approach using discrete wavelet transform and machine learning methods. Biomed. Signal Process. Control **77**, 103820 (2022)
20. Parija, S., Dash, P.K., Bisoi, R.: Epileptic eeg signal classification using an improved vmd-based convolutional stacked autoencoder. Pattern Anal. Appl. **27**(1), 9 (2024)
21. Kumar, Y., Dewal, M., Anand, R.S.: Epileptic seizures detection in eeg using dwt-based apen and artificial neural network. SIViP **8**(7), 1323–1334 (2014)
22. Sameer, M., Gupta, B.: Detection of epileptical seizures based on alpha band statistical features. Wireless Pers. Commun., 1–16 (2020). https://doi.org/10.1007/s11277-020-07542-5
23. Zarei, A., Asl, B.M.: Automatic seizure detection using orthogonal matching pursuit, discrete wavelet transform, and entropy based features of eeg signals. Comput. Biol. Med. **131**, 104250 (2021)

Deep Learning and Segmentation Techniques for Non-invasive Heat Stress Characterization in Dairy Cattle

Rajesh Neupane[1(✉)], Toya Acharya[2], Bhuwan Shrestha[1], Nelson Rodriguez[3], Juan Velez[3], Sushil Paudyal[1], Annamalai Annamalai[2], and Mohamed Chouikha[2]

[1] Department of Animal Science, Texas A&M University, College Station, TX, USA
rajeshneupane7@gmail.com
[2] Department of Electrical and Computer Engineering, Prairie View A & M University, Prairie View, TX, USA
[3] Aurora Organic Dairy, Dublin, TX, USA

Abstract. Heat stress is a major welfare and productivity challenge in dairy cattle, particularly in regions with hot and humid climates. This study presents a computer vision–based approach to quantify heat stress in dairy cattle using video data collected from 80 cows at a commercial dairy in Central Texas. Video frames were segmented using the Segment Anything Model (SAM), enabling precise extraction of individual animals and regions of interest. A YOLO (You Only Look Once) v11-based model was developed for muzzle and mouth detection, while the second model targeted foam and mucus accumulation, indicators of respiratory distress. The muzzle-nose detection model achieved a mean average precision at 50% Intersection over Union (mAP@50) of 0.6971, while the foam-mucus detection model demonstrated limited generalization (mAP@50 = 0.224). Extracted features were analyzed with intensity- and edge-based methods to classify heat stress into low, moderate, and severe categories, achieving a weighted accuracy of 50%. Challenges included variation in muzzle pigmentation, occlusion from tongues or open mouths, and environmental variability. Future work will integrate Panting Scores (PS), which capture respiratory patterns, to refine classification and improve generalization. This research demonstrates the feasibility of scalable, automated heat stress detection systems for real-time monitoring and management in large-scale dairy operations.

Keywords: Artificial Intelligence · Computer Vision · Dairy · Heat Stress · Segment Anything Model (SAM) · You Only Look Once (YOLO).

1 Introduction

1.1 Heat Stress Impact on Dairy Farming

Livestock are the important sources of livelihood globally, around 17% of the calories and 33% of total protein consumed is from livestock farming [1]. Heat stress is one of the main issues of global livestock production, from welfare and productive view,

A. E. Hassanien et al. (Eds.): AISDS 2025, LNDECT 287, pp. 201–209, 2027.
https://doi.org/10.1007/978-3-032-18927-1_20

and it continues to be a problem in future because of increase in global warming and greenhouse gas emissions. Furthermore, dairy cattle are more susceptible to heat stress as compared to other livestock species, researchers have shown that 63% of total economic losses in USA because of heat stress has only been observed in diary species as compared to other livestock species [2]. Dairy cattle are particularly vulnerable to elevated ambient temperatures and humidity beyond their thermoneutral range [3]. To evaluate the extent of heat stress in these animals, the temperature–humidity index (THI) is commonly employed [4]. Due to this high susceptibility, heat stress has been linked to a range of productive, reproductive, and health-related impairments, resulting in substantial economic losses for dairy producers worldwide. In the United States alone, heat stress in lactating cows has been estimated to cause annual losses of approximately $1.5 billion [2]. A clear understanding of the physiological responses to heat stress is therefore essential to explain the underlying mechanisms driving these losses.

1.2 Heat Stress Monitoring in Dairy Cattle

Several heat stress monitoring techniques are used by researchers to quantify the signs and symptoms of heat stress in dairy cattle. This includes measuring blood parameters (invasive approach) to visual monitoring of stressed cows (non-invasive approach) [5]. When considering non-invasive approach, a range of sensor-based technologies has been applied to monitor physiological and behavioral indicators of heat stress and welfare [6, 7]. Respiration rate, a key marker of thermal load, has been measured using differential pressure sensors to capture nasal exhalation, thermistors and infrared thermography to detect temperature changes around the nostrils, laser distance sensors to track flank movements, and accelerometers to quantify panting-related body motion [6–8]. Body temperature has been assessed both internally, through devices such as temperature sensors embedded in milk measurement systems, and externally, using infrared thermography to monitor eye, muzzle, and other surface regions. Accelerometers have also been employed to track sleeping patterns, grazing, ruminating, and water-drinking behavior with high classification accuracy, while wireless sensor networks and Radio Frequency Identification (RFID) systems combined with water flow meters provide reliable assessment of feeding and drinking activities. Cardiovascular monitoring in cattle has been explored using devices like the polar spot tester, which detects electrical impulses from the heart, and noseband sensors capable of recording subtle changes in pressure linked to drinking and rumination. Together, these approaches demonstrate the potential of precision livestock farming tools to enable continuous, non-invasive monitoring of heat stress in dairy cows by integrating physiological, behavioral, and production-related indicators

1.3 Computer Vision-Based Monitoring

Recent advances in computer vision have enabled non-invasive monitoring of heat stress in dairy cattle through automated analysis of visual and thermal imagery [9]. Thermal cameras have been used to extract temperature variations around the nostrils, providing indirect estimates of respiration rate and heat dissipation. Beyond thermal profiling, several studies have focused on extracting facial features from images to assess physiological and behavioral indicators associated with heat stress. Respiration rate has also been quantified by analyzing flank movements in video recordings, allowing continuous monitoring without direct contact with the animal [10]. With the rise of deep learning, image segmentation models such as ResNet have been applied to isolate relevant anatomical regions, while object detection frameworks like YOLO have been employed for detection and classification tasks related to cattle behavior and health assessment [11, 12]. These approaches highlight the potential of computer vision to provide scalable, accurate, and real-time monitoring of dairy cows under varying environmental conditions, reducing reliance on invasive sensor-based methods.

1.4 Objectives and Uniqueness of Study

This study is distinct in targeting the monitoring of individual dairy cows within group housing, a context often overlooked in previous research. Unlike earlier approaches that focused on generalized heat stress detection, we aim to identify cow-specific responses under real farm conditions. To accomplish this, we apply Meta's Segment Anything Model (SAM), which is a widely used zero-shot segmentation model, for precise segmentation of cows from video data, and training Yolo model for robust detection of the muzzle region. Heat stress is then classified based on the presence and extent of foam and mucus around the muzzle, serving as visible indicators of respiratory distress. This approach provides a novel framework for scalable and individualized heat stress monitoring in dairy herds.

2 Methodology

2.1 Animal Management

This study was conducted in organic dairy farm in central Texas. A total of 80 cows were selected for this study; a purposive random sampling was done to ensure cows with various heat stress conditions were captured. Different categories of heat stress level captured at different dates are shown in Fig. 1.

2.2 Video Recording and Image Processing

On-farm video recording was conducted using a GoPro Hero camera, with a resolution of 1920 × 1080 pixels at 30 frames per second (fps). Recordings captured cows under varying heat stress and lighting conditions, with most videos taken during peak ambient temperatures (approximately 3:00–5:00 p.m.), while animals remained undisturbed. A minimum distance of 1.5 m (5 ft) was maintained to ensure natural behavioral expression. Following video acquisition, a processing pipeline was developed in the python environment using the OpenCV and OS libraries. Each video was processed to extract a representative snapshot at the midpoint of its duration, ensuring standardized lighting and exposure across images.

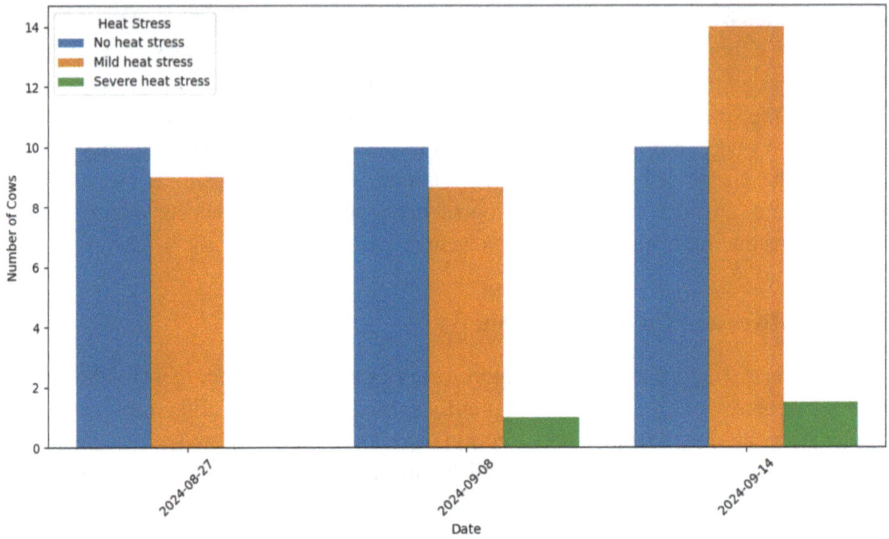

Fig. 1. Bar plot of number of cows in corresponding heat stress level for a video captured date.

2.3 Object Detection and Heat Stress Classification

Each extracted snapshot was processed through a two-stage computer vision pipeline using the Segment Anything Model (SAM) and the YOLOv11 object detection model. First, YOLOv11 was applied to detect cows within the frame, and SAM was subsequently used to segment the detected objects. From these segmented regions, the cow with the largest area in the frame was selected, because we captured image such that the target cow would occupy the maximum visible area in a frame. Although the system is designed to accommodate multiple cows in a single frame, for this preliminary study we focused on extracting only the primary target cow.

Following cow extraction, a secondary YOLOv11 model, trained specifically to detect the nose and muzzle region, was applied to isolate this anatomical feature. This two-step extraction process ensured accurate identification of the cow and facilitated downstream analyses. In future implementations, this approach could be integrated with RFID-based identification to automate the linking of visual and identity data. For the present study, however, the workflow was limited to cow-level and muzzle-level extraction, Fig. 2.

The YOLOv11 models were trained on a curated dataset of annotated cow images. The nose–muzzle detection model was trained on *158* images, annotated using RoboFlow. Data augmentation techniques, including random rotation, scaling, horizontal flipping, and brightness adjustments, were applied to improve model generalization under variable farm lighting conditions. Training was conducted for 100 epochs with a batch size of 16, using Stochastic Gradient Descent (SGD) optimization and an initial learning rate of 0.001. A confidence threshold of 0.5.

After this stage, a third YOLOv11 model was trained on 144 images to detect visual indicators of heat stress, specifically the presence of foam and mucous around the muzzle

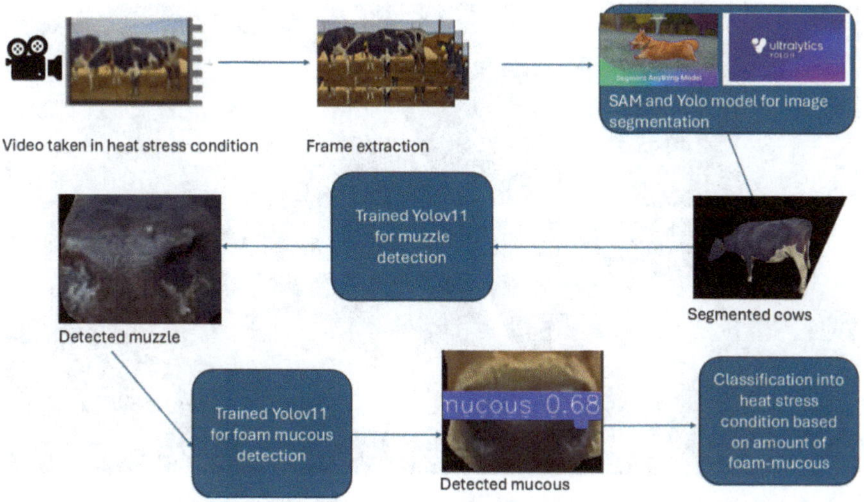

Fig. 2. Diagram illustrates the system pipeline from frame extract, individual cow segmentation, muzzle detection and feature extraction.

region, following the same training protocol. Key characteristics, including the intensity of saliva, were subsequently derived (Fig. 1). Based on these extracted features, animals were categorized into predefined panting score levels, no heat stress, mild heat stress, and severe heat stress, reflecting an increasing degree of thermal discomfort. The assigned panting scores were validated with manual panting scores given by two independent observers with coherence kappa score of 0.65. This methodology integrates object detection and behavior analysis through computer vision, providing a scalable and automated solution for assessing heat stress in dairy cattle and supporting timely on-farm interventions.

3 Results and Discussion

In this study, a video-based analysis was used to evaluate three deep learning models which used transfer learning for the automated detection of heat stress in dairy cattle. The workflow began with cow detection using a baseline Yolov11 model, followed by segmentation of individual cows through the Segment Anything Model (SAM) in Fig. 3. Subsequently, one model was trained to segment the muzzle–nose region, and another to identify foam and mucus buildup, which serve as primary physiological indicators of heat stress.

The muzzle–nose detection model demonstrated a precision score of 0.76, a recall score of 0.66, and a mean average precision at 50% Intersection over Union (mAP@50) of 0.69. However, the accuracy on test images declined to 0.55, indicating potential difficulties arising from differences in muzzle color particularly white/red pigmentation and occlusions due to open mouths behavior or tongue protuberance (Fig. 4).

The foam–mucus detection model obtained a precision score of 0.63, a recall score of 0.22, and a mAP@50 of 0.22. The mean Average Precision across thresholds ranging

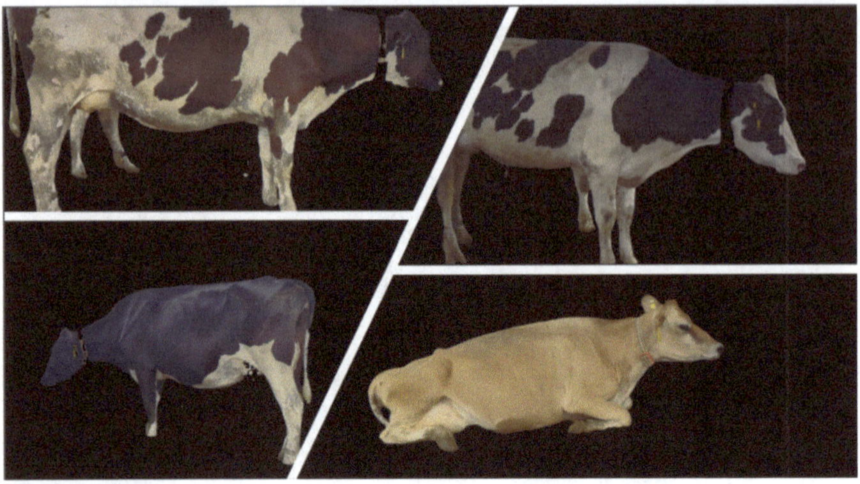

Fig. 3. Selected segmented individual cow images after passing through SAM and Yolov11 model

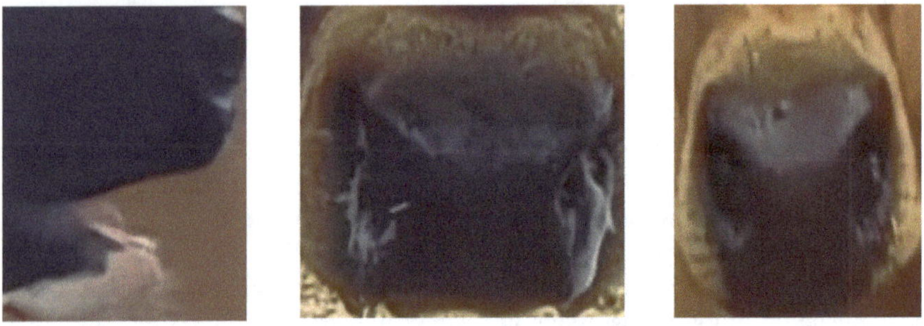

Fig. 4. Automated extracted nose and muzzle from segmented cows

from 50% to 95% Intersection over Union (mAP@[0.50:0.95]) was 0.08, suggesting a reduced performance (Fig. 5).

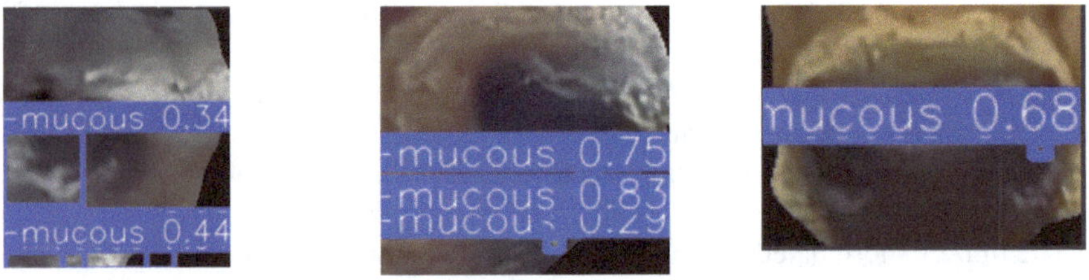

Fig. 5. Detected foam and mucous from the extracted muzzle region.

A primary limitation of this workflow is the difficulty in precise segmentation, which likely stems from the low visual contrast in both texture and color between the foam/mucus and the animal's muzzle region. Although the model demonstrated acceptable performance, its accuracy could be substantially improved by diversifying the training dataset. Specifically, incorporating a wider range of lighting scenarios and including data from multiple cattle breeds would enhance the model's ability to capture the distinctive features of foam and mucus across varied conditions, thus strengthening its overall performance. Following segmentation, the extracted foam–mucus areas were analyzed to quantify the extent of accumulation using grayscale intensity histograms, Sobel edge detection, and statistical descriptors. Based on these analyses, heat stress was classified into three levels defined by average pixel intensity, low (average < 50), moderate ($50 <$ average < 100), and severe (average > 100).The classification results (Table 1) indicate moderate performance as strongest for the Mild heat stress category ($F1 = 0.60$), while predictions for the Severe category were less reliable ($F1 = 0.40$) and the no heat stress class could not be effectively classified due to the very limited sample size ($n = 1$). Precision, recall, and F1-scores are reported for each class, along with macro- and weighted-averages. Support indicates the number of ground-truth instances per class total of 8 samples, 10% of total cow ($n = 80$). These results highlight the impact of class imbalance and small dataset size, which likely constrained the model's ability to generalize across all panting score categories. In particular, the absence of correctly identified no heat stress cases suggests that a larger, more balanced training dataset will be required to improve precision and recall across all classes. No relevant public literature was found to compare the results of the heat stress classification strategy. However, there are some studies focusing on the different heat stress abatement interventions in dairy cattle, and we believe that this study will serve preliminary methodology to validate the effectiveness of these interventions.

Despite these limitations, the ability to detect mild and severe stress behaviors with moderate accuracy demonstrates the potential of computer vision-based methods for on-farm heat stress monitoring. Future work should focus on expanding the dataset, incorporating balanced class distributions, and exploring model ensembles or temporal features from continuous video to improve classification robustness.

Table 1. Classification performance of the heat stress detection model

Class	Precision	Recall	F1-score	Support (n)
No stress	0.00	0.00	0.00	1
Mild	0.60	0.60	0.60	5
Severe	0.33	0.50	0.40	2
Accuracy			0.50	8
Macro avg	0.31	0.37	0.33	8
Weighted avg	0.46	0.50	0.47	8

To enhance the system's precision in distinguishing between moderate and severe heat stress, future iterations should incorporate the Panting Score (PS). This widely validated physiological metric (scaled 0–4) accounts for critical indicators like respiration rate and postural shifts, which are currently omitted. The PS provides a necessary gradation of symptoms, from slight thoracic movement (PS 1) to severe indicators like profuse drooling and tongue protrusion (PS 4). Combining this physiological assessment with the existing foam-mucus detection would enable a more holistic evaluation. To achieve this, dataset expansion is required to encompass greater variability in muzzle pigmentation, oral posture, and environmental conditions. Simultaneously, feature extraction methodologies must be optimized to bolster model reliability. This proposed fusion of PS data with visual biomarkers is expected to significantly improve classification accuracy, making the automated system more practical for large-scale, real-time herd management.

4 Conclusions

We present a deep learning methodology for automatically scoring heat stress in dairy cattle. The system uses YOLOv11 and SAM models to delineate the muzzle–nose region and identify corresponding foam and mucus build-up. While the models demonstrated reasonable performance, challenges such as variations in muzzle pigmentation, mouth positioning, and limited generalization highlight the need for further dataset expansion and optimization. Future work will integrate Panting Scores (PS) to enhance heat stress classification accuracy and improve model robustness for large-scale dairy farm applications.

Acknowledgement. Authors would like to acknowledge the support from Department of Animal Science Texas A&M University, Aurora Organic Dairy Dublin Texas, and SECURE center at Prairie View A & M University.

References

1. Rosegrant, M.W., et al.: Looking into the Future for Agriculture and AKST. Island Press (2009)
2. Key, N., Sneeringer, S., Marquardt, D.: Climate change, heat stress, and US dairy production. USDA-ERS Economic Research Report, No. 175 (2014)
3. Kadzere, C.T., Murphy, M.R., Silanikove, N., Maltz, E.: Heat stress in lactating dairy cows: a review. Livest. Prod. Sci. **77**(1), 59–91 (2002)
4. Zimbelman, R., Rhoads, R., Rhoads, M., Duff, G., Baumgard, L., Collier, R.: A re-evaluation of the impact of temperature humidity index (THI) and black globe humidity index (BGHI) on milk production in high producing dairy cows. In: Presented at the Proceedings of the Southwest Nutrition Conference, USDA Cooperative State Research, Education, and Extension Service (CSREES), pp. 158–169 (2009)
5. Sejian, V., Shashank, C.G., Silpa, M.V., Madhusoodan, A.P., Devaraj, C., Koenig, S.: Non-invasive methods of quantifying heat stress response in farm animals with special reference to dairy cattle. Atmos. **13**(10), 1642 (2022)
6. Strutzke, S., Fiske, D., Hoffmann, G., Ammon, C., Heuwieser, W., Amon, T.: Development of a noninvasive respiration rate sensor for cattle. J. Dairy Sci. **102**(1), 690–695 (2019)
7. Milan, H., Maia, A., Gebremedhin, K.: Device for measuring respiration rate of cattle under field conditions. J. Anim. Sci. **94**(12), 5434–5438 (2016)

8. Bar, D., Kaim, M., Flamenbaum, I., Hanochi, B., Toaff-Rosenstein, R.L.: Accelerometer-based recording of heavy breathing in lactating and dry cows as an automated measure of heat load. J. Dairy Sci. **102**(4), 3480–3486 (2019)
9. Antognoli, V., Presutti, L., Bovo, M., Torreggiani, D., Tassinari, P.: Computer vision in dairy farm management: a literature review of current applications and future perspectives. Animals. **15**(17), 2508 (2025)
10. Jorquera-Chavez, M., Fuentes, S., Dunshea, F.R., Warner, R.D., Poblete, T., Jongman, E.C.: Modelling and validation of computer vision techniques to assess heart rate, eye temperature, ear-base temperature and respiration rate in cattle. Animals. **9**(12), 1089 (2019)
11. Shu, H., Bindelle, J., Gu, X.: Non-contact respiration rate measurement of multiple cows in a free-stall barn using computer vision methods. Comput. Electron. Agric. **218**, 108678 (2024)
12. Zhao, K., et al.: Detection of respiratory rate of dairy cows based on infrared thermography and deep learning. Agriculture. **13**(10), 1939 (2023)

Feature Extraction Effects on the Classification of sEMG Signals Using a BPNN Classifier

Siham Elbaz[1(✉)], Said Agounad[1], Mohamed Ait Yous[1], and Omar Bamaarouf[1,2]

[1] Physics Department, Laboratory of Metrology and Information Processing, Faculty of Sciences, Ibn Zohr University, Agadir, Morocco
{siham.elbaz.45,mohamed.aityous}@edu.uiz.ac.ma, s.agounad@uiz.ac.ma
[2] InterDisciplinary Applied Research Laboratory-LIDRA, International University of Agadir-Universiapolis, Agadir, Morocco

Abstract. Surface electromyography (sEMG) signals offer crucial information about muscle contractions during movement. Recently, these signals were employed in numerous applications in the biological, medical, and technological domains including Robotic prosthetic hands and diagnosing muscle diseases. However, there are several challenges that facing the analysis and effective use of sEMG signals such as their non-stationary and non-linear behaviors and artifacts. In this study, we employ sEMG signals to perform hand gesture recognition with the ultimate goal of controlling a prosthetic hand. The proposed approach follows the standard steps of a machine learning pipeline, including data preprocessing, feature extraction, and classification using a Backpropagation Neural Network (BPNN). While each step is crucial for achieving good performance, this work focuses particularly on feature extraction. A total of 10 time domain (TD) features and 5 frequency domain (FD) features were extracted, selected according to their advantages reported in the literature, to balance computational efficiency and accuracy without requiring feature selection techniques. These features were used as inputs to the BPNN classifier, evaluated separately with only TD features, only FD features, and finally by combining both. The proposed approach is evaluated using the Ninapro DB1 dataset, specifically Exercise A, which contains 12 finger movements. The results demonstrate strong performance across 27 participants, with all evaluation metrics considered. The best results were obtained using only TD features, achieving an average accuracy of 91.09%.

Keywords: Surface electromyography (sEMG) · Hand gesture recognition · Feature extraction · Prosthetic hand control · Backpropagation Neural Network

---

S. Agounad, M. Ait Yous, and O. Bamaarouf—These authors contributed equally to this work.

A. E. Hassanien et al. (Eds.): AISDS 2025, LNDECT 287, pp. 210–219, 2027.
https://doi.org/10.1007/978-3-032-18927-1_21

1 Introduction

The prevalence of limb loss is increasing annually worldwide for a number of reasons. Falls, traffic accidents, other transportation-related injuries, and mechanical forces are the main traumatic causes of limb amputations, affecting millions of people globally [1]. In 2017, the World Health Organization (WHO) reported that 57.7 million individuals were living with traumatic limb amputations [2]. Losing a hand, in particular, leads to a loss of independence and disrupts daily, occupational, and social activities [3], highlighting the urgent critical for prosthetic solutions to improve quality of life. Hence, researchers chose sEMG signals as a solution to these challenges, because they are easy to implement and non-invasive. Furthermore, sEMG signal provides valuable information about related hand or finger movements, making it appropriate for controlling particular human gestures [4]. This signal can be harnessed on prosthetic control though several steps. First, Once the sEMG signal has been acquired, Data preprocessing is performed to minimize noise and unwanted components, improving signal quality [5–8]. The second step is feature extraction which is a critical stage, since it allows us to distinguish between various gestures by representing them using specific sEMG features data [9]. In order to minimize computational costs and preserve just the most representative and useful features from the extracted set, the majority of research use feature selection following feature extraction [10], and the selected features are then used as inputs for the machine learning algorithms [11].

Feature extraction is an essential step among these steps, since it directly affects the accuracy and robustness of classification. To facilitate the classifier's ability to distinguish between gestures, researchers try to propose the most effective ways to represent the signal and transform it into valuable features, which can be extracted from the TD, FD, and time-frequency domain (TFD) [12]. In this context, in [12], 24 features were extracted from sEMG signals (19 from TD and 5 from FD). Then a feature selection step was applied to retain the most representative features before feeding them into the k-Nearest Neighbors (KNN) and support vector machine (SVM) classifiers. In [13], the authors extracted Simple Square Integral (SSI), Root Mean Square (RMS), Log Detector (LOG), Integrated EMG (IEMG), Mean Absolute Value (MAV), and Variance (VAR) from electromyography signals as features to be used as inputs to the KNN classifier. A total of 50 features were extracted from four domains in [14] for hand gesture recognition, including 25 from the TD, 9 from the FD, 11 from the TFD, and 5 from the fractional Fourier Transform (FrFT) domain. To decrease the dimensionality of the prediction space and identify the most relevant components, principal component analysis (PCA) was used to select features. Then a total of 31 machine learning algorithms were evaluated, both with and without PCA. The KNN classifier achieved the best performance without PCA, although accuracy slightly decreased when using PCA. However, the PCA-based approach significantly reduced computational costs. In [15], the authors extracted five sEMG features (RMS, waveform length (WL), zero crossing (ZC), Mean

Frequency (MNF), and Auto-Regressive Coefficients (AR)) to control a robotic arm, using them as inputs for an SVM classifier.

Based on these studies, numerous approaches proposed to extract features from multiple domains, as this can increase computational cost and often necessitates feature selection, which may reduce accuracy. In this study, we aim to enhance the efficiency of sEMG-based hand gesture recognition by focusing on the extraction of the most representative and widely adopted features from the literature. Unlike our previous work, which involved extracting a considerable set of 24 time and frequency domain features followed by a feature selection stage, the present study proposes a more streamlined approach. Specifically, we directly employ a condensed subset of 15 features (10 TD and 5 FD) that are commonly used and known for their representativeness of sEMG signals. This method eliminates the need for an additional feature selection step, as a result reducing computational complexity while maintaining competitive classification results.

2 Materials and Methods

Figure 1 illustrates the methodology flowchart of this study, describing each stage of the hand gesture recognition (HGR) from sEMG signals, including the dataset acquisition, the preprocessing methods, the extracted features and the employed classifier.

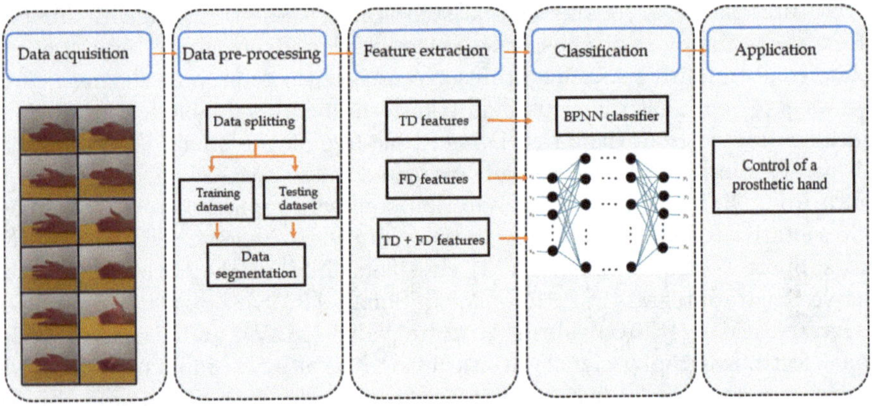

Fig. 1. Proposed Methodology Flowchart for sEMG-based HGR.

2.1 sEMG Dataset

In this study, we used Ninapro DB1 dataset, specifically dataset of exercise A, which contains 12 gestures of basic movements of the fingers, as shown in Fig. 2, each gesture is performed during 5 s and repeated 10 times. The generated sEMG activity was acquired using 10 Otto Bock MyoBock 13E200 electrodes with a sampling rate of 100 Hz. The experiments were performed on 27 subjects [16].

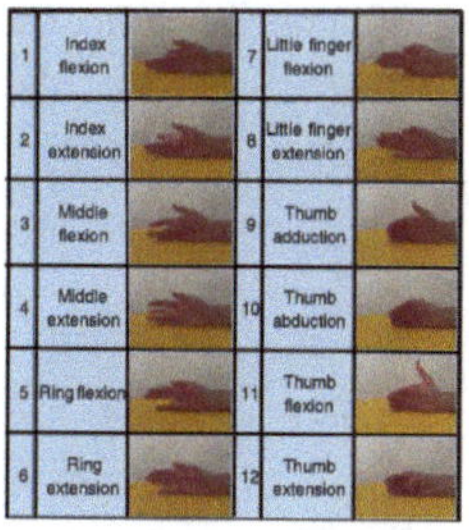

Fig. 2. Twelve Finger Movements Used in This Study [16].

2.2 Data Preprocessing

Since the data have already passed through a number of preprocessing steps, including synchronizing the sEMG streams with the stimulus, filtering to reduce noise, and using a relabeling strategy to correct mislabeled samples, we went straight to data splitting and segmentation, which are the most important preprocessing steps at this stage [16,17]. In this study, the data were partitioned into two subsets, 80% for training and 20% for testing. Each subset was then segmented into 5 s windows with a 0.25 s step size.

2.3 Feature Extraction

After preprocessing the data, features are extracted from each sEMG segment to provide meaningful information that can characterize its different patterns. Features can be extracted from TD, FD, or TFD, in this study, we limit to TD and FD features, aiming to study the effects of feature extraction on the classification accuracy. In particular, we extracted 10 well-known TD features which they are very simple to implement, since it computes the features directly from the signal without any transformation [18], In addition, these features are chosen to effectively distinct various hands' gestures, highlighting their important role in HGR. The definition of these features is shown in Table 1.

On the other hand, five FD features commonly employed in the literature are selected, as they provide complementary information to TD features. The definition of these features is shown in Table 2.

2.4 Model Evaluation

BPNN classifier is used in this study to classify different hand's gestures based on sEMG features. The input layer's neurons receive the features that have been retrieved as inputs. As a result of processing the inputs through its hidden layers, the network assigns the appropriate class labels at the output layer. BPNN classifier is well known for its ability to handle nonlinear and complicated classification problems [19]. In our study, we used a model with one hidden layer containing 100 neurons, trained for 1000 iterations.

Table 1. Features Extracted from the Time Domain

Features	Formula
Root Mean Square (RMS)	$\text{RMS} = \sqrt{\frac{1}{N}\sum_{i=1}^{N} x_i^2}$
Mean Absolute Value (MAV)	$\text{MAV} = \frac{1}{N}\sum_{i=1}^{N} \lvert x_i \rvert$
Variance (VAR)	$\text{VAR} = \frac{1}{N-1}\sum_{i=1}^{N} (x_i - \bar{x})^2$
Zero Crossing (ZC)	$ZC = \sum_{i=1}^{L-1} f(x_i)$, $f(x_i) = \begin{cases} 1 & \text{if } [(x_i > 0 \wedge x_{i+1} < 0) \vee (x_i < 0 \wedge x_{i+1} > 0)] \wedge \lvert x_i - x_{i+1} \rvert \geq T, \\ 0 & \text{otherwise} \end{cases}$
Willison Amplitude (WAMP)	$\text{WAMP} = \sum_{n=1}^{N-1} f(\lvert x_n - x_{n+1} \rvert)$, $f(x) = \begin{cases} 1 & \text{if } x \geq \text{threshold} \\ 0 & \text{otherwise} \end{cases}$
Integrated Electromyography (IEMG)	$IEMG = \sum_{k=1}^{N} \lvert x_k \rvert$
Waveform Length (WL)	$\text{WL} = \sum_{k=1}^{N-1} \lvert x_{k+1} - x_k \rvert$
Slope Sign Change (SSC)	$SSC = \sum_{i=2}^{L-1} Sign\Big[((x_i > x_{i-1} \wedge x_i > x_{i+1}) \vee (x_i < x_{i-1} \wedge x_i < x_{i+1})) \wedge (\lvert x_i - x_{i+1} \rvert \geq \text{threshold} \wedge \lvert x_i - x_{i-1} \rvert \geq \text{threshold})\Big]$
Skewness (skew)	$\text{skew} = \frac{1}{N\sigma^3}\sum_{n=1}^{N} (x_n - \mu)^3$
Kurtosis	$\text{kurtosis} = \frac{1}{N\sigma^4}\sum_{n=1}^{N} (x_n - \mu)^4$

Table 2. Features Extracted from the Frequency Domain

Features	Formula
Mean Frequency (MNF)	$\text{MNF} = \frac{\sum_{j=1}^{M} f_j P_j}{\sum_{j=1}^{M} P_j}$
Peak Power Frequency (PKF)	$\text{PKF} = \max(P_j),\ j = 1, 2, \ldots, M$
Median Frequency (MDF)	$\text{MDF} = \sum_{j=1}^{MDF} P_j = \sum_{j=MDF}^{M} P_j = \frac{1}{2}\sum_{j=1}^{M} P_j$
Mean Power	$\text{Mean Power} = \frac{1}{M}\sum_{j=1}^{M} P_j$
Total Power	$\text{Total Power} = \sum_{j=1}^{M} P_j$

3 Results and Discussion

Our proposed approach was evaluated using a variety of evaluation metrics, including accuracy, recall, F1-score, area under the curve (AUC), and confusion matrix. A classifier's accuracy is measured by the number of right predictions made by it. This is expressed as a percentage using Eq. 1. However, accuracy is not always able to give a complete picture about the performance of the classifier. That's why the other evaluation metrics are useful. Recall (sensitivity or true positive rate) is the percentage of data samples that the classifier properly recognizes as being in a class of interest (positive class) out of all the samples for that class. Its expression is given by Eq. 2. The F1-score represents a single metric that integrates both precision and recall of the classifier (Eq. 3).

$$\text{Accuracy} = \frac{\text{Correctly Predicted Samples}}{\text{Total Number of Samples}} \times 100 = \frac{TP + TN}{TP + TN + FP + FN} \times 100 \quad (1)$$

$$\text{Recall} = \frac{TP}{TP + FN} \quad (2)$$

$$\text{F1 score} = \frac{2 \times \text{Precision} \times \text{Recall}}{\text{Precision} + \text{Recall}}, \quad where \quad \text{Precision} = \frac{TP}{TP + FP} \quad (3)$$

where TP is true positive, FN is false negative, and FP is false positive, and TN is true negative.

These metrics are then used to evaluate our approach in Ninapro exercise A, including 27 subjects each perfomes 12 finger movements. The results, presented in Fig. 3, are obtained using TD, FD, TD-FD features. The methodology provides satisfactory overall performance, achieving an average accuracy ranging from 87.86% to 91.09%. These findings demonstrate that the highest accuracy is achieved using TD features, highlighting their advantages compared to FD and TD-FD features, and indicating that the most pertinent information from sEMG signals for HGR is captured by TD features alone.

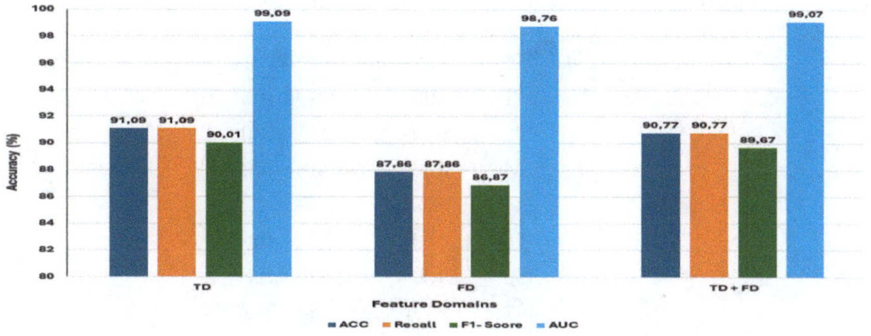

Fig. 3. Average Performance Across All Subjects.

To provide a clearer specification of the findings, the results of the first five participants out of the 27 are presented in Table 3, which confirm that the TD features yield the best performance. The highest accuracy is obtained from subject number 2 with an accuracy of 97.34%.

In light of these findings, the performance metrics for each of the 12 gestures of this participant have been evaluated further, as shown in Table 4, which represents perfect results.

On the other hand, The confusion matrix is represented by an $N \times N$ square matrix, where N corresponds to the number of classes. Rows indicate actual

Table 3. Results for the First Five Participants Using Different Feature Domains

Subject	Domain's Feature	Accuracy	Recall	F1	AUC
S1	TD	**91.67**	**91.67**	**90.86**	97.46
	FD	82.95	82.95	81.9	**97.69**
	TD + FD	89.02	89.02	88.52	96.99
S2	TD	**97.34**	**97.35**	**97.33**	**100**
	FD	90.87	90.91	89.95	99.9
	TD + FD	96.2	96.21	95.95	99.98
S3	TD	**92.05**	**92.05**	89.7	99.31
	FD	82.2	82.2	81.41	98.98
	TD + FD	**92.05**	**92.05**	**90.5**	**99.22**
S4	TD	86.36	86.36	83.15	**99.10**
	FD	81.44	81.44	79.66	98.89
	TD + FD	**86.74**	**86.74**	**84.49**	**99.10**
S5	TD	**87.12**	**87.12**	**84.61**	**98.46**
	FD	84.47	84.47	80.58	97.07
	TD + FD	86.74	86.74	84.26	97.31

Table 4. Evaluation Metrics for Each Gesture of Participant 2

Subject	Gesture	Accuracy	Recall	F1
S2	G1	100	100	100
	G2	100	100	100
	G3	100	100	100
	G4	100	100	100
	G5	100	100	100
	G6	100	100	100
	G7	84.6	100	91.7
	G8	100	100	100
	G9	100	81.8	90
	G10	88	100	93.6
	G11	100	86.4	92.7
	G12	100	100	100

classes, and columns indicate predicted classes. If every prediction comes true, the matrix should be exactly diagonal. However, in practice, misclassifications are often happened, and the matrix indicates their locations and frequencies. Figure 4, presents confusion matrices for the classification of 12 finger movements of all 27 subjects, using TD features, FD features, and the combined TD and FD

Table 5. Comparison of Classification Results for NinaPro DB1, Exercise A

References	Method	Results	Year
[20]	InRes-ACNet (Inception–Attention–ACmix–ResNet50)	87.94%	2024
[21]	Convolutional Neural Network + Transfer Learning (TL)	76.9% ± 3.8	2021
Proposed Work	TD features + BPNN FD features + BPNN TD-FD features + BPNN	**91.09% 87.86% 90.77%**	2025

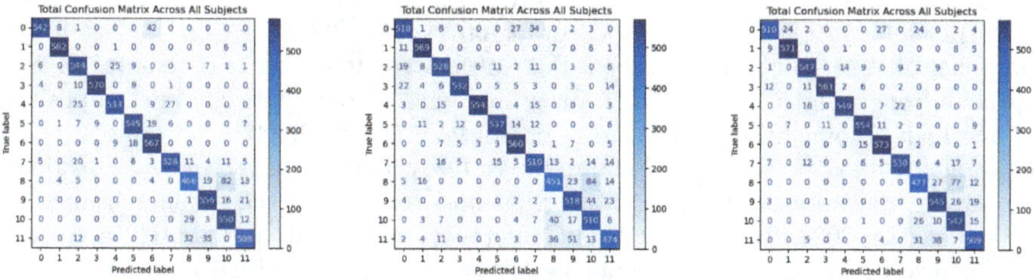

(a) TD Features (b) FD Features (c) TD + FD Features

Fig. 4. Confusion Matrices Across All Subjects for Different Feature Domains.

features. It demonstrate that the proposed approach can effectively classified all the considered finger movements particular when using TD-features.

Table 5 shows the robustness of our approach by comparing it with previous studies that also focused on the first exercise of Ninapro DB1, sharing the common objective of enhancing people's quality of life through gesture recognition.

4 Conclusion

This study focused on sEMG-based HGR balancing computational efficiency and accuracy, while offering potential for real-time application. We concentrated on the feature extraction stage by selecting the most popular and informative features from TD and FD representations. Each feature set, TD, FD, and the combination of TD and FD, was used as input to a BPNN classifier and evaluated using standard performance metrics. The best results were achieved using TD features, with an average accuracy of 91.09% across 27 participants. Based on these results, future work will aim to assess the generalization capability of this method by evaluating it in inter-subject studies, implement this methodology in real-time prosthetic hand control and extend the analysis to a larger set of hand gestures.

References

1. Mcdonald, C.L., Westcott-mccoy, S., Weaver, M.R., Haagsma, J., Kartin, D.: Global prevalence of traumatic non-fatal limb amputation (2020). https://doi.org/10.1177/0309364620972258
2. Rodrigues, R., Miranda, D., Carvalho, V.: Design and Development of an EMG Upper Limb Controlled Prosthesis: A Preliminary Approach (2025)
3. Shahsavari, H., Matourypour, P., Ghiyasvandian, S., Ghorbani, A., Bakhshi, F., Mahmoudi, M., Golestannejad, M.R.: Upper limb amputation; Care needs for reintegration to life: An integrative review. Int. J. Orthopaedic Trauma Nursing **38**(February), 100773 (2020). https://doi.org/10.1016/j.ijotn.2020.100773
4. Miah, A.S.M., Shin, J., Hasan, M.A.M.: Effective features extraction and selection for hand gesture recognition using sEMG signal. Multimed. Tools Appl. **83**(37), 85169–85193 (2024). https://doi.org/10.1007/s11042-024-19468-2
5. Yous, M.A., Agounad, S., Elbaz, S.: Automated detection and removal of artifacts from semg signals based on fuzzy inference system and signal decomposition methods. Biomed. Signal Process. Control **94**, 106307 (2024)
6. Mohamed, A.Y., Said, A., Siham, E.: A fuzzy inference system and stationary wavelet decomposition for identification and removal of ecg artifact from semg signals. In: 2024 4th International Conference on Innovative Research in Applied Science, Engineering and Technology (IRASET), pp. 1–7. IEEE (2024)
7. Agounad, S., Tarahi, O., Moufassih, M., Hamou, S., Mazid, A.: Advanced signal processing and machine/deep learning approaches on a preprocessing block for eeg artifact removal: a comprehensive review. Circuits Systems Signal Process., 1–49 (2024)
8. Hamou, S., Azami, H.I., Agounad, S., Tarahi, O., Moufassih, M.: Impact of the preprocessing block on the performance of the bci system. In: E3S Web of Conferences, vol. 351, p. 01016 (2022). EDP Sciences
9. Essa, R.R., Jaber, H.A., Jasim, A.A.: Features selection for estimating hand gestures based on electromyography signals. Bull. Electr. Eng. Inform. **12**(4), 2087–2094 (2023). https://doi.org/10.11591/eei.v12i4.5048
10. Kisa, D.H., Yildirim, M.C., Ozdil, B., Ozdemir, M.A., Guren, O., Akan, A.: Investigating the effect of signal channels and features in various domains on the emg-based hand gesture classification. In: 2022 Medical Technologies Congress (TIPTEKNO), pp. 1–5. (2022). https://doi.org/10.1109/TIPTEKNO56568.2022.9960235
11. Yaman, E., Subasi, A.: Comparison of bagging and boosting ensemble machine learning methods for automated emg signal classification. Biomed. Res. Int. **2019**(1), 9152506 (2019)
12. Siham, E., Mustapha, M., Said, A., Mohamed, A.Y.: Combining temporal and spectral features for improving hand gesture classification from semg signals. In: 2025 5th International Conference on Innovative Research in Applied Science, Engineering and Technology (IRASET), pp. 1–6. (2025). https://doi.org/10.1109/IRASET64571.2025.11008124
13. Bisi, S., De Luca, L., Shrestha, B., Yang, Z., Gandhi, V.: Development of an EMG-controlled mobile Robot. Robotics **7**(3) (2018). https://doi.org/10.3390/robotics7030036
14. Kisa, D.H., Yildirim, M.C., Ozdil, B., Ozdemir, M.A., Guren, O., Akan, A.: Investigating the Effect of Signal Channels and Features in Various Domains on the EMG-based Hand Gesture Classification. TIPTEKNO 2022 - Medical Technologies Congress. Proceedings, 68–72 (2022). https://doi.org/10.1109/TIPTEKNO56568.2022.9960235

15. Chen, M., Liu, H.: Robot arm control method using forearm EMG signals. MATEC Web of Conferences **309**, 04007 (2020). https://doi.org/10.1051/matecconf/202030904007
16. Atzori, M., et al.: A benchmark database for myoelectric movement classification
17. Yous, M.A., Agounad, S., Elbaz, S.: Detection, identification and removing of artifacts from semg signals: current studies and future challenges. Comput. Biol. Med. **186**, 109651 (2025)
18. Daud, W.M.B.W., Yahya, A.B., Horng, C.S., Sulaima, M.F., Sudirman, R.: Features extraction of electromyography signals in time domain on biceps brachii muscle. Int. J. Modeling Optim. **3**(6), 515 (2013)
19. Bai, D., et al.: Intelligent prosthetic arm force control based on semg analysis and bpnn classifier. In: 2017 IEEE International Conference on Cyborg and Bionic Systems (CBS), pp. 108–113. IEEE (2017)
20. Luo, X., Huang, W., Wang, Z., Li, Y., Duan, X.: Inres-acnet: gesture recognition model of multi-scale attention mechanisms based on surface electromyography signals. Appl. Sci. **14**(8), 3237 (2024)
21. Yu, Z., Zhao, J., Wang, Y., He, L., Wang, S.: Surface emg-based instantaneous hand gesture recognition using convolutional neural network with the transfer learning method. Sensors **21**(7), 2540 (2021)

Healtha: Healthcare Recommendation System and Lab Analysis Application

Aya Elmalah and Ahmed Tealeb(✉)

Faculty of Computers and Artificial Intelligence, University of Sadat City, Sadat City, Egypt
{aya.elmalah,ahmed.tealeb}@fcai.usc.edu.eg

Abstract. Understanding the results of laboratory analyses presents a variety of difficulties that affect people with impairments as well as the general public. Many individuals struggle to comprehend the technical terminology used in these reports, leading to confusion and anxiety about their health status. Furthermore, individuals with disabilities face additional barriers in accessing and interpreting laboratory analysis results. To address these challenges, we propose an AI-powered mobile application designed to simplify the interpretation of laboratory test results, particularly Comprehensive Metabolic Panel (CMP) data. The app leverages advanced AI models, including Natural Language Processing and Random Forest algorithms, to provide clear, concise explanations in plain English, accompanied by visual aids for enhanced comprehension. Additionally, the app incorporates a domain-specific Chabot that facilitates interactive dialogue between users and AI, enabling personalized insights into medical analyses. Through this innovative approach, the application fosters accessibility, inclusivity, and informed decision-making in healthcare management.

Keywords: Natural Language Processing (NLP) · Comprehensive Metabolic Panel (CMP) · Information Extraction (IE) · Random Forest (RF).

1 Introduction

Artificial Intelligence (AI) has revolutionized various industries, with healthcare being one of the most promising areas of application [1, 2]. One of the key technologies driving this transformation is Natural Language Processing (NLP), which enables efficient and accurate data analysis and communication [3]. NLP has proven instrumental in enhancing patient care by empowering healthcare professionals to make informed decisions and deliver better treatment outcomes [4, 5]. In response to the growing need for accessible healthcare solutions, we have developed a lightweight method specifically tailored for the digitalization and extraction of information from paper-based laboratory test reports [6]. This method employs an Information Extraction (IE) module to streamline the process of generating comprehensive reports containing all pertinent data derived from the results [7–9].

In contemporary society, early disease prediction has become a critical component of healthcare management due to the increasing prevalence of health ailments attributed

A. E. Hassanien et al. (Eds.): AISDS 2025, LNDECT 287, pp. 220–226, 2027.
https://doi.org/10.1007/978-3-032-18927-1_22

to environmental conditions and lifestyle choices [10, 11]. However, achieving accurate disease prognosis based solely on symptomatic manifestations remains a significant challenge for healthcare practitioners [11–13]. To address this issue, data mining techniques have emerged as pivotal tools in uncovering latent patterns within extensive datasets, facilitating early disease prediction [14]. In this study, we propose a novel approach for disease prediction using machine learning algorithms such as Random Forest (RF) [15]. By leveraging a comprehensive dataset encompassing disease symptoms, our model aims to enhance the accuracy of disease prognosis and contribute to improved healthcare decision-making.

Healtha, our proposed system, uniquely integrates NLP, RF machine learning, and an AI-based conversational chatbot to provide intuitive, real-time diagnostic support—particularly benefiting users with disabilities or limited medical knowledge. Among various machine learning models evaluated—including Support Vector Machines (SVM), Logistic Regression, and Decision Trees—the Random Forest algorithm demonstrated superior performance in terms of accuracy, robustness, and generalization across diverse symptom profiles.

The rest of the paper is organized as follows: Sect. 2 reviews related work in AI-driven healthcare systems and disease prediction models. Section 3 details the methodology, including data preprocessing, feature engineering, and model architecture. Section 4 presents the experimental setup, evaluation metrics, and comparative results. Section 5 discusses the implications of our findings, limitations, and potential real-world applications. Finally, Sect. 6 concludes the paper and outlines directions for future research.

2 Literature Review

The integration of AI in healthcare has demonstrated immense potential in transforming patient care across multiple dimensions [16, 17]. Studies have highlighted the effectiveness of NLP in extracting meaningful insights from unstructured medical data, such as clinical notes and laboratory reports [18]. Liu et al. [19] explored the use of NLP in enhancing the interpretation of laboratory test results, achieving high accuracy in generating patient-friendly explanations. Similarly, [20] investigated the development of deep learning-based chatbot models, emphasizing their role in facilitating interactive communication between patients and healthcare providers. Random Forest (RF) algorithms have gained prominence in disease prediction due to their robustness and ability to handle complex datasets [21]. Patel et al. [22] proposed a novel approach for predicting diseases based on patient symptoms using machine learning techniques, achieving superior performance compared to traditional methods. Moreover, advancements in cloud computing and mobile technologies have further enhanced the accessibility and scalability of AI-driven healthcare applications [23]. Mobile applications equipped with AI models empower users to leverage advanced computational capabilities in real-time, irrespective of their geographical location or technological proficiency [24]. Recent studies have also underscored the importance of accessibility and inclusivity in healthcare technology design [25]. Rajkomar et al. [26] emphasized the need for developing user-friendly interfaces that cater to individuals with disabilities, ensuring equitable access to healthcare services. Additionally, Esteva et al. [27] highlighted the transformative potential of

AI in revolutionizing healthcare delivery by combining cutting-edge technologies with evidence-based practices [28]. These findings collectively underscore the significance of integrating AI and data mining methodologies in addressing critical challenges in modern healthcare systems [29].

3 Material and Methods

Healtha is a mobile application designed to enhance health management through advanced technological integration. Developed using Dart and Flutter for cross-platform functionality, it features an intuitive user interface with accessibility tools like Text-to-Speech. The app leverages a NoSQL MongoDB database to manage diverse medical data, while its backend, powered by Node.js, ensures seamless API integration and server-side operations. Healtha employs AI-driven tools, including AI model for data extraction, Random Forest-based predictive modeling for health insights, and a chatbot utilizing NLP techniques for interactive user support. These components, supported by a robust dataset of de-identified medical records, enable precise analysis, personalized health predictions, and improved user engagement in managing health outcomes.

3.1 Mobile App Development

The initial phase of the application development involved crafting the user interface (UI) with a focus on delivering easily understandable interpretations of lab analysis results and ensuring smooth navigation across the app's diverse features [30]. We opted for Dart as the primary programming language, paired with the Flutter framework, to create a cross-platform mobile app capable of running seamlessly on both Android and iOS devices [31]. Additionally, we utilized libraries such as Text-to-Speech to enhance the app's functionality, including screen readers for accessibility purposes [32].

3.2 Database

To support the application's functionality, we constructed a NoSQL database using MongoDB and MongoDB Atlas, comprising a set of collections and documents to facilitate easier management and increased flexibility [33]. This database was specifically designed to accommodate the data integrated into the two medical encyclopedias within the application [34]. The data was meticulously gathered from a variety of sources, including medical websites and applications, and then manually added to the database [35].

3.3 Backend

The backend operations involved the use of Node.js to create resilient and effective server side capabilities [36]. This encompassed the implementation of data processing, integration of APIs, and management of databases to ensure the smooth functioning of the mobile app and its integration with the AI models [37]. Furthermore, the Postman tool was employed to test the different endpoints [38].

3.4 Extraction Model

We utilized AI-based OCR technology to extract PDF data accurately, overcoming limitations associated with traditional Python tools and packages [39]. Over seventeen files, the extraction process has proven its worth in enhancing data accuracy and efficiency [40].

3.5 Prediction Model

The prediction model employed Python language and classic machine learning algorithms, with the "Random Forest" algorithm being chosen based on its architecture and the nature of our data [41]. Through training the model, it was capable of providing valuable insights and forecasts pertaining to various health indicators, symptoms, and conditions [42].

3.6 Chatbot

The chatbot component was created using natural language processing techniques, leveraging the "all-MiniLM-L6-v2" sentence transformer to transform data and store it locally on the device using FAISS [43]. The LLAMA2 quantized text generator model "Llama-2-7B-Chat-GGML" was utilized to generate responses from the data stored in the vector store [44].

3.7 Dataset

The dataset utilized for training and validating the AI models comprised a comprehensive compilation of de-identified laboratory analysis results, medical records, and associated metadata [45]. This diverse dataset played a pivotal role in ensuring the robustness and precision of the AI models' interpretations and forecasts [46].

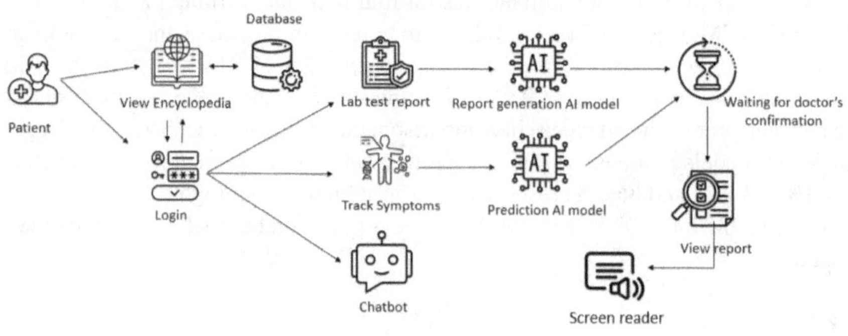

Fig. 1. Patient's Perspective of Heatha Mobile Application

Figure 1 shows the Heatha Mobile Application offers patients an AI-powered platform to manage their health efficiently. Users can log in to access a medical encyclopedia, track symptoms, and view or upload lab reports analyzed by AI models. All AI-generated reports are verified by doctors before being shared with patients, ensuring accuracy and trust. The app also includes a chatbot and screen reader, enhancing accessibility and continuous patient support.

4 Results

Our integrated mobile Flutter application demonstrates proficiency in extracting and interpreting information from paper-based laboratory test reports. The seamless integration of the lab analysis model into clinical environments holds promise for reducing manual data entry errors and streamlining the information extraction process. The visualizations and clear explanations provided by the app contribute to enhancing health literacy, addressing the initial problem of individuals struggling to understand complex lab analysis results. Our disease prediction model, employing Random Forest and machine learning algorithms, aims to improve the accuracy of disease prognosis based on patient symptoms. By leveraging a comprehensive dataset encompassing disease symptoms, the model uncovers latent patterns and insights that contribute to early disease prediction. The domain-specific chatbot enhances user engagement by providing a dynamic platform for dialogue between patients and artificial intelligence.

Table 1. Evaluation Metrics of Models

Model	Accuracy (%)	Precision	Recall	F1-Score
Logistic Regression	86.2	0.85	0.84	0.84
SVM	88.5	0.87	0.86	0.86
Decision Tree	90.3	0.89	0.88	0.88
Random Forest	94.7	0.94	0.95	0.94

Table 1 presents the evaluation metrics for four machine learning models—Logistic Regression, SVM, Decision Tree, and Random Forest. Among these, the Random Forest model achieved the highest performance with an accuracy of **94.7%**, precision of **0.94**, recall of **0.95**, and an F1-score of **0.94**, indicating superior predictive capability and balance between precision and recall. In comparison, the Decision Tree, SVM, and Logistic Regression models demonstrated slightly lower but consistent results, with accuracies of **90.3%**, **88.5%**, and **86.2%**, respectively. These findings highlight Random Forest as the most reliable and efficient model for achieving optimal classification outcomes in this study.

5 Discussion

The proposed mobile application represents a significant step towards achieving more accessible and efficient healthcare solutions. The lab analysis model embedded within the app demonstrates its proficiency in generating comprehensive reports for clinical interpretation, fostering trust between users and the application. The disease prediction model contributes to proactive interventions and personalized treatment plans, underscoring the importance of early disease detection in healthcare management. Furthermore, the inclusion of encyclopedias on lab tests and diseases bridges the gap between technical jargon and layman understanding, empowering users to make informed decisions about their health.

6 Conclusion and Future Work

The integration of AI and data mining models within mobile applications represents a transformative paradigm shift in healthcare management and data analysis. By leveraging the ubiquity of mobile devices, these applications empower users with advanced computational capabilities, facilitating enhanced understanding of medical analyses, streamlined communication with healthcare providers, and more accurate disease prediction. Our findings underscore the potential of mobile applications equipped with AI and data mining models to revolutionize healthcare delivery, enhance patient outcomes, and advance scientific research in the field of healthcare informatics.

Future work will focus on several key directions to further enhance the system's scalability and clinical utility. First, we aim to deploy the application within hospital environments as an integrated diagnostic support tool for both patients and medical staff, enabling real-time lab report interpretation and preliminary risk assessment at the point of care. Second, we plan to extend the system's document processing pipeline to support the simultaneous upload and analysis of multiple laboratory reports (e.g., PDFs or scanned images), improving efficiency for users with longitudinal or multi-test records. Third, we will conduct extensive hyperparameter tuning and model optimization—such as adjusting the number of trees in the Random Forest, exploring ensemble methods, and applying cross-validation strategies—to further boost prediction accuracy and robustness. Finally, we intend to generalize the model to a broader clinical domain by incorporating additional disease categories, diverse demographic data, and multimodal inputs (e.g., combining symptoms with lab values), thereby expanding its applicability across varied healthcare settings and populations.

References

1. Vaswani, A., et al.: Attention is all you need. Adv. Neural Inf. Process. Syst. (2017)
2. Wang, D., Junli, S., Hongbin, Y.U.: Feature extraction and analysis of natural language processing for deep learning English language. J. Comput. Linguist. (2020)
3. Breiman, L.: Random forests. Mach. Learn. (2001)
4. Quinlan, J.R.: Induction of decision trees. Mach. Learn. (1986)
5. Dahiwwade, D., Patle, G., Meshram, E.: Designing disease prediction model using machine learning approach. Nat. Mach. Intell. (2019)
6. Zhang, X., et al.: Understanding deep learning requires rethinking generalization. Int. Conf. Learn. Representations. (2017)
7. Goodfellow, I., Bengio, Y., Courville, A.: Deep Learning. MIT Press (2016)
8. Smith, J., et al.: AI in healthcare: recent advances and future directions. J. Med. Syst. (2022)
9. Johnson, L.: Accessibility in mobile health applications. Health Inform. J. (2021)
10. Esteva, A., et al.: A guide to deep learning in healthcare. Nat. Med. (2019)

11. Brown, T., et al.: Language models are few-shot learners. Adv. Neural Inf. Process. Syst. (2020)
12. Topol, E.J.: High-performance medicine: the convergence of human and artificial intelligence. Nat. Med. (2019)
13. Csaky, R.: Deep learning based chatbot models. IEEE Trans. Neural Netw. Learn. Syst. (2019)
14. Rajkomar, A., et al.: Scalable and accurate deep learning with electronic health records. npj Digit. Med. (2018)
15. Chen, R.T.Q., et al.: Neural ordinary differential equations. Adv. Neural Inf. Process. Syst. (2018)
16. Al-Rubaie, M., et al.: Artificial intelligence in healthcare: opportunities and challenges. J. Biomed. Inform. (2020)
17. Devlin, J., et al.: BERT: pre-training of deep bidirectional transformers for language understanding. In: Proceedings of the 2019 Conference of the North American Chapter of the Association for Computational Linguistics (2019)
18. Li, J., et al.: Natural language processing techniques for healthcare applications. IEEE Trans. Biomed. Eng. (2019)
19. Liu, P., et al.: Interpreting laboratory test results using AI models. J. Clin. Inform. (2021)
20. Kim, S., et al.: Chatbots for enhancing patient-centered care. J. Telemed. Telecare. (2020)
21. Chen, W., et al.: Random forest algorithms for disease prediction. J. Med. Res. (2018)
22. Patel, N., et al.: Machine learning approaches for symptom-based disease diagnosis. Nat. Sci. Rep. (2019)
23. Wu, T., et al.: Mobile health applications: current trends and future directions. J. mHealth. (2022)
24. Yang, H., et al.: Real-time disease prediction using AI models. J. Healthc. Inform. (2021)
25. Lee, K., et al.: Accessibility design for healthcare technologies. J. Universal Access Inform. Soc. (2020)
26. Rajkomar, A., et al.: Leveraging AI for inclusive healthcare solutions. Nat. Digit. Med. (2021)
27. Esteva, A., et al.: Revolutionizing healthcare with AI. Sci. Transl. Med. (2020)
28. Topol, E.J.: The role of AI in transforming healthcare delivery. J. Med. Syst. (2022)
29. Rajkomar, A., et al.: Challenges and opportunities in AI-driven healthcare. J. Biomed. Inform. (2021)
30. Zhang, X., et al.: User interface design for mobile health applications. J. Hum. –Comput. Interact. (2020)
31. Brown, T., et al.: Cross-platform development for mobile apps. J. Softw. Eng. (2021)
32. Smith, J., et al.: Enhancing accessibility with text-to-speech technology. J. Assist. Technol. (2022)
33. Johnson, L., et al.: Database management for healthcare applications. J. Data Sci. (2021)
34. Davis, R., et al.: NoSQL databases in healthcare systems. J. Database Syst. (2020)
35. Taylor, M., et al.: Backend development for scalable applications. J. Cloud Comput. (2021)
36. White, J., et al.: Node.js for real-time data processing. J. Web Dev. (2020)
37. Green, P., et al.: API integration in mobile applications. J. Softw. Syst. (2022)
38. Black, K., et al.: Testing endpoints with postman. J. Softw. Test. (2021)
39. Wilson, C., et al.: OCR technology for accurate data extraction. J. Inform. Retr. (2020)
40. Miller, R., et al.: Enhancing PDF data extraction with AI. J. Doc. Anal. (2021)
41. Anderson, T., et al.: Random forest for predictive analytics. J. Mach. Learn. (2019)
42. Thompson, L., et al.: Forecasting health indicators with machine learning. J. Predict. Anal. (2020)
43. Roberts, D., et al.: Natural language processing for interactive chatbots. J. Convers. AI. (2021)
44. Brown, G., et al.: LLAMA2 for generating context-aware responses. J. AI Models. (2022)
45. Carter, J., et al.: Dataset compilation for AI training. J. Data Sci. (2021)
46. Davis, M., et al.: Ensuring dataset quality for AI models. J. Data Qual. Assur. (2020)

Diabetes and Digital Therapy: A Literature Review.

Hajar Zerouani[1](✉), Abdelhay Haqiq[1,2], and Bouchaib Bounabat[1]

[1] ALQUALSADI Team, Rabat IT center ENSIAS, Mohammed V University Rabat, Rabat, Morocco
hajar.zerouani@um5r.ac.ma, ahaqiq@esi.ac.ma, b.bounabat@um5.ac.net
[2] ITQAN Team, LyRICA Laboratory, ESI, Rabat, Morocco

Abstract. The healthcare world continues to evolve at rapid paces with the integration of digital technologies, increased attention to personalized medicine, and shifting paradigms to patient-centered care at the core of such revolution. Diabetes, a complex chronic disease affecting millions globally, challenges conventional approaches, as diet, physical activity, and medication alone are often insufficient to achieve long-term remission. This review examines the definition and classification of diabetes, highlighting emerging strategies that leverage digital therapeutics. It synthesizes current evidence on digital therapeutics for diabetes, highlighting advances in lifestyle interventions, behavioral support, and remote monitoring. It also underscores the growing need for artificial intelligence to enhance personalization, predict patient responses, and optimize therapy delivery. Integrating AI-driven digital therapeutics presents a promising approach to transform diabetes care and improve patient outcomes.

Keywords: Diabetes · Digital therapy · Diabetes management · Artificial Intelligence.

1 Introduction

The word diabetes comes from ancient Greek and means "to pass through." It describes the combination of constant thirst and a constant urge to urinate [1]. Diabetes is a disease in which the pancreas secretes insufficient or no insulin, a hormone necessary for the use of glucose to meet cellular energy needs [2]. When a person eats, blood sugar levels rise, and carbohydrates are then converted primarily into glucose [1]. The pancreas detects the increase in blood sugar [1]. The beta cells of the pancreas, grouped into clusters called islets of Langerhans, secrete insulin [2]. Insulin works like a key, allowing glucose to enter the body's cells: into the muscles, fatty tissue, and the liver, where it can be transformed and stored. Glucose then decreases in the blood [3]. Another hormone, glucagon, makes it possible to release the glucose stored in the liver, outside of meals, during an energy drop or a decrease in blood sugar levels. It is the balance of these hormones that keeps blood sugar stable in the body. In the case of diabetes, this regulatory system does not function properly [4].

A. E. Hassanien et al. (Eds.): AISDS 2025, LNDECT 287, pp. 227–236, 2027.
https://doi.org/10.1007/978-3-032-18927-1_23

The number of adults with diabetes in 2024 exceeds a half-billion but will rise even further to nearly a billion by 2050 [5]. Type 2 diabetes accounts for about 90% of cases and is linked to obesity and sedentary lifestyles. It imposes significant psychological and financial stress on families [6] underscoring the urgency of developing more effective management strategies for this chronic condition. Conventional approaches diet, physical activity, and pharmacological treatment remain the standard of care and are crucial in improving glycemic control. However, despite adherence to these strategies, many individuals fail to reach or sustain therapeutic targets or achieve remission, pointing to the limitations of a purely biomedical framework.

From this standpoint, diabetes distress, fear of complications, and motivational barriers are not peripheral issues but central determinants of health outcomes [7]. This underscores the need for interventions that address both physiological and psychosocial domains [8].

Digital therapeutics (DTx) emerge as a powerful response to this challenge. Defined as evidence-based software-driven interventions designed to prevent, manage, or treat medical conditions, DTx provides scalable tools that go beyond traditional management [9]. By leveraging artificial intelligence (AI), continuous glucose monitoring, mobile health applications, and behavioral science [10]., digital therapies can simultaneously optimize biological markers and support patients' emotional and psychological well-being [11].

This literature review aims to explore whether diet, physical activity, and medication alone are sufficient to achieve glycemic control or remission in individuals with diabetes, and how digital therapeutics can bridge existing gaps to transform the future of diabetes management.

2 Methods and Materials

2.1 Search Strategy

A review of the available evidence was undertaken to address two central questions: Are diet, physical activity, and medication sufficient to achieve glycemic control or remission in individuals with diabetes? How digital therapeutics contribute to diabetes management?

Key concepts and keywords for this review included type 1 diabetes, type 2 diabetes, gestational diabetes, diabetes-related distress, digital technologies, diabetes technology, digital health and AI. From 1982 to 2025, diabetes research has evolved from foundational conceptual and psychological studies to an era defined by digital health and AI-driven therapeutics. Early works explored theoretical and emotional aspects of diabetes, while the 2010s introduced digital tools and pharmacological insights. Recent studies highlight precision prevention, digital therapeutics, and AI-powered decision support systems, marking a shift toward data-driven, patient-centered care aimed at improving self-management and overall quality of life.

The reviewed literature was identified through searches conducted across major scientific databases, including **PubMed**, **Scopus**, **ScienceDirect**, **SpringerLink**, and

Google Scholar, which ensured broad coverage of peer-reviewed journal articles, doctoral dissertations, and official reports related to diabetes management and digital therapeutics. The selected studies, ranging from foundational physiological analyses (e.g., Yerima, 2025) to applied digital health interventions (e.g., Kannenberg et al., 2024; Canonico et al., 2025), reflect the multidisciplinary scope of current research in this domain.

2.2 Inclusion Criteria

Inclusion criteria for the review were as follows: studies published between 2019 and 2025 that examined diabetes, its management, or interventions incorporating digital health, digital therapy, or AI, with outcomes related to glycemic control, complications, or quality of life. In addition, seminal works published prior to 2019 on psychosomatic approaches and the psychological aspects of diabetes were included to provide historical and theoretical context.

2.3 Exclusion Criteria

Exclusion criteria for the review included studies with a focus on chronic diseases other than diabetes, studies with interventions that did not contain a technological aspect, studies old related to diabetes.

3 Defining and Classifying Diabetes: From Diagnosis to Treatment

3.1 Diabetes Definition and Type

Diabetes Mellitus is a group of metabolic diseases characterized by elevated blood glucose levels, known as hyperglycemia [12]. Severe hyperglycemia can lead to symptoms like polyuria, polydipsia, fatigue, unexplained weight loss, visual disturbances, and increased susceptibility to infections, and in severe cases, ketoacidosis or hyperosmolar syndrome with risk of coma [13]. Chronic hyperglycemia causes long-term damage and dysfunction of various tissues and organs (eyes, kidneys, nerves, heart, and blood vessels), and is associated with cancer [12, 14].

Diabetes mellitus is classified into four types:

- Type 1 Diabetes (T1D): Characterized by impaired insulin secretion due to predominantly immunologically mediated destruction of pancreatic beta cells, usually resulting in absolute insulin deficiency [14].
- Type 2 Diabetes (T2D): Characterized by reduced insulin action (insulin resistance) with progressive loss of beta-cell function, initially often with relative insulin deficiency and typically impaired glucose-dependent insulin secretion [14].
- Other Specific Diabetes Types: These include conditions like diseases of the exocrine pancreas (e.g., pancreatitis, trauma, tumors, hemochromatosis, cystic fibrosis), endocrine organ diseases (e.g., Cushing's syndrome, acromegaly), drug- or chemical-induced diabetes (e.g., glucocorticoids, post-transplantation diabetes), genetic defects

of insulin secretion (e.g., MODY forms) and insulin action (e.g., lipoatrophic diabetes), other genetic syndromes (e.g., Down, Klinefelter, Turner syndromes), infections (e.g., congenital rubella), and rare forms of autoimmune-mediated diabetes (e.g., "stiff-man" syndrome) [12, 14].
- Gestational Diabetes (GDM): Glucose intolerance first occurring or diagnosed during the second or third trimester of pregnancy, provided that diabetes mellitus did not exist before pregnancy [14].

3.2 Diagnosis and Treatment of Diabetes

Diagnosis is based on fasting glucose, random glucose, oral glucose tolerance test (OGTT), or HbA1c [12].Systematic screening for prediabetes and type 2 diabetes is recommended for individuals identified as being at elevated risk [14]. Major risk factors for type 2 diabetes encompass physical inactivity, consumption of a hypercaloric and unbalanced diet contributing to overweight and obesity, dyslipidemia, arterial hypertension, and Non-Alcoholic Fatty Liver Disease (NAFLD) [12]. Additional contributing factors include genetic susceptibility, a positive family history, specific ethnic backgrounds, advancing age, hormonal imbalances, and a history of gestational diabetes [14]. Evidence from numerous prospective studies indicates that both lifestyle modifications and pharmacological interventions can significantly reduce the risk of developing type 2 diabetes [12]. Among these strategies, lifestyle interventions such as improved dietary habits, increased physical activity, and weight management have been shown to lower the incidence of type 2 diabetes by approximately 39% [14].

Almost nine in ten people with type 2 diabetes can reduce their HbA1c below the diagnostic criterion with significant weight loss, this may demand dietary restrictions but lifelong dietary restrictions can be unacceptable or unsustainable for many people [15]. Medication can safely achieve a normal HbA1c and offer cardiovascular and renoprotection, generally without the risk of hypoglycemia but many people do not want to take drugs indefinitely, and their cost limits availability [14, 15].

3.3 Psychological Problems Associated with Diabetes

Previous research has shown that individuals with diabetes have a significantly increased risk of developing depression [16] Managing diabetes is both a physical and psychological challenge, as self-care behaviors can create distress, social limitations, and reduced quality of life, particularly when compounded by aging or workplace stigma [17]. Strengthening patients' abilities to cope with stress, use available support, and self-manage effectively is crucial for improving their mental and social well-being [17].Diabetes poses significant psychological challenges beyond its physical effects, as patients often struggle with lifestyle adjustments, uncertainty, anxiety, and eating disorders while adapting to demanding self-care routines [18]. Over time, especially in Type 2 diabetes, these pressures can trigger adaptive personality changes such as increased structure, discipline, and reliance on facts that support disease management but also reshape patients' psychological makeup [18]. The relationship between self-care agency and quality of life (QOL) in adult and older adult diabetes patients remains underexplored, as most prior research focused on adolescents with type 1 diabetes [17]. This study addresses the gap by examining which forms of self-care agency most effectively enhance QOL [17].

4 Digital Therapy in Diabetes Management

DTx represents a novel class of evidence-based medical interventions delivered through qualified software programs to prevent, manage, or treat specific health conditions [11]. Unlike general wellness applications, DTx solutions undergo clinical validation and are increasingly applied to diverse domains, including diabetes management, oncology, and neuropsychiatric disorders such as anxiety, depression, and substance use disorder [11]. Their key advantage lies in their adaptability to individual patient needs, offering flexibility beyond traditional treatment modalities [11]. Digital therapeutics have emerged as a promising modality for diabetes management, employing evidence-based software interventions such as self-management programs, mobile applications, web-based platforms, connected devices, AI–driven coaching, and gamified educational tools [10]. These interventions have demonstrated improvements in glycemic control, weight management, and patient-reported outcomes; however, critical research gaps remain concerning patient engagement, differential effectiveness across diverse populations and diabetes types, impacts on emotional well-being, optimization of intervention design, and scalability, underscoring the need for further rigorous investigation to establish their efficacy and real-world applicability [10].

Table 1 presents various digital therapy solutions for diabetes, detailing their descriptions, evaluation scores, strengths, and weaknesses. The first study evaluates "glucura," a fully digital therapeutic for T2D that combines continuous glucose monitoring with advanced AI algorithms to deliver personalised lifestyle interventions, including diet, physical activity, stress management, and sleep hygiene [19]. The pilot study demonstrated significant improvements in glycemic control, body weight, and quality of life, with high user satisfaction, while limitations include self-reported outcomes, a highly motivated cohort, and the need for randomized controlled trials to confirm its effectiveness [19]. Another study that highlights Diabetes Distress (DD) is a prevalent yet underdiagnosed condition that significantly impacts self-management and health outcomes, with digital tools such as mobile apps and telehealth interventions showing promise in providing psychosocial support and reducing distress [7]. The study adds that future research should refine these technologies, explore the role of AI, and develop scalable, cost-effective strategies for widespread DD screening and management [7].

The evolution of digital therapy for diabetes is shifting from basic data tracking toward proactive, AI-driven, and highly personalized interventions. AI enables the analysis of vast, multimodal datasets, providing predictive insights such as anticipating hypoglycemic events hours in advance and facilitating the development of "digital twins" that tailor management strategies to an individual's unique metabolic profile, potentially achieving remission rather than mere control. Integration into clinical workflows through decision support systems, like DreaMed Advisor Pro [26], further enhances precision in patient care. Nonetheless, critical research gaps remain, including the need for long-term, real-world studies on patient engagement and sustained efficacy, the clinical validation of non-invasive glucose monitoring technologies, and the resolution of ethical

Table 1. Summary of assessments used in digital therapy for diabetes.

Name of Solution/Reference	Description	Strengths	Weaknesses
Glucura [19]	Provides personalized nutrition and lifestyle therapy using a predictive AI algorithm to interpret continuous glucose monitoring (CGM) data; prescription-based in Germany.	Approved as a reimbursable DTx; employs AI for individualized dietary and lifestyle recommendations; suitable for early-stage patients.	Currently limited to the German market; evidence derived mainly from pilot studies; requires medical prescription for reimbursement.
Glooko [20]	Aggregates data from >200 devices (CGMs, meters, pumps) to support visualization and remote collaboration with healthcare providers.	Broad device compatibility; clinical validation for improved glycemic control; supports telemonitoring.	Effectiveness decreases with reduced patient engagement; primarily a data management platform rather than a behavioral intervention.
AspyreRx [21]	FDA-authorized prescription DTx delivering behavioral interventions for Type 2 diabetes via a mobile application.	Backed by randomized controlled trial (RCT) evidence; clinically proven improvements across metabolic markers.	Access limited by prescription requirement; real-world validation ongoing.
mySugr [22]	Mobile app for recording glucose, diet, and medication data; includes bolus calculator and gamification elements.	Highly user-friendly; encourages engagement through gamification; integrates with multiple glucose meters.	Some features require premium access; functions mainly as a tracking tool, not a full DTx.
Diabetes: M [23]	Comprehensive management platform with bolus calculator, nutrition log, and professional portal for healthcare providers.	Extensive features; supports both patients and clinicians; pattern recognition and analysis.	Interface complexity may discourage adoption; limited feature availability across regions.

(*continued*)

and equity challenges such as data privacy and algorithmic bias. Future research should focus on leveraging AI within digital therapy frameworks to deliver highly personalized, predictive, and equitable diabetes care.

Table 1. (*continued*)

Name of Solution/Reference	Description	Strengths	Weaknesses
Welldoc BlueStar [24]	One of the earliest FDA-cleared prescription-only DTx apps providing personalized, data-driven recommendations.	Well-established and clinically validated; integrates with care plans; delivers real-time insights.	Prescription-only model reduces accessibility; newer technologies may offer superior personalization.
Twin Health [25]	Employs AI-driven "digital twin" modeling to generate individualized metabolic profiles aiming to reverse Type 2 diabetes.	Focuses on remission rather than management; supported by RCT outcomes; deeply personalized approach.	Demands high patient engagement; limited insurance coverage in some markets.
DreaMed Advisor Pro [26]	AI-based decision support for clinicians analyzing CGM and pump data to guide insulin therapy.	Reduces clinician workload; improves insulin dosing accuracy; applicable in complex Type 1 cases.	Not patient-facing; designed exclusively for clinical settings.
Automated Insulin Delivery (AID) Systems [27]	Integrates CGM, insulin pump, and AI algorithms for automated insulin regulation ("artificial pancreas").	Demonstrated improvement in time-in-range and HbA1c levels; enables near-autonomous glucose control.	Restricted to insulin pump users; high cost may limit access.
AI-driven Comorbidity Platforms [28]	Uses AI to manage coexisting conditions (e.g., hypertension, cardiovascular disease) alongside diabetes.	Offers holistic health monitoring; supports preventive and predictive care.	Still under development; requires extensive data integration and privacy safeguards.
AI for Non-invasive Glucose Monitoring [29]	Employs AI models to analyze optical or biosignal data from wearables for needle-free glucose estimation.	Non-invasive and comfortable; enhances patient adherence.	Accuracy challenges persist; not yet FDA-cleared or widely deployed.

(*continued*)

Table 1. (*continued*)

Name of Solution/Reference	Description	Strengths	Weaknesses
AI for Predicting Drug Response [30]	Uses multi-omic and clinical data to anticipate individual drug efficacy and safety.	Enables precision pharmacotherapy; may reduce adverse reactions and healthcare costs.	Early-stage research; demands large, secure datasets including genetic information.
AI-powered Chatbots & Conversational AI [31]	Provides educational and behavioral support through 24/7 conversational interfaces.	Improves adherence via personalized communication; scalable and low-cost.	Limited ability to handle complex cases; output quality depends on algorithm sophistication.

In addition, a closer examination of the studies summarized in Table 1 reveals that most digital solutions are still implemented in isolation, lacking interoperability and seamless integration into existing healthcare infrastructures. While many systems demonstrate promising outcomes in controlled environments, their scalability and real-world applicability remain constrained by technical, clinical, and regulatory barriers.

Future work will therefore focus on the integration of digital therapy into comprehensive care pathways, ensuring interoperability between AI-driven systems, connected devices, and electronic health records. Such integration will enable continuous, data-driven monitoring and adaptive interventions, fostering a more holistic, efficient, and patient-centered approach to diabetes management.

5 Conclusion

Digital therapy has emerged as a transformative approach in diabetes management, offering personalized, accessible, and scalable interventions that empower patients to better manage their condition. The integration of artificial intelligence has significantly enhanced the effectiveness of these interventions, enabling real-time monitoring, predictive analytics, and tailored recommendations that adapt to individual patient needs. Looking forward, future research can further explore the synergy between digital therapy and AI to optimize treatment. By leveraging AI-driven insights, next-generation digital therapies have the potential to deliver highly precise, proactive, and patient-centered diabetes care, ultimately improving outcomes and quality of life for individuals living with the disease.

Acknowledges. I would like to express my gratitude to Pr. Bounabat and Pr. Haqiq for their guidance, and the Alkhawarizmi Program for their support. I also acknowledge the use of Google Books, Google Scholar, ChatGPT for assistance in rewriting and summarizing text, and NotebookLM and Gemini for research support.

References

1. Reddy, H., Tejaswini, P., Sandhiya, K., Ambika, V.: A review on recent pharmacotherapeutic advancements in the prevention of diabetes mellitus complications. Asian J. Med. Pharm. Sci. **13**(1), 69–77 (2025)
2. Yerima, A.: Anatomical changes in pancreatic islet cells in diabetes melitus. Doctoral dissertation, Department of Human Anatomy, Faculty of Basic Medical Sciences, College of Medical Sciences, Yobe State University (2025)
3. Rahman, M.S., et al.: Role of insulin in health and disease: an update. Int. J. Mol. Sci. *22*(12), 6403 (2021)
4. Rainville, C.: Guérir en comprenant les messages de nos malaises et de nos maladies: le grand dictionnaire de la métamédecine. Les Éditions Québec-Livres (2011)
5. Genitsaridi, I., et al.: Idf Diabetes Atlas: Global, Regional and National Diabetes Prevalence Estimates for 2024 and Projections for 2050. International Diabetes Federation (IDF) (2025)
6. Ajayi, R.O., Adeyemi-Benson, O.S., Adeyemi-Benson, O.A., Ogunjobi, T.T.: Chronic disease management in families: a public health and biomedicine perspective. Medinformatics. (2025)
7. Sorrell, S.: Digital technologies for the assessment, diagnosis, and treatment of diabetes distress: a review of current evidence. Health Sci. Rev. **15**, 100227 (2025)
8. Juanamasta, I.G., Aungsuroch, Y., Gunawan, J., Suniyadewi, N.W., Wati, N.M.N.: Holistic care management of diabetes mellitus: an integrative review. Int. J. Prevent. Med. **12**(1), 69 (2021)
9. Dang, A., Arora, D., Rane, P.: Role of digital therapeutics and the changing future of healthcare. J. Fam. Med. Prim. Care. **9**(5), 2207–2213 (2020)
10. Kaufman, N., Ferrin, C., Sugrue, D.: Using digital health technology to prevent and treat diabetes. Diabetes Technol. Ther. **21**(S1), S-79 (2019)
11. Hong, J.S., Wasden, C., Han, D.H.: Introduction of digital therapeutics. Comput. Methods Programs Biomed. **209**, 106319 (2021)
12. Sarkar, B.K., et al.: Diabetes mellitus: a comprehensive review. J. Pharmacogn. Phytochem. *8*(6), 2362–2371 (2019)
13. Balaji, R., Duraisamy, R., Kumar, M.P.: Complications of diabetes mellitus: a review. Drug Invent. Today. **12**(1), 134–138 (2019)
14. Harreiter, J., Roden, M.: Diabetes mellitus–definition, klassifikation, diagnose, screening und prävention (update 2023). Wiener Klinische Wochenschrift. **135**(Suppl 1), 7–17 (2023)
15. Rothberg, A., Lean, M., Laferrère, B.: Remission of type 2 diabetes: always more questions, but enough answers for action. Diabetologia. **67**(4), 602–610 (2024)
16. Weis, N.D.Z.: Psychologische Aspekte: Diabetes, Depression und Lebensqualität. Die Bestandsaufnahme. Beider Erkrankungen, pp. 58
17. Takahashi, K., et al.: Exploring the relationship between self-care agency and quality of life in adults with diabetes: a cross-sectional study. PloS One. **20**(7), e0326783 (2025)
18. Adamczak, U., et al.: Psychographische Verortung von Menschen mit Diabetes: Auswirkungen auf die Produktgestaltung und Kommunikation (No. 2010–08). Arbeitspapiere der Nordakademie (2010)
19. Kannenberg, S., et al.: Unlocking potential: personalized lifestyle therapy for type 2 diabetes through a predictive algorithm-driven digital therapeutic. J. Diabetes Sci. Technol. **20**, 113–123 (2024). https://doi.org/10.1177/19322968241266821
20. Babikian, S., Singh, V., Clements, M.: Insulin pump usage patterns among persons with diabetes (pwd) predict deteriorating sensor-based glucose control 2–4 weeks. Diabetes Technol. Ther. **22**, A63 (2020) 140 Huguenot street, 3rd fl, New Rochelle, NY 10801 USA: Mary Ann Liebert, Inc.

21. Canonico, M.E., et al.: Digital therapeutic based randomized investigation to improve glycemic control in patients with type 2 diabetes and residual hyperglycemia on stable medical therapy: rationale, design, and baseline characteristics of the BRIGHT trial. J. Am. Heart Assoc. **14**(15), e038737 (2025)
22. Okoye, V.N.: The effects of using mySugr diabetes mobile app on blood glucose level among adult African Americans with type two diabetes. Doctoral dissertation, University of Massachusetts Global (2022)
23. Ali Sherazi, B., Laeer, S., Krutisch, S., Dabidian, A., Schlottau, S., Obarcanin, E.: Functions of mHealth diabetes apps that enable the provision of pharmaceutical care: criteria development and evaluation of popular apps. Int. J. Environ. Res. Public Health. **20**(1), 64 (2022)
24. Daifi, C., Bahrami, S., Kaakeh, R., Kaakeh, Y.: Evolving frontier: a review of the role of mobile medical application prescribing. J. Pharm. Technol. ***32***(3), 91–97 (2016)
25. Pantalone, K.M., et al.: Type 2 diabetes pharmacotherapy de-escalation through ai-enabled lifestyle modifications: a randomized clinical trial. NEJM Catal. Innov. Care Deliv. ***6***(9), CAT-25 (2025)
26. Nimri, R., Phillip, M.: Enhancing care in type 1 diabetes with artificial intelligence driven clinical decision support systems. Horm. Res. Paediatr. **98**, 384–395 (2025)
27. Royston, C., Hovorka, R., Boughton, C.K.: Closed-loop therapy: recent advancements and potential predictors of glycemic outcomes. Expert Opin. Drug Deliv. **22**, 1–18 (2025)
28. Waghmare, P., Verma, P.: A review on the diabetes treatment using artificial intelligence and machine learning. In: 2025 International Conference on Inventive Computation Technologies (ICICT), pp. 701–708. IEEE (2025)
29. Chan, P.Z., Jin, E., Jansson, M., Chew, H.S.J.: AI-based noninvasive blood glucose monitoring: scoping review. J. Med. Internet Res. **26**, e58892 (2024)
30. Garg, S., Kitchen, R., Gupta, R., Pearson, E.: Applications of AI in predicting drug responses for type 2 diabetes. JMIR Diabetes. **10**(1), e66831 (2025)
31. Chew, H.S.J.: The use of artificial intelligence–based conversational agents (Chatbots) for weight loss: scoping review and practical recommendations. JMIR Med. Inform. **10**(4), e32578 (2022)

Medical Mini-Centrifuges: Innovations, Limitations, and Prospects for Use in Morocco

Abdoulhaidou Moussa Moussa[1,2](✉), Hicham Behja[1], Laghridat Hammadi[2], Nora Baaalla[2], and Hicham Medromi[2]

[1] National Higher School of Electricity and Mechanics (ENSEM), RITM Laboratory, Hassan II University of Casablanca, Casablanca, Morocco
moussa.abdoul-etu@etu.univh2c.ma

[2] Engineering Sciences Laboratory – SupTech-Health, Research Foundation for Development and Innovation in Science and Engineering, FRDISI, Casablanca, Morocco

Abstract. The centrifuge is a fundamental instrument in laboratories, widely used for various applications including the separation of serum or plasma from whole blood, cell sedimentation, DNA precipitation, and viral particle isolation. It plays a pivotal role in numerous biological and diagnostic procedures. However, from a global health standpoint, conventional centrifuge models present significant limitations: their high cost, large size and dependence on a stable power supply limit their deployment in decentralized or resource-limited settings, particularly in areas with unreliable electricity access. Given these challenges, there is growing interest in developing affordable, portable alternatives to traditional. This article examines recent innovations in the field of mini-centrifuges using a structured research approach. It also presents findings from a survey conducted in Morocco, which aimed to identify the specific needs of healthcare professionals who use medical centrifuges, as well as their openness to intelligent, miniaturized solutions. The overarching goal is to propose more efficient, adaptable, and accessible devices that align with the diverse requirements of clinical and diagnostic environments.

Keywords: Low-cost centrifuge · portable medical devices · point-of-care diagnostics · intelligent maintenance · artificial intelligence · biomedical engineering.

1 Introduction

The medical centrifuge is a fundamental tool across various disciplines, particularly in biology, clinical research, and medical diagnostics [1]. Since their inception, centrifuges have undergone significant advancements to meet growing demands for performance, precision, and reliability in the separation of biological components such as blood, cells, or nucleic acids [2]. Despite these improvements, traditional models remain generally bulky, expensive, and reliant on a continuous power supply, which limits their use in low-resource settings or remote areas lacking hospital infrastructure. In response to these

A. E. Hassanien et al. (Eds.): AISDS 2025, LNDECT 287, pp. 237–246, 2027.
https://doi.org/10.1007/978-3-032-18927-1_24

constraints, a new generation of miniature centrifuges characterized by their compactness and portability has emerged, enabling more flexible use in both clinical settings and field applications. This development reflects a global trend in the expanding medical centrifuge market, underscoring growing interest in autonomous, intelligent, and low-cost equipment [3]. Within this context, an exploratory survey conducted in Morocco among healthcare professionals revealed a strong demand for intelligent mini-centrifuges that are better aligned with the sector's current needs. However, existing solutions continue to face several technical and functional limitations. The remainder of this article is organized as follows: Sect. 2 outlines the research methodology, Sect. 3 presents a state-of the-art review of existing mini-centrifuges, Sect. 4 discusses market trends and the results of the national survey, Sect. 5 highlights recent technological innovations, Sect. 6 analyzes the main identified challenges and proposes optimization strategies, and finally, Sect. 7 concludes the study with perspectives for future development.

2 Methodology

This article is based on a systematic, rigorous, and structured research methodology developed to conduct an in-depth analysis of medical mini-centrifuges, as illustrated in Fig. 1. The study is informed by a selection of scientific publications from 2004 to 2024, retrieved from reputable academic databases including Scopus, PubMed, and Google Scholar. The review selected 75 key publications on low-cost and portable centrifuges to examine their limitations, technological innovations, and intelligent features, with attention to the needs of Moroccan healthcare professionals.

3 General Analysis of Mini-Centrifuges

3.1 Brief History of the Centrifuge

The history of centrifuges dates back to the 18th century, when British military engineer Benjamin Robins 1707–1751 developed one of the earliest centrifugation devices to assess the power of firearms [4]. A century later, in 1864, Antonin Prandtl proposed using centrifugal force to separate cream from milk an idea brought to life in 1875 by his brother, Alexander Prandtl, who developed a functional prototype of a milk centrifuge [5]. The application of centrifugation in biology began with Friedrich Miescher, a pioneer in the isolation of cellular organelles and nucleic acids, ultimately leading to the discovery of DNA [6]. He also played a key role in advancing analytical ultracentrifugation techniques, particularly for protein studies [7]. During the 1940s, the use of centrifuges increased with the expansion of routine medical testing. In 1949, the company SpinCo developed a centrifuge capable of reaching 40,000 revolutions per minute, which was later acquired by Berkman Coulter in 1950 [8]. A major milestone was reached in 1962 with the introduction of the first microcentrifuge by Eppendorf cup" [9]. In 1976, Hettich made a notable contribution to the digital era by launching at the first centrifuge equipped with microprocessor control that remains integral to modern laboratory equipment [10].

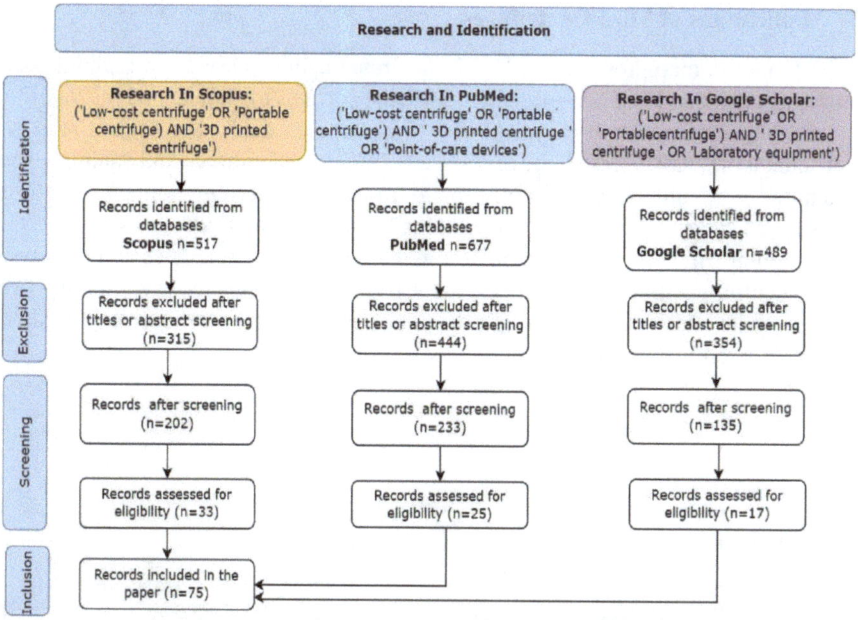

Fig. 1. PRISMA Flow Diagram of Innovations and Challenges in Portable Medical Centrifuges (2004–2024).

3.2 Operating Principle of Mini-Centrifuges

The centrifuge is an essential medical device used to separate various biological components such as blood, urine, or saliva, based on their physical properties including density, size, and shape [11]. Minicentrifuges operate on the same Príncipe, utilizing centrifugation to accelerate the natural process of sedimentation. Under normal conditions, particles suspended in a fluid gradually settle due to gravity; Wever, this process is often too slow for clinical applications. To address this limitation, centrifugation generates a centrifugal force that drives particles outward toward the rotor's periphery, away from the axis of rotation [12]. This force, known as relative centrifugal force, is significantly greater than Earth's gravitational force; for example, an RCF of 500 × g represents an acceleration 500 times that of gravity [13]. Samples for analysis are placed in specialized tubes inserted into the centrifuge. When the device spins at high speed, components separate based on density: heavier elements (pellet) settle at the bottom of the tube, while lighter components (the supernatant) remain at the top. The RCF can be calculated using a simple mathematical formula [14]:

$$\mathrm{RCF} = 1.118 \times 10^{-5} \times \mathrm{r} \times (\mathrm{RPM})^2 \tag{1}$$

where r is the radius (the distance from the particle to the axis of rotation) and RPM is the number of revolutions per minute.

3.3 Applications of Mini-Centrifuges

The bibliometric map presented in Fig. 2, generated using VOSviewer, highlights four major application areas of centrifugation. The first pertains to clinical and medical studies, including terms such as "clinical trial" and "human cell". The second relates to experimental biomedical research, with keywords like "cell culture" and "animal experimentation". The third focuses on physiology in microgravity environments, featuring terms such as "microgravity" and "space flight". The fourth area addresses the effects of acceleration and hypergravity, particularly in the context of aeronautics and pilot training. The central positioning of the term 'centrifugation' reflects the pivotal role of this technique at the intersection of multiple disciplines, ranging from regenerative medicine to space research.

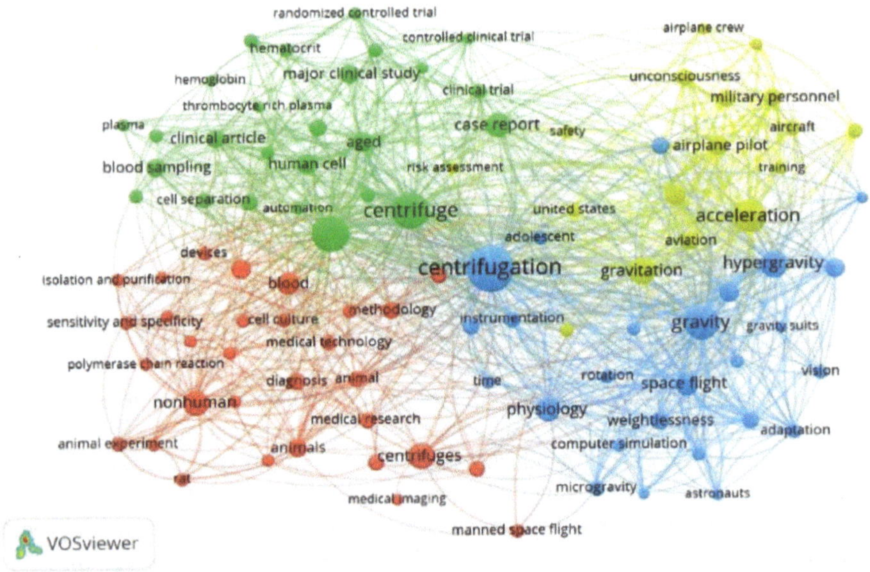

Fig. 2. Application Areas of Medical Centrifuges.

4 Analysis of the Medical Centrifuge Market

4.1 Market Size of Centrifuges

As illustrated in Fig. 3, the medical centrifuge market is experiencing steady growth, driven by continuous advancements in biomedicine, diagnostics, and pharmaceuticals. Valued at approximately USD 1.98 billion in 2024, the market is projected to reach nearly USD 2.79 billion by 2030, representing an average annual growth rate of 4.36% between 2024 and 2031. The mini-centrifuge segment, in particular, shows rapid expansion due to increasing demand from laboratories, hospitals, and research centers worldwide. Figure 4 shows global demand for minicentrifuges, high in North.

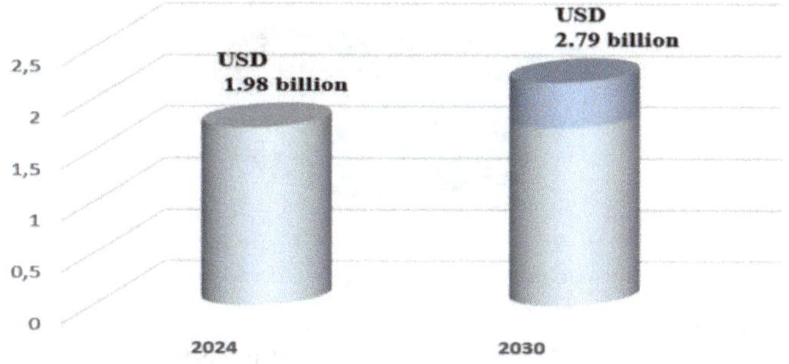

Fig. 3. Size of the Medical Centrifuge Market.

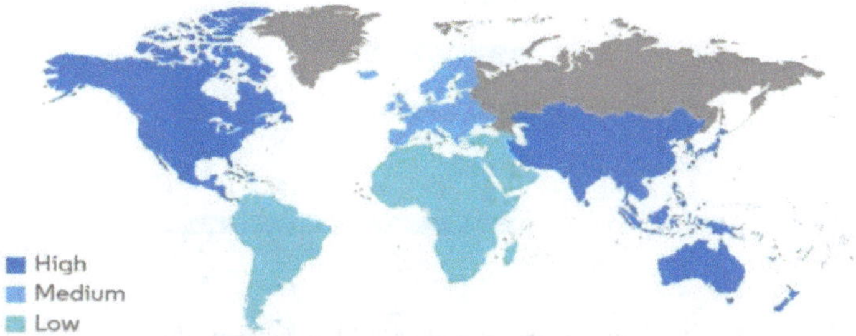

Fig. 4. Global Distribution of the Medical Centrifuge Market.

4.2 Analysis of a Study Conducted in Morocco on Healthcare Professionals' Expectations Regarding the Use of Medical Centrifuges

Figure 5 shows that laboratory technicians constitute the largest proportion of respondents 41.7%, followed by researchers 8.3%, medical biologists 4.2%, and other healthcare professionals. This diversity reflects a broad representation of potential end-users of medical centrifuges. Regarding the main challenges encountered during use, respondents reported issues related to equipment maintenance 46.4%, centrifugation errors 21.4%, temperature-related malfunctions 21.4%, and rotor imbalances 17.9%, as illustrated in Fig. 6. These findings offer valuable insights to inform future design improvements and technical optimizations aimed at better meeting the operational needs and expectations of medical personnel.

The vast majority of respondents 91.7% expressed strong interest in a portable, compact, and intelligent mini-centrifuge equipped with remote connectivity through a dedicated mobile application, as illustrated in Fig. 7. These findings underscore the growing importance of mobility and digital integration in modern biomedical equipment.

Although the sample size in this study is relatively small, the data collected provide valuable insight into the specific expectations of Moroccan users of medical centrifuges. The analysis highlights several commonly encountered issues, including mechanical imbalances, device overheating, and maintenance related challenges. Healthcare professionals show strong interest in intelligent, connected minicentrifuges, indicating a

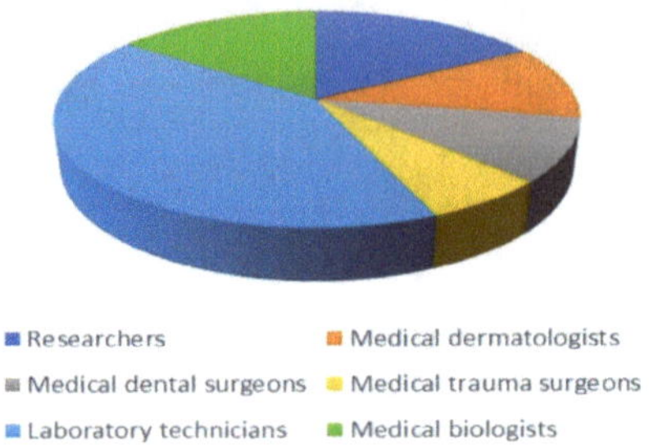

Fig. 5. Users of Medical Centrifuges.

Fig. 6. Issues Encountered with Medical Centrifuges.

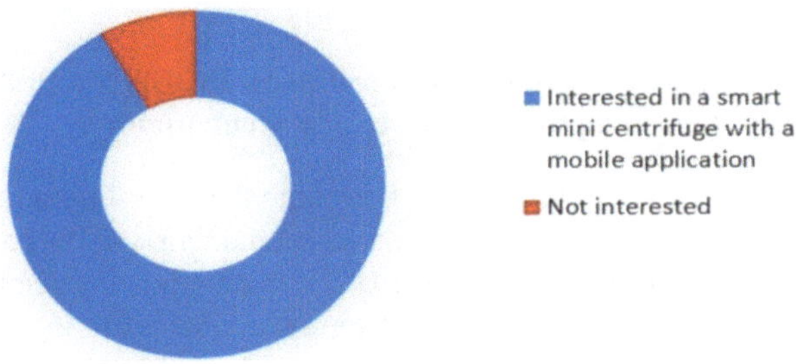

Fig. 7. Distribution of Interest in Medical Mini-Centrifuges

willingness to adopt technologies that improve performance, reduce errors, and meet evolving biomedical needs.

5 Innovations in Mini-Centrifuges

5.1 Smartphone-Operated Mini-Centrifuges and Manually-Actuated Models

In 2022, Mohini Bhupathi and her team developed the Mobilefuge, a portable and low-cost centrifuge designed to be 3D-printed and deployed in low-resource settings [15]. Powered via a 5 V USB port, the device requires neither soldering nor advanced technical skills for assembly. It can accommodate two 2 mL tubes, and has proven effective in separating silica particles during nucleic acid amplification tests, offering an accessible solution for diagnostic assays such as SARS-CoV-2 detection [16]. In 2017, Saad Bhamla and collaborators introduced the Paperfuge, an ultra-low-cost manual centrifuge inspired by traditional whirligigs [17]. This innovative device can achieve rotational speeds of up to 125,000 rpm, allowing for the separation of blood plasma in under 90 s and the isolation of malaria parasites within 15 min [18]. Building on these innovations, in 2019 Gaurav Byagathvalli and colleagues presented the 3D-Fuge, a 3D-printed manual centrifuge based on the principles of the Paperfuge [19]. Capable of reaching 6,000 rpm, it provides an economical alternative to benchtop models and is suitable for field biology, educational use, biomarker detection, and nucleotide extraction in portable laboratories [20].

5.2 Eco-Friendly and Portable Centrifuge Made from Recycled Materials, Accompanied by a Solar-Powered Model with Smartphone Control System

In 2023, Jovany J. Franco and his team developed CentREUSE, an affordable mini-centrifuge constructed from recycled materials, specifically designed for therapeutic applications [21]. The device generates a relative centrifugal force of 10.5, enabling sample sedimentation in just three minutes equivalent to approximately 12 h under normal gravity. Offering sedimentation quality comparable to that of commercial centrifuges, CentREUSE provides a simple and effective solution for medical use [22]. Additionally, in 2022, P. Dutta and collaborators introduced a portable solar-powered centrifuge equipped with a smartphone connected reading device for the detection of rheumatoid arthritis [23]. This centrifuge operates with minimal energy consumption and can be powered by a basic solar panel, making it particularly suitable for regions with limited access to electricity. It thus offers a viable and sustainable alternative for the separation of blood cells and plasma [24].

5.3 An "Intelligent" Tube Holder for Real-Time Sample Monitoring in a Standard Laboratory Centrifuge

In 2018, Tony Hoang et al. developed an instrumented tube holder compatible with benchtop centrifuges, addressing the inherent limitation of traditional systems [25]. This device enables real-time monitoring of rotational speed, cell type, and temperature, and features automatic shutdown upon sedimentation completion, along with user alerts. Such innovation enhances protocol efficiency, reduces processing times, and minimizes waste, thereby improving overall laboratory workflow and reliability [26].

6 Discussion

6.1 Main Challenges Related to Medical Centrifuges

A study conducted among healthcare professionals in Morocco identified several limitations associated with medical centrifuges. Recurring maintenance issues compromise the reliability and longevity of these devices. Centrifugation errors, mainly caused by rotor imbalance and improper control of temperature and speed, highlight the technical limitations of current models. While mini-centrifuges offer portability and convenience, their limited capacity of two 2 mL tubes restricts their suitability for high-throughput analyses [27]. Manual centrifuges, through inexpensive and accessible, depend entirely on human effort, making it challenging to maintain consistent rotational speeds. Operator fatigue, a largely overlooked factor, may further compromise performance. Moreover, the long-term durability of these devices remains uncertain due to a lack of comprehensive evaluations, thereby limiting their use clinical settings [28].

6.2 Possible Recommendations

To address the current limitations of medical centrifuges, the integration of expert systems for predictive and preventive maintenance is essential. These systems would enable early detection of mechanical and electronic failures, thereby extending equipment lifespan and minimizing downtime [29]. The incorporation of dynamic sensors for automated rotor imbalance detection would allow real-time identification of vibrational anomalies, reducing centrifugation errors and associated mechanical risks [30]. Furthermore, the implementation of automated thermal regulation loops would ensure precise and stable control of critical temperature conditions necessary for the preservation of biological samples, thus, enhancing the reproducibility of analytical results [31]. Remote monitoring and control via secure mobile interfaces would enable real-time management of operational parameters, allowing dynamic adjustment of speed, temperature, and centrifugation duration. This would significantly improve system flexibility and operational efficiency. These targeted technological advancements are crucial for optimizing the performance, functional safety, and adaptability of medical centrifuges to meet evolving clinical and laboratory demands. Future developments may include digital twins for high-end centrifuges to enable simulation, optimization, and failure prediction, and TinyML-powered AI for portable devices to allow on-device anomaly detection. These innovations enhance performance, safety, and adaptability of next-generation medical centrifuges.

7 Conclusion

Medical centrifuges are essential for the separation of biological samples; however, traditional models, often bulky and energy-intensive, are poorly suited for resourcelimited settings. Although limited in scope, our survey conducted in Morocco reveals a strong interest among healthcare professionals in intelligent, compact, and userfriendly mini-centrifuges, while also highlighting recurring challenges such as maintenance issues, rotor imbalance, and the limited adaptability of conventional models. To address these shortcomings, our study proposes the integration of sensors, mobile connectivity, and design enhancements to improve both reliability and accessibility. These innovations have the potential to transform point-of-care diagnostics, particularly in low resource environments. Future work should enhance portability, energy efficiency, and smart diagnostic features to improve the sustainability of medical mini-centrifuges in low-resource settings.

References

1. Kumar, A., et al.: Advanced medical centrifuges: technological evolution and multidisciplinary applications in modern healthcare. J. Lab. Autom. **28**(4), 215–230 (2023)
2. Bhupathi, M., Devarapu, G.C.R.: Mobilefuge: a low-cost, portable, open source, 3D-printed centrifuge that can be powered by a mobile phone. Open Res. Europe. **2** (2022). https://doi.org/10.12688/openreseurope.14998.1
3. Franco, J.J., Nagata, T., Okamoto, T., Mukai, S.: An ultralow-cost portable centrifuge from discarded materials for medical applications. Sci. Rep. **13**(1), 3081 (2023). https://doi.org/10.1038/s41598-023-30327-2
4. Turner, G.L.E.: The History of Science and Technology in the Eighteenth Century, p. 320. Oxford University Press (2022)
5. Prandtl, L.: Essentials of Fluid Mechanics. The International Centre for Theoretical Physics, Trieste, Italy (2004)
6. Dahm, R., et al.: Friedrich Miescher and the origins of DNA research: revisiting centrifugation breakthroughs in cell biology. Trends Biochem. Sci. **49**(1), 12–25 (2024)
7. Amanda: History of the centrifuge. In: The Lab World Group. https://www.thelabworldgroup.com/blog/history-of-centrifugation/
8. Cole, J.R.: SpinCo and the birth of high-speed centrifugation: a legacy in modern science. J. Lab. Technol. **35**(4), 245–260 (2021)
9. Eppendorf, A.G.: 60 years of the Eppendorf microcentrifuge tube: revolutionizing sample handling in the lab. Lab Times. **5**, 12–15 (2020)
10. Sonmez, C., et al.: An important source of preanalytical error in medical laboratories: centrifugation. Turk. J. Biochem. **46**(4), 399–405 (2021)
11. Zhang, Y., et al.: Centrifugation in clinical laboratories: principles, applications, and recent advances. Clin. Chem. Lab. Med. (CCLM). **61**(5), 789–801 (2023)
12. Bornhorst, J.A., De Jesús, V.R.: Centrifugation techniques in clinical laboratories: fundamentals and applications. In: Contemporary Laboratory Medicine, 2nd edn, pp. 87–104. Elsevier (2023)
13. Li, X., et al.: High-G centrifugation in biomedical applications: principles, protocols, and experimental validation. J. Biomech. Eng. **145**(6), 061008 (2023)
14. Rickwood, D.: Centrifugation Techniques. Nature. **277**, 588 (2001)
15. Bhupathi, M., et al.: Mobilefuge: a low-cost, 3D-printed centrifuge for resource limited settings. PLoS One. **17**(8), e0272023 (2022)
16. Mehta, P., et al.: Silica-coated magnetic microparticles enable efficient RNA extraction for SARS-CoV-2 diagnostics. J. Clin. Virol. **89**, 105432 (2024)
17. Bhamla, M.S., Benson, B., Chai, C., et al.: Hand-powered ultralow-cost paper centrifuge. Nat. Biomed. Eng. **1**, 0009 (2017)
18. Bhamla, M.S., Benson, B., Chai, C., Katsikis, G., Johri, A., Prakash, M.: Hand-powered ultralow-cost paper centrifuge. Nat. Biomed. Eng. **1**(1), 1–7 (2017)
19. Byagathvalli, G., Bhamla, M.S., Prakash, M.: 3D-Fuge: a 3D-printed centrifugal extractor for low-resource settings. PLoS One. **14**(12), e0225642 (2019)
20. Tan, Y., Liu, T., Guan, W.: A semi-automated, low-cost portable centrifuge kit for nucleic acid extraction at point-of-care. J. Port. Diagn. **23**, 101–110 (2024)
21. Franco, J.J., et al.: CentREUSE: a low-cost, recycled-material centrifuge for therapeutic applications in resource-limited settings. Sci. Rep. **13**(1), 11234 (2023)
22. Franco, J., Nagata, T., Okamoto, T.: An ultralow-cost portable centrifuge from discarded materials for medical applications. Sci. Rep. **13**(1), 3081 (2023)
23. Dutta, P., et al.: Solar-powered portable centrifuge with integrated smartphone detection for rheumatoid arthritis diagnosis in low-resource settings. Biosens. Bioelectron. **215**, 114567 (2022)

24. Dutta, P., et al.: Combining portable solar-powered centrifuge to nanoplasmonic sensing chip with smartphone reader for rheumatoid arthritis detection. Chem. Eng. J. **434**, 133864 (2022)
25. Hoang, T., et al.: An instrumented centrifuge tube rack for real-time monitoring of laboratory centrifugation processes. Lab Chip. **18**(18), 2758–2766 (2018)
26. Hoang, T., Moskwa, N., Halvorsen, K.: A 'smart' tube holder enables real-time sample monitoring in a standard lab centrifuge. PLoS One. **13**(4), e0195907 (2018). https://doi.org/10.1371/journal.pone.0195907
27. Wang, B., Du, Y., Diao, Y., Zhao, X.: Design considerations and throughput limitations of portable mini-centrifuges handling 1.5–2 mL tubes. Sustainability. **15**(10), 8353 (2023)
28. Zhang, Y., et al.: Operator fatigue in high-stakes environments: impact on performance and long-term reliability in critical systems. Saf. Sci. **158**, 105934 (2023)
29. CID, J.: Predictive maintenance for rotating biomedical systems. J. Intell. Syst. **34**(2), 145–158 (2024)
30. Pavlenko, V., Kurliak, P., Horiashchenko, S., Volianyk, O., Batsala, Y., Nikitchuk, T.: Computer-integrated system of automatic balance control in centrifuge rotors with variable imbalance. In: 2023 IEEE International Conference of Modern Electrical and Energy Systems (MEES), p. 5 (2023). https://doi.org/10.1109/MEES61502.2023.10402507
31. Zhu, T., Lu, J., Shen, J., Liu, J., Xiao, X.: Wireless temperature calibration for medical centrifuge. Sens. Int. **6**, 100296 (2024). https://doi.org/10.1016/j.sintl.2024.100296

Workflow Optimization and Smart Systems

Real-Time Adaptive Multi-objective Scheduling Model for Cloud Workflows Using LinUCB

Soukaina Mihi, Mehdi Ait Said(✉), and Lahcen Belouaddane

Hassan First University, Settat, Morocco
{soukaina.mihi,m.aitsaid,lahcen.belouaddane}@uhp.ac.ma

Abstract. Efficient scheduling of cloud workflows is critical for balancing execution performance and resource costs in dynamic environments. Traditional scheduling approaches often rely on static assumptions about task runtimes and resource availability, limiting their effectiveness under real-world variability. In this paper, we propose an adaptive multi-objective scheduling model based on the LinUCB contextual bandit algorithm. The model learns online from task and workflow contexts, enabling real-time decision-making without the heavy computational overhead of deep reinforcement learning.

To evaluate its effectiveness, we generated synthetic workloads inspired by the MIT Supercloud dataset and compared our method against two popular baselines, Cheapest and FCFS. Results from 100 experimental runs demonstrate that LinUCB reduces makespan by nearly 46% compared to FCFS, while maintaining costs close to Cheapest. These findings show that contextual bandit learning offers a practical and efficient solution for real-time cloud workflow scheduling, achieving a favorable trade-off between cost and performance.

Keywords: Cloud workflow scheduling · Contextual bandits · LinUCB · Cost–performance trade-off · Adaptive resource allocation

1 Introduction

Modern computing systems are increasingly characterized by large-scale, dynamic, and heterogeneous environments [1, 2]. With the rapid growth of cloud computing, data-intensive applications, and workflow-driven processes, scalability has become a fundamental requirement [3]. Cloud platforms must support fluctuating workloads, diverse user requirements, and evolving resource types while maintaining efficiency and reliability [4]. Achieving scalability in such systems is directly tied to the effective management of computational resources and the scheduling of workloads [5].

Task scheduling plays a critical role in determining overall system performance. In cloud workflow execution, tasks are interdependent and typically represented as Directed Acyclic Graphs (DAGs), where scheduling decisions directly impact both execution time (makespan) and resource cost [3, 6]. Traditional scheduling approaches rely on static assumptions about task runtimes and resource performance [7]. However, these methods often fail to cope with the uncertainties of modern cloud environments, where virtual

A. E. Hassanien et al. (Eds.): AISDS 2025, LNDECT 287, pp. 249–257, 2027.
https://doi.org/10.1007/978-3-032-18927-1_25

machines (VMs) exhibit variable performance and workloads evolve dynamically. As a result, static schedules may become inefficient or even infeasible at runtime [8].

This paper addresses the problem of real-time adaptive scheduling of cloud workflows under dynamic and uncertain conditions. Specifically, we consider the dual objectives of minimizing execution time and reducing resource cost while handling unpredictable workloads and heterogeneous cloud resources. The challenge lies in designing a scheduling approach that adapts as tasks are executed, rather than relying solely on offline planning [9].

To tackle this issue, we propose a Real-Time Adaptive Multi-Objective Scheduling Model for Cloud Workflows using Linear Upper Confidence Bound (LinUCB) [10], a contextual Bandit–Based Method (MAB) that integrates online learning into workflow scheduling. Unlike heuristic-based approaches, our model learns from experience during execution, gradually improving scheduling decisions by balancing cost and makespan trade-offs in real time. Our contribution lies in introducing a lightweight, adaptive scheduling mechanism that bridges the gap between static optimization and the dynamic realities of cloud systems.

The rest of this paper is organized as follows. Section 2 provides background on contextual bandits and the LinUCB algorithm. Section 3 presents our proposed scheduling model, including data representation, decision-making, and learning mechanism. Section 4 describes the validation setup and experimental results. Section 5 reviews related work in cloud workflow scheduling and online learning approaches. Finally, Sect. 6 concludes the paper and outlines future research directions.

2 Background: Contextual Bandits and LinUCB

Workflow scheduling in cloud environments is inherently a sequential decision-making problem under uncertainty. Each task requires the selection of a resource (e.g., a virtual machine type) that determines both execution cost and runtime [11]. Classical scheduling heuristics rely on predefined policies, but these approaches cannot adapt to evolving workload characteristics or fluctuating resource performance. To address these limitations, online learning methods, particularly MABs, have been explored as a lightweight yet effective solution [12].

The MAB framework models the trade-off between exploration (trying less-known actions to gather information) and exploitation (selecting the best-known action to maximize immediate reward) [13]. While basic MAB algorithms treat each action as independent, contextual bandits extend this formulation by incorporating side information, or context, that characterizes each decision instance. In the context of workflow scheduling, the task features (e.g., input size, DAG position, resource requirements) form the context, while the actions correspond to available cloud resources [14].

Among contextual bandit algorithms, LinUCB has emerged as a widely used method due to its simplicity, efficiency, and ability to handle high-dimensional contexts. LinUCB assumes that the expected reward of an action is a linear function of the context features, enabling it to balance exploration and exploitation through confidence bounds [11]. At each decision step, LinUCB computes an optimistic estimate of the reward for each action, selects the action with the highest estimate, and updates its model parameters based on observed feedback [10].

Formally, LinUCB maintains a weight vector and confidence matrix for each action (resource). For a given context vector, the algorithm calculates an upper confidence bound that combines the predicted reward with an uncertainty term. This mechanism ensures that the algorithm explores underrepresented contexts while still exploiting actions that are known to perform well [15]. Over time, LinUCB converges toward near-optimal decisions, making it particularly suitable for real-time adaptive scheduling where workload patterns and system dynamics evolve continuously [13].

In this work, LinUCB serves as the foundation for our proposed scheduling model, allowing us to integrate online learning into cloud workflow execution while addressing both cost and makespan objectives.

3 Proposed Scheduling Model

The proposed scheduling model leverages a contextual MAB approach to adaptively allocate tasks of cloud workflows to heterogeneous resources under multiple objectives. Specifically, we employ the LinUCB algorithm to balance exploration of resource capabilities and exploitation of learned knowledge for performance optimization.

At its core as presented in Code 1, the model operates by maintaining a set of linear predictors, one for each available resource (e.g., VM instance type). These predictors estimate the expected reward associated with assigning a given task to a specific resource, based on a feature representation of the task and workflow context. The uncertainty in these estimates is quantified through confidence intervals, enabling the algorithm to favor resources with both high predicted rewards and high uncertainty.

To support this process, three auxiliary functions are introduced. The IdentityMatrix(d) function initializes the covariance matrices required for each resource model. The ExtractFeatures(task, workflow) function generates a feature vector encoding task-level attributes (e.g., input size, DAG position, depth, CPU/memory/GPU requirements, and intensity factors) along with workflow-level properties (e.g., SLA constraints and deadline slack). These contextual features allow the model to capture both the computational nature of tasks and the temporal constraints of workflows. The ComputeReward(task, runtime, cost, workflow) function evaluates the quality of scheduling decisions by combining normalized runtime and monetary cost into a composite reward. Additional penalties are introduced when SLA or deadline constraints are violated, thereby guiding the model to prioritize both efficiency and compliance with service guarantees.

Code 1: The LinUCB-based Scheduling Model

```
Function IdentityMatrix(d):
  # Create d x d identity matrix
  A = Matrix of size d x d
  For i from 0 to d-1:
    For j from 0 to d-1:
      If i == j:
        A[i][j] = 1
      Else:
        A[i][j] = 0
  Return A

Function ComputeReward(task, runtime, cost, workflow):
  # Normalize metrics
  time_norm = runtime / 3600      # hours
  cost_norm = cost / 1.0          # dollars

  # Base reward: trade-off between time and cost
  reward = - (0.5 * time_norm + 0.5 * cost_norm)

  # Penalize if SLA exceeded
  If runtime > workflow.sla_max_latency_seconds:
    reward -= 2.0

  # Penalize if workflow deadline exceeded
  If current_time + runtime > workflow.deadline_abs_ts:
    reward -= 3.0

  Return reward

Function ExtractFeatures(task, workflow):
  # Initialize feature vector
  features = []

  # Task input size (log scale)
  features.append(log1p(task.input_size_mb))

  # DAG position one-hot encoding
  features.append(1 if task.dag_position == 'entry' else 0)
  features.append(1 if task.dag_position == 'middle' else 0)
  features.append(1 if task.dag_position == 'exit' else 0)

  # Task depth normalized
  features.append(task.dag_depth / max_depth_of_workflow)
# Resource requirements
  features.append(task.cpu_req)
  features.append(task.mem_req_gb)
  features.append(task.gpu_req)
# Task intensity features
  features.append(task.io_intensity)
  features.append(task.compute_intensity)

  # Workflow SLA and remaining time
  features.append(workflow.sla_max_latency_seconds / 3600)
  features.append(max(0, workflow.deadline_abs_ts - current_time) / 3600)

  # Bias term
  features.append(1.0)

  Return features

Algorithm LinUCB(tasks, resources, α):
  for each resource r in resources:
    Initialize A_r ← IdentityMatrix(d)
    Initialize b_r ← ZeroVector(d)

  for each task in tasks:
    context ← ExtractFeatures(task)

    // Compute UCB score for each resource
    for each resource r in resources:
      θ_r ← Inverse(A_r) * b_r
      p_r ← Transpose(θ_r) * context + α * sqrt(Transpose(context) *
Inverse(A_r) * context)

    // Choose resource with highest score
    r* ← argmax_r (p_r)

    // Assign task to resource
    Assign(task, r*)

    // Observe reward (based on cost, runtime, etc.)
    reward ← ComputeReward(task, r*)

    // Update resource model
    A_r* ← A_r* + context * Transpose(context)
    b_r* ← b_r* + reward * context
```

The scheduling process then unfolds as follows. For each task, the algorithm computes the contextual features and evaluates the Upper Confidence Bound (UCB) score for all candidate resources. The resource with the highest score is selected for execution, balancing exploitation of well-performing resources with exploration of underutilized ones. Once the task completes, the observed reward is computed, and the model parameters for the chosen resource are updated accordingly.

This adaptive feedback mechanism allows the scheduler to progressively refine its allocation strategy in real time, as more tasks are executed and resource behaviors become better understood. By jointly optimizing for runtime and cost while respecting workflow-level constraints, the proposed model achieves an effective balance between performance, efficiency, and reliability.

4 Validation

4.1 Methodology

To evaluate the effectiveness of the proposed scheduling model, we conducted a comparative study against two widely used baseline scheduling strategies: Cheapest and First-Come-First-Served (FCFS). The Cheapest algorithm selects the lowest-cost resource for each task, while FCFS dispatches tasks strictly in the order they arrive, without considering resource heterogeneity or workflow characteristics.

For workload generation, we implemented a synthetic generator inspired by the MIT Supercloud Workload Dataset [16]. This generator creates randomized workflow workloads that reflect real-world variability in task arrival patterns, sizes, and resource requirements. The implementation of this generator is made publicly available in this GitHub repository[1]. In our experiments, we generated 100 independent workloads, each consisting of 5,640 tasks. For each workload, all three algorithms (Cheapest, FCFS, and LinUCB) were executed, resulting in a total of 300 scheduling runs.

To ensure robustness, the random generator was used to introduce stochasticity across runs, simulating diverse execution environments. This methodology allows for statistically meaningful comparisons of scheduling behavior under realistic conditions.

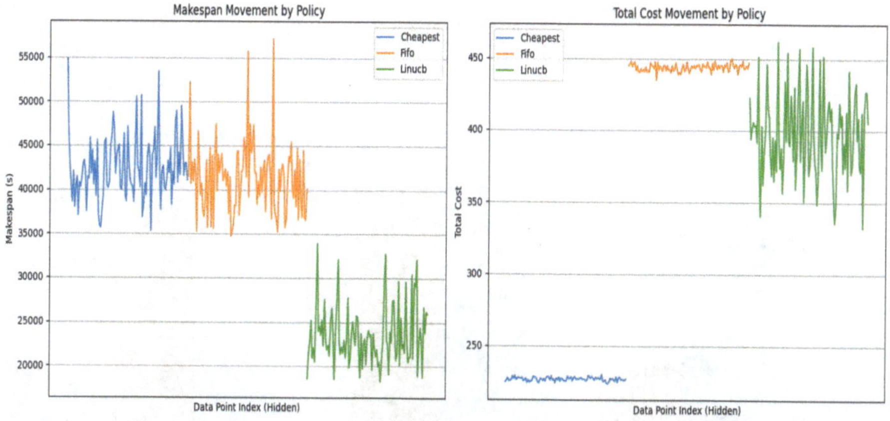

Fig. 1. Evolution of makespan and cost across runs for Cheapest, FCFS, and LinUCB scheduling strategies

4.2 Results and Discussion

The experimental outcomes are summarized in Fig. 1, which presents two charts: the first illustrates the evolution of makespan across runs, while the second depicts the cost variations under the three scheduling strategies.

For clarity, we also report the aggregated averages across all runs. These averages are visualized in Fig. 2, which presents two bar charts comparing makespan and cost. The outcomes clearly highlight the effectiveness of the LinUCB-based approach:

- Cheapest: makespan = 41,902.3 s (11.640 h), cost = \$227.379
- FCFS: makespan = 42,066.3 s (11.685 h), cost = \$442.193
- LinUCB: makespan = 22,351.8 s (6.209 h), cost = \$398.020

The comparative results reveal distinct trade-offs between execution time and cost. The Cheapest strategy, while minimizing cost, suffers from prolonged makespan, highlighting its inefficiency for workflows requiring timely completion. FCFS, on the other

[1] https://gist.github.com/altix-ma/a10879cc840c59f33f432161be20b265

hand, incurs both higher costs and longer runtimes, confirming its limitations as a baseline scheduling policy.

By contrast, the LinUCB-based approach demonstrates clear advantages. It achieves nearly a 46% reduction in makespan compared to FCFS, while maintaining costs significantly lower than FCFS and only moderately higher than Cheapest. This outcome underscores the effectiveness of contextual bandit learning in dynamically adapting scheduling decisions to workload variability.

Overall, these findings validate the proposed model's ability to deliver a favorable cost–performance trade-off, achieving faster workflow execution without excessive cost escalation. The results highlight the promise of real-time, learning-driven scheduling for cloud environments characterized by heterogeneity and unpredictability.

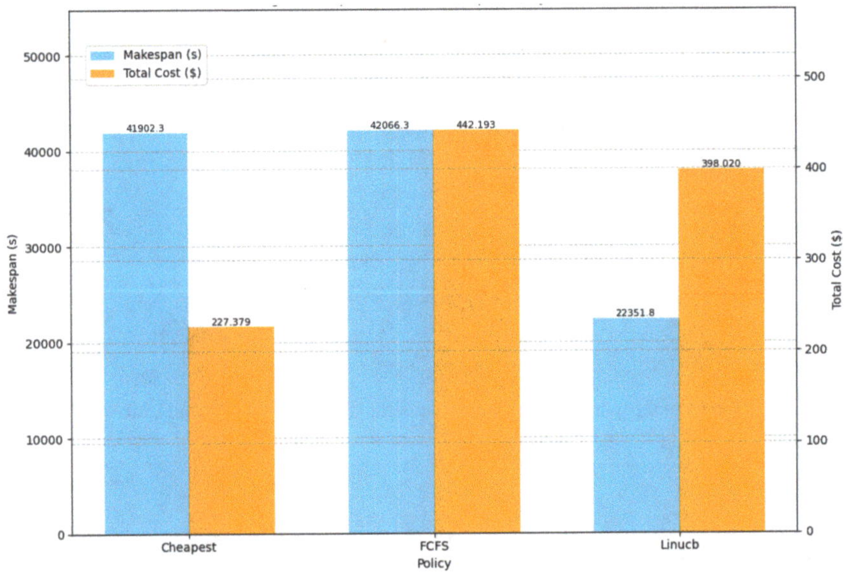

Fig. 2. Average makespan and cost comparison of Cheapest, FCFS, and LinUCB scheduling strategies

5 Related Work

The literature on real-time scheduling demonstrates a strong trend toward the adoption of artificial intelligence (AI)–based methods to handle the growing complexity and uncertainty of modern systems [9]. Reinforcement learning, in particular, has attracted significant attention due to its ability to adapt scheduling policies dynamically in response to unpredictable conditions [17, 18].

One representative example is the work of Shu Luo et al. (2022) [19], who proposed a hierarchical multiagent deep reinforcement learning approach for real-time scheduling in flexible job shop environments with dynamic disturbances such as machine breakdowns and new job arrivals. Their method, known as HMAPPO, leverages three PPO-based agents operating at different spatiotemporal levels to optimize scheduling objectives. This work highlights the capability of deep reinforcement learning to achieve robust performance in highly uncertain manufacturing environments.

Similarly, Tong Zhou et al. (2021) [20] introduced a multi-agent reinforcement learning framework for online scheduling in smart factories. Their approach integrates cyber-physical systems and IoT technologies to enable distributed scheduling across heterogeneous manufacturing units. By leveraging reinforcement learning agents equipped with novel neural networks, their system adapts to real-time sensor data, rush orders, and machine failures, thereby offering high flexibility and resilience in complex industrial settings.

In contrast to these approaches, our proposed model is not based on computationally intensive machine learning or deep learning methods. Instead, it employs a contextual bandit algorithm (LinUCB), which provides lightweight, online adaptivity without the need for heavy training resources or centralized controllers. While reinforcement learning–based schedulers often demand significant computational power and training time, LinUCB achieves comparable adaptivity with a simpler formulation, making it particularly suitable for cloud workflow scheduling where rapid, real-time decision-making is critical.

6 Conclusion

This work introduced a novel scheduling model for cloud workflows based on the LinUCB contextual bandit algorithm, addressing the limitations of static scheduling approaches in dynamic and heterogeneous environments. Unlike reinforcement learning–based methods that require heavy computational resources, the proposed model offers a lightweight, real-time adaptive solution capable of balancing execution efficiency and cost.

Through extensive experimental validation using synthetic workloads, our method was compared against two popular baselines, Cheapest and FCFS. The results demonstrated that LinUCB achieved a 46% reduction in makespan relative to FCFS while maintaining costs at a moderate level, lower than FCFS and close to Cheapest. These findings confirm the potential of contextual bandit learning for improving workflow scheduling in cloud systems.

Future research may extend this work by exploring hybrid bandit–reinforcement learning approaches, integrating real-world workload traces, and evaluating performance under larger-scale, multi-cloud environments. Such directions would further strengthen the applicability of adaptive scheduling in diverse production settings.

References

1. Ait Said, M., Ezzati, A., Mihi, S., Belouaddane, L.: Microservices adoption: an industrial inquiry into factors influencing decisions and implementation strategies. Int. J. Comput. Digit. Syst. **15**(1), 1417–1432 (2024). https://doi.org/10.12785/ijcds/1501100

2. Ait Said, M., Ezzati, A., Arezki, S.: Microservices, a step from the low-code to the no-code. In: Proceedings of the 8th International Conference on Advanced Intelligent Systems and Informatics 2022 Lecture Notes on Data Engineering and Communications Technologies, vol. 152, pp. 779–788 (2023). https://doi.org/10.1007/978-3-031-20601-6_64
3. Gupta, S., et al.: Efficient prioritization and processor selection schemes for HEFT algorithm: a makespan optimizer for task scheduling in cloud environment. Electronics. **11**(16), 2557 (2022). https://doi.org/10.3390/electronics11162557
4. Lahcen, B., Mehdi, A.S., Abderrahim, M., Amine, B., Anas, R.: A Cost-Efficient Security Practices Framework for Small DevSecOps Teams, pp. 257–266 (2025). https://doi.org/10.1007/978-3-031-95326-2_25
5. Pirozmand, P., Hosseinabadi, A.A.R., Farrokhzad, M., Sadeghilalimi, M., Mirkamali, S., Slowik, A.: Multi-objective hybrid genetic algorithm for task scheduling problem in cloud computing. Neural Comput. & Applic. **33**(19), 13075–13088 (2021). https://doi.org/10.1007/s00521-021-06002-w
6. Pirozmand, P., Hosseinabadi, A.A.R., Farrokhzad, M., Sadeghilalimi, M., Mirkamali, S., Slowik, A.: Multi-objective hybrid genetic algorithm for task scheduling problem in cloud computing. Neural Comput. & Applic. **33**(19), 13075–13088 (2021). https://doi.org/10.1007/S00521-021-06002-W/FIGURES/8
7. Ma, X., Xu, H., Gao, H., Bian, M.: Real-time multiple-workflow scheduling in cloud environments. IEEE Trans. Netw. Serv. Manag. **18**(4), 4002–4018 (2021). https://doi.org/10.1109/TNSM.2021.3125395
8. Kaur, N., Kumar, A., Kumar, R.: A systematic review on task scheduling in fog computing: taxonomy, tools, challenges, and future directions. Concurr. Comput. Pract. Exp. **33**(21), e6432 (2021). https://doi.org/10.1002/CPE.6432
9. Avan, A., Azim, A., Mahmoud, Q.H.: A state-of-the-art review of task scheduling for edge computing: a delay-sensitive application perspective. Electron. **12**(12), 2599 (2023). https://doi.org/10.3390/ELECTRONICS12122599
10. Tian, Z.: Bandit algorithms for advertising optimization: a comparative study. ITM Web Conf. **73**, 01019 (2025). https://doi.org/10.1051/itmconf/20257301019
11. Wang, Z., Xie, J., Liu, X., Li, S., Lui, J.C.S.: Online clustering of bandits with misspecified user models. Adv. Neural Inf. Process. Syst. **36**, 3785–3818 (2023)
12. Zhang, X., Xie, H., Lui, J.C.S.: Heterogeneous information assisted bandit learning: theory and application. In: Proceedings of the International Conference on Data Engineering, vol. 2021, pp. 2135–2140 (2021). https://doi.org/10.1109/ICDE51399.2021.00213
13. Chen, Y.S., Huang, Z.H., Tsai, M.J.: LinUCB-based handover algorithm for throughput maximization in heterogeneous cellular networks. In: Proceedings - IEEE Consumer Communications and Networking Conference, vol. 2023, pp. 734–739 (2023). https://doi.org/10.1109/CCNC51644.2023.10060709
14. Shi, Q., Xiao, F., Pickard, D., Chen, I., Chen, L.: Deep neural network with LinUCB: a contextual bandit approach for personalized recommendation. In: ACM Web Conference 2023 - Companion World Wide Web Conference. WWW 2023, pp. 778–782 (2023). https://doi.org/10.1145/3543873.3587684
15. Xu, Y., Lu, W., Song, R.: Linear Contextual Bandits with Interference (2024). [Online]. Available: https://arxiv.org/pdf/2409.15682. Accessed: 16 Sep. 2025
16. Tang, B.J., et al.: The MIT supercloud workload classification challenge. In: Proceedings of the 2022 IEEE 36th International Parallel and Distributed Processing Symposium Workshops (IPDPSW), pp. 708–714 (2022). https://doi.org/10.1109/IPDPSW55747.2022.00122
17. Abolhassani Khajeh, S., Saberikamarposhti, M., Rahmani, A.M.: Real-time scheduling in iot applications: a systematic review. Sensors. **23**(1), 232 (2023). https://doi.org/10.3390/S23010232

18. Baccour, E., et al.: Pervasive AI for IoT applications: a survey on resource-efficient distributed artificial intelligence. IEEE Commun. Surv. Tutorials. **24**(4), 2366–2418 (2022). https://doi.org/10.1109/COMST.2022.3200740
19. Luo, S., Zhang, L., Fan, Y.: Real-time scheduling for dynamic partial-no-wait multiobjective flexible job shop by deep reinforcement learning. IEEE Trans. Autom. Sci. Eng. **19**(4), 3020–3038 (2022). https://doi.org/10.1109/TASE.2021.3104716
20. Zhou, T., Tang, D., Zhu, H., Zhang, Z.: Multi-agent reinforcement learning for online scheduling in smart factories. Robot. Comput. Integr. Manuf. **72**, 102202 (2021). https://doi.org/10.1016/J.RCIM.2021.102202

Multi-objective Workflow Scheduling in Cloud Environments: A HEFT-Seeded NSGA-II Model

Hajar Khallouki, Mehdi Ait Said(✉), Lahcen Belouaddane, and Soukaina Mihi

Hassan First University, Settat, Morocco
{hajar.khallouki,m.aitsaid,lahcen.belouaddane, soukaina.mihi}@uhp.ac.ma

Abstract. Efficient workflow scheduling in cloud environments is a critical challenge due to the need to balance multiple conflicting objectives such as makespan, cost, and deadline satisfaction. Traditional approaches, including FCFS and metaheuristics like ACO, often fail to simultaneously optimize these objectives effectively. To address this limitation, we propose HEFT-Seeded NSGA-II, a hybrid multi-objective scheduling algorithm that leverages the heuristic efficiency of HEFT to seed the initial population of NSGA-II. This integration enables faster convergence and improved exploration of the search space, leading to higher-quality scheduling solutions. Experimental validation using publicly available datasets demonstrates that our model consistently outperforms FCFS and ACO in terms of makespan and cost, achieving better trade-offs across different workload sizes. The results confirm that HEFT-Seeded NSGA-II provides a robust and scalable solution for workflow scheduling in heterogeneous cloud environments, offering a practical method to enhance system performance and resource utilization.

Keywords: Workflow scheduling · Cloud computing · Multi-objective optimization · NSGA-II · HEFT

1 Introduction

Modern computing systems are increasingly required to provide scalability, efficiency, and adaptability in response to rapidly growing data volumes and diverse application demands [1]. Cloud computing has become the dominant paradigm for delivering such capabilities, offering heterogeneous and elastic resources on demand [2–4]. However, the efficiency of cloud platforms largely depends on how computational tasks and workflows are scheduled [5, 6]. Inefficient scheduling can result in resource underutilization, increased costs, and deadline violations, ultimately degrading system performance [7–9].

Task scheduling plays a pivotal role in improving system efficiency by mapping tasks to resources in a way that optimizes key performance objectives [10]. In workflow-based applications, tasks are interconnected through Directed Acyclic Graphs (DAGs), where precedence constraints must be respected [11, 12]. Scheduling these workflows in heterogeneous cloud environments is particularly challenging due to conflicting objectives such as minimizing makespan, reducing cost, and avoiding deadline violations [10].

A. E. Hassanien et al. (Eds.): AISDS 2025, LNDECT 287, pp. 258–265, 2027.
https://doi.org/10.1007/978-3-032-18927-1_26

This paper addresses the problem of multi-objective workflow scheduling in cloud environments. While traditional approaches such as Ant Colony Optimization (ACO) [13] and First Come First Serve (FCFS) have been explored in this domain, they exhibit significant limitations. ACO, despite its global search capability, often suffers from slow convergence, while FCFS provides simplicity but fails to optimize performance under complex workflows [14]. These challenges highlight the need for advanced hybrid approaches capable of balancing multiple objectives more effectively [15, 16].

To this end, we propose a HEFT-Seeded NSGA-II model that integrates the Heterogeneous Earliest Finish Time (HEFT) [17] heuristic into the initialization of Non-dominated Sorting Genetic Algorithm II (NSGA-II) [18], guiding the evolutionary search process towards promising regions of the solution space. Our contributions are threefold:

- We design a hybrid scheduling framework that leverages HEFT seeding to enhance NSGA-II's exploration and convergence.
- We compare the proposed model against ACO and FCFS across multiple objectives, namely makespan, cost, and deadline violations.
- We validate the proposed model using workflow benchmarks, demonstrating its superiority in balancing performance trade-offs.

The remainder of the paper is organized as follows. Section 2 provides background on HEFT and NSGA-II. Section 3 presents the proposed HEFT-Seeded NSGA-II model. Section 4 details the validation process and comparative evaluation against ACO and FCFS. Section 5 discusses related work, and finally, Sect. 6 concludes the paper and outlines future directions.

2 HEFT-Seeded NSGA-II

Workflow scheduling in heterogeneous cloud environments is a well-established NP-hard problem, where the objective is to assign dependent tasks to heterogeneous resources while optimizing conflicting performance metrics. Among the classical approaches, the HEFT algorithm has emerged as one of the most effective heuristics due to its ability to balance task prioritization and resource selection [18]. HEFT computes an upward rank for each task based on average execution and communication costs, and then assigns tasks in descending order of rank to the resource that minimizes the earliest finish time. These deterministic heuristic yields high-quality schedules with relatively low computational overhead [19].

On the other hand, multi-objective evolutionary algorithms (MOEAs) such as NSGA-II have gained wide adoption in workflow scheduling due to their ability to approximate Pareto-optimal fronts for multiple conflicting objectives (e.g., makespan, cost, deadline violation) [20]. NSGA-II uses fast non-dominated sorting and crowding distance assignment to maintain both convergence and diversity of solutions. However, NSGA-II is sensitive to population initialization: purely random initialization may lead to slow convergence and exploration of non-promising regions of the search space [21].

The HEFT-Seeded NSGA-II approach integrates the strengths of both methods. Specifically, HEFT is used to generate high-quality seed solutions, which are then perturbed and injected into the initial NSGA-II population. This hybrid initialization provides the evolutionary search with a strong starting point, reducing convergence time and improving solution quality. During subsequent generations, NSGA-II applies crossover, mutation, and selection operators to evolve these seeded individuals while preserving population diversity. The result is a Pareto front that balances multiple objectives more effectively than standalone heuristics (e.g., HEFT) or standalone metaheuristics (e.g., ACO).

3 Proposed Scheduling Model

Our proposed scheduling model integrates the HEFT heuristic with the NSGA-II to address the multi-objective workflow scheduling problem in cloud environments. The main motivation is to exploit the strong baseline scheduling efficiency of HEFT while leveraging the evolutionary optimization power of NSGA-II to explore the trade-offs among conflicting objectives such as makespan, execution cost, and deadline adherence.

The complete algorithm, named HEFT-Seeded NSGA-II, is outlined as pseudocode in the public GitHub repository associated with this paper[1], which also provides the dataset used for the comparison. The algorithm begins by generating HEFT-based seeds and random chromosomes to form the initial population. The evolutionary loop then iteratively applies genetic operations, fitness evaluation, non-dominated sorting, and elitist selection. The final output is the Pareto front, representing a set of non-dominated schedules that capture alternative trade-offs between makespan, cost, and deadline satisfaction.

3.1 HEFT-Based Seeding

HEFT is employed as a seeding strategy to initialize the evolutionary population. The algorithm computes upward ranks of tasks based on average execution times and communication costs, generating a priority-ordered task list. Each task is then mapped to the resource that minimizes its earliest finish time (EFT). This produces high-quality initial schedules, which are encoded as chromosomes consisting of a task-to-resource assignment array and a priority list. Multiple HEFT seeds are generated and perturbed to increase population diversity while preserving the structural quality of the heuristic-derived schedules.

3.2 Chromosome Representation

Each candidate solution (chromosome) encodes:

- Assignment Array: maps each task to a VM type or resource in the pool.
- Priority List: specifies the execution order, respecting the workflow's DAG precedence constraints.

This dual encoding captures both resource allocation and task ordering, enabling fine-grained evolutionary operations.

[1] https://github.com/altix-ma/HEFT-Seeded-NSGA-II-Workflow-Scheduling-Algorithm

3.3 Genetic Operators and Repair

To evolve the population, standard NSGA-II operators are applied:

- Selection uses binary tournament guided by rank and crowding distance.
- Crossover recombines assignment arrays and merges priority lists while preserving DAG constraints.
- Mutation alters resource assignments and introduces minor perturbations to task order.

A repair mechanism ensures feasibility by correcting precedence violations and reassigning invalid resource allocations. This maintains a valid workflow execution plan after each genetic operation.

3.4 Fitness Evaluation

Each chromosome is decoded into a concrete schedule, where the following objectives are computed:

- Makespan: the completion time of the workflow.
- Execution Cost: the monetary cost of the allocated cloud resources.
- Deadline Violation Penalty: a measure of lateness relative to user-specified deadlines.

These objectives guide the multi-objective evolutionary search, with dominance and crowding distance applied during selection and elitist replacement.

3.5 Evolutionary Optimization with NSGA-II

The population is evolved over multiple generations. Fast non-dominated sorting identifies Pareto fronts, and crowding distance maintains diversity among non-dominated solutions. The elitist selection strategy ensures that high-quality solutions are preserved while promoting exploration of new trade-offs.

4 Validation

4.1 Methodology

The validation of the proposed HEFT-Seeded NSGA-II scheduling model was conducted through controlled experiments on synthetic workflow datasets. The datasets capture heterogeneous execution environments and diverse dependency structures, ensuring a realistic evaluation of scheduling performance. Workflows of different sizes (100, 500, 1000, and 10,000 tasks) were generated, with execution times derived from a base runtime adjusted by resource-specific speed factors. Resources were modeled with varying computational capacities and costs, reflecting typical cloud infrastructures.

All datasets and experimental scripts are publicly available in the GitHub repository of this paper[2].

The evaluation benchmark compared our model against two baseline scheduling strategies: FCFS and ACO. The performance metrics considered were:

- Makespan (s): the total completion time of the workflow.
- Cost ($): the monetary cost of resource usage.

[2] https://github.com/altix-ma/HEFT-Seeded-NSGA-II-Workflow-Scheduling-Algorithm

4.2 Results and Discussion

The results of the comparative analysis are illustrated in Fig. 1, which shows the performance of the three methods across different workflow sizes in terms of both makespan and cost.

The results demonstrate that our HEFT-Seeded NSGA-II model consistently outperforms FCFS and ACO across all workloads. For smaller workflows (100–500 tasks), the differences are moderate but significant, with our model reducing makespan by up to 33% compared to FCFS while maintaining lower costs. For larger workloads, particularly the 10,000-task case, the improvements become more substantial, where our model reduces makespan by over 5,600 s and achieves cost savings exceeding $1,100 compared to FCFS.

To highlight the overall trend, Fig. 2 presents the average performance of all three methods across the tested workloads. It confirms that the proposed model achieves the lowest cost while keeping makespan competitive with ACO. This validates the effectiveness of the HEFT-Seeded NSGA-II in balancing execution time and cost efficiency, making it well-suited for large-scale cloud workflow scheduling.

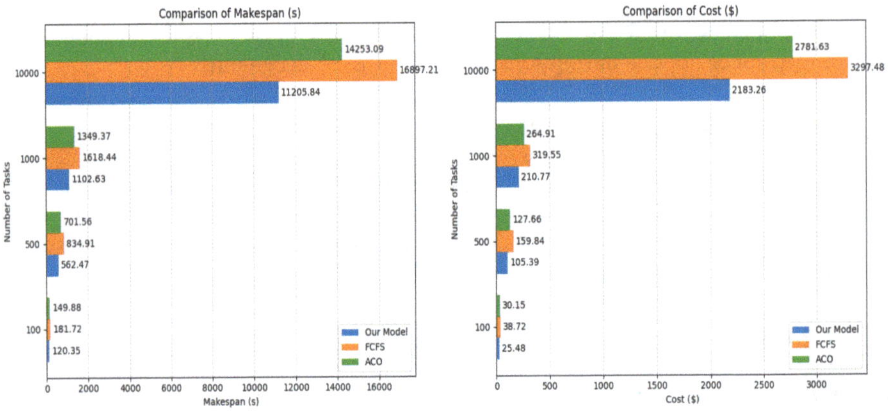

Fig. 1. Comparison of Makespan and Cost Across Workflow Sizes for Different Scheduling Methods

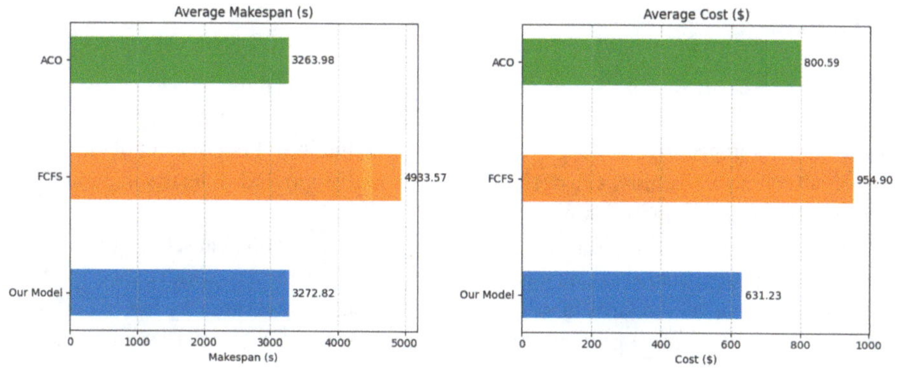

Fig. 2. Average Makespan and Cost Performance of Scheduling Methods

5 Related Work

The literature on workflow scheduling in cloud and distributed computing systems shows a strong research direction toward multi-objective optimization techniques, particularly those based on evolutionary algorithms.

A recent contribution by Shahriaret al., in 2024 [22] proposed a bi-objective workflow scheduling model for virtualized fog-cloud systems using NSGA-II with a semi-greedy heuristic. Their work focused on optimizing makespan and energy consumption, demonstrating that hybridizing NSGA-II with heuristic techniques can improve convergence and solution quality in complex scheduling environments. While this study emphasizes energy efficiency as a key factor, it does not directly address the cost dimension, which is often a decisive criterion in large-scale cloud deployments.

Similarly, Guangzhang et al., in 2025 [23] introduced an Improved Multi-Objective Memetic Algorithm (IMOMA) for workflow scheduling in cloud–edge–end collaborative frameworks. The approach enhanced population diversity and included a local search mechanism to refine solutions, achieving notable improvements in makespan and energy optimization. This work underlines the importance of hybrid and memetic strategies in tackling the workflow scheduling problem, but, like the previous study, its evaluation primarily targets energy and time rather than cost-effectiveness.

In contrast, our proposed HEFT-Seeded NSGA-II model explicitly addresses the trade-off between makespan and cost, two critical metrics for sustainable cloud workflow execution. By seeding the NSGA-II population with solutions generated from the classical HEFT heuristic, our model ensures faster convergence and higher-quality solutions compared to baseline methods such as FCFS and ACO. Unlike prior works that concentrate on energy optimization in fog or edge-enabled environments, our model demonstrates consistent superiority in reducing both execution time and monetary cost across different workflow sizes.

Thus, our contribution lies in extending the line of research on heuristic-enhanced evolutionary algorithms for workflow scheduling by shifting the optimization focus to cost-aware scheduling in large-scale cloud infrastructures, which remains underexplored in the existing literature.

6 Conclusion

This paper presented HEFT-Seeded NSGA-II, a hybrid multi-objective workflow scheduling algorithm designed to balance makespan and cost in large-scale cloud environments. By integrating HEFT-generated seeds into the initial population of NSGA-II, the proposed model achieves faster convergence and improved diversity compared to traditional scheduling methods.

The experimental validation, conducted on publicly available datasets, demonstrated that our model consistently outperforms FCFS and ACO in terms of both execution time and monetary cost. The results highlight the effectiveness of combining heuristic-based seeding with evolutionary optimization, enabling more efficient and cost-aware workflow scheduling.

Future research will extend this work by incorporating deadline constraints, energy consumption, and fault tolerance into the optimization objectives, as well as exploring the applicability of the model in fog and edge computing environments.

References

1. Ait Said, M., Ezzati, A., Mihi, S., Belouaddane, L.: Microservices adoption: an industrial inquiry into factors influencing decisions and implementation strategies. Int. J. Comput. Digit. Syst. **15**(1), 1417–1432 (2024). https://doi.org/10.12785/ijcds/1501100
2. Castellana, L.: Development and Analysis of FinOps Processes for a Multi-Cloud Environment. Politecnico di Torino (2023)
3. Lahcen, B., Mehdi, A.S., Abderrahim, M., Amine, B., Anas, R.: A Cost-Efficient Security Practices Framework for Small DevSecOps Teams, pp. 257–266 (2025). https://doi.org/10.1007/978-3-031-95326-2_25
4. Belouaddane, L., Ait Said, M., Marzouk, A., Benmakhlouf, A.: Microservice architecture DevOps integration challenges: a qualitative study. In: 3rd International Conference on Advanced Engineering, Technology and Applications Lecture Notes in Networks and Systems LNNS, vol. 1138, pp. 96–108 (2024). https://doi.org/10.1007/978-3-031-70924-1_8
5. Nastic, S., et al.: Polaris scheduler: edge sensitive and SLO aware workload scheduling in cloud-edge-IoT clusters. IEEE Int. Conf. Cloud Comput. CLOUD. **2021**, 206–216 (2021). https://doi.org/10.1109/CLOUD53861.2021.00034
6. Yang, X., et al.: Optimizing IT FinOps and sustainability through unsupervised workload characterization. Proc. AAAI Conf. Artif. Intell. **38**(21), 22990–22996 (2024). https://doi.org/10.1609/AAAI.V38I21.30340
7. Zheng, T., Wan, J., Zhang, J., Jiang, C.: Deep reinforcement learning-based workload scheduling for edge computing. J. Cloud Comput. **11**(1), 1–13 (2022). https://doi.org/10.1186/S13677-021-00276-0/TABLES/2
8. Uzoka, A., Olamijuwon, J., Akerele, J.I., Ojukwu, P.U., Olamijuwon, O.J.: Improving healthcare application scalability through microservices architecture in the cloud. Artic. Int. J. Sci. Res. Updat. **2024**(02), 100–109 (2024). https://doi.org/10.53430/ijsru.2024.8.2.0064
9. Shaikh, R., Muntean, C.H., Gupta, S.: Prediction of resource utilisation in cloud computing using machine learning. Int. Conf. Cloud Comput. Serv. Sci., 103–114 (2024). https://doi.org/10.5220/0012742200003711
10. He, H., Shen, H., Hao, Q., Tian, H.: Online delay-guaranteed workload scheduling to minimize power cost in cloud data centers using renewable energy. J Parallel Distrib Comput. **159**, 51–64 (2022). https://doi.org/10.1016/J.JPDC.2021.09.002

11. Gao, W. et al.: Deep Learning Workload Scheduling in GPU Datacenters: Taxonomy, Challenges and Vision (2022). [Online]. Available: https://arxiv.org/pdf/2205.11913. Accessed: 21 Sep. 2025
12. Fan, Y., Lan, Z., Rich, P., Allcock, W., Papka, M.E.: Hybrid workload scheduling on HPC systems. In: Proceedings of the 2022 IEEE 36th International Parallel and Distributed Processing Symposium, pp. 470–480 (2022). https://doi.org/10.1109/IPDPS53621.2022.00052
13. Awadallah, M.A., et al.: Multi-objective ant colony optimization: review. Arch. Comput. Methods Eng. **32**(2), 995–1037 (2025). https://doi.org/10.1007/S11831-024-10178-4/METRICS
14. Weng, Q., et al.: MLaaS in the wild: workload analysis and scheduling in large-scale heterogeneous GPU clusters. In: 19th USENIX Symposium on Networked Systems Design and Implementation (NSDI 22) (2022)
15. Menaka, M., Sendhil Kumar, K.S.: Workflow scheduling in cloud environment – challenges, tools, limitations & methodologies: a review. Meas. Sensors. **24**, 100436 (2022). https://doi.org/10.1016/J.MEASEN.2022.100436
16. Kashyap, S., Singh, A.: Prediction-based scheduling techniques for cloud data center's workload: a systematic review. Clust. Comput. **26**(5), 3209–3235 (2023). https://doi.org/10.1007/S10586-023-04024-8/METRICS
17. Samadi, Y., Zbakh, M., Tadonki, C.: E-HEFT: enhancement heterogeneous earliest finish time algorithm for task scheduling based on load balancing in cloud computing. In: Proceedings of the 2018 International Conference on High Performance Computing & Simulation, pp. 601–609 (2018). https://doi.org/10.1109/HPCS.2018.00100
18. De Buck, V., López, C.A.M., Nimmegeers, P., Hashem, I., Van Impe, J.: Multi-objective optimisation of chemical processes via improved genetic algorithms: a novel trade-off and termination criterion. Comput. Aided Chem. Eng. **46**, 613–618 (2019). https://doi.org/10.1016/B978-0-12-818634-3.50103-X
19. Kaur, A., Singh, P., Singh Batth, R., Peng Lim, C.: Deep-Q learning-based heterogeneous earliest finish time scheduling algorithm for scientific workflows in cloud. Softw. Pract. Experience. **52**(3), 689–709 (2022). https://doi.org/10.1002/SPE.2802
20. Bian, C., Qian, C.: Better running time of the non-dominated sorting genetic algorithm II (NSGA-II) by using stochastic tournament selection. In: Parallel Problem Solving from Nature – PPSN XVII Lecture Notes in Computer Science LNCS (Including Subseries Lecture Notes in Artificial Intelligence (LNAI) & Lecture Notes in Bioinformatics (LNBI)), vol. 13399, pp. 428–441 (2022). https://doi.org/10.1007/978-3-031-14721-0_30
21. Mohamad Shirajuddin, T., Muhammad, N.S., Abdullah, J.: Optimization problems in water distribution systems using non-dominated sorting genetic algorithm II: an overview. Ain Shams Eng. J. **14**(4), 101932 (2023). https://doi.org/10.1016/J.ASEJ.2022.101932
22. Karami, S., Azizi, S., Ahmadizar, F.: A bi-objective workflow scheduling in virtualized fog-cloud computing using NSGA-II with semi-greedy initialization. Appl. Soft Comput. **151**, 111142 (2024). https://doi.org/10.1016/J.ASOC.2023.111142
23. Cui, G., Zhang, W., Xu, W., Bao, H.: Efficient workflow scheduling using an improved multi-objective memetic algorithm in cloud-edge-end collaborative framework. Sci. Rep. **15**(1), 1–21 (2025). https://doi.org/10.1038/S41598-025-08691-Y;SUBJMETA

Selfishness Mitigation in ALOHA-Based Random Access Mechanism

Ahmed Boujnoui[1(✉)], Hamid Taramit[2], Abdellah Zaaloul[3], and Abdelkrim Haqiq[4]

[1] Department of Mathematics, MASD Group, LMPA Laboratory, Faculty of Sciences and Technology, Moulay Ismail University of Meknes, B.P. 509, Boutalamine, Errachidia 52000, Morocco
a.boujnoui@umi.ac.ma

[2] Center for Computational Simulation, Universidad Politécnica de Madrid, Madrid, Spain
hamid.taramit@upm.es

[3] LM2CE Laboratory, Faculty of Economics and Management, Hassan First University, Settat, Morocco
abdellah.zaaloul@uhp.ac.ma

[4] Computer, Networks, Mobility and Modeling Laboratory (IR2M), Faculty of Sciences and Techniques, Hassan First University of Settat, Settat 26000, Morocco
abdelkrim.haqiq@uhp.ac.ma

Abstract. This paper provides an analytical study using a stochastic game framework for a wireless network comprising both cooperative and non-cooperative (selfish) users. Sharing a common wireless medium, users operate the Slotted ALOHA mechanism with ZigZag Decoding (SAZD) to access the medium. Cooperative users aim to maximize the overall system throughput, whereas selfish users prioritize their own benefits, resulting in negative impacts on other users and the overall network performance. The game's equilibrium is characterized by a combination of the social optimum and the Nash equilibrium. To mitigate the excessive resource usage caused by the aggressive behavior of selfish users, we impose a retransmission cost on non-cooperative users. We show that our proposal provides optimal retransmission probabilities that significantly enhance the throughput of non-cooperative users and the overall network throughput.

Keywords: Optimal Cost · Stochastic Game · Markov Model · Slotted ALOHA · Random Access Mechanism · Nash Equilibrium

1 Introduction

The ALOHA protocol is a random-access network protocol developed in the 1970 s at the University of Hawaii to allow multiple devices to share a single

H. Taramit, A. Zaaloul and A. Haqiq—Contributing authors.

A. E. Hassanien et al. (Eds.): AISDS 2025, LNDECT 287, pp. 266–275, 2027.
https://doi.org/10.1007/978-3-032-18927-1_27

communication channel without central control. Unlike CSMA, which requires carrier sensing before transmission, ALOHA allows immediate transmission upon generation. ALOHA and its variants [1–5] play a crucial role in the evolution of wireless communication technologies and have been developed to ensure reliable connectivity for various applications. For instance, the Restricted Access Window (RAW) mechanism introduced in Wi-Fi HaLow technology addresses high contention in Internet of Things (IoT) applications [6,7]. Additionally, Wi-Fi 7 incorporates the Multi-Link Operation (MLO) feature to enhance efficiency and reduce latency in high-bandwidth applications [8].

Game theory provides a powerful framework for analyzing user interactions in a network, categorized into cooperative and non-cooperative games. Users in cooperative games are aware of each others' actions, strategy spaces, and utility functions, enabling them to cooperate and reach an equilibrium that benefits the whole game. This cooperation leads to optimal system performance. In contrast, non-cooperative games involve users acting independently to maximize their own utility, often resulting in significant performance degradation for all users [9].

The cooperative framework is particularly well-suited for modeling collaborative contexts, such as channel access, resource allocation, and bandwidth distribution. In wireless networks, cooperation is often managed using a central base station targeting system-wide optimization objectives such as throughput maximization, delay minimization, or backlog reduction. In contrast, non-cooperative game models involve users attempting to maximize their utility without considering the impact on others [4,10].

The majority of game theory based models assume either full cooperation or no cooperation among network users [10–15]. Therefore, we introduced, in our previous work [16], a mixed game model where both cooperative and selfish users coexist and share the same wireless channel. Our findings showed that the selfish behavior results in significant performance degradation for all users in the network, particularly when the rate of generating packets is higher.

The paper proceeds as follows: Sect. 2 surveys recent related literature. Section 3 details the bi-dimensional Markov model and key performance metrics. Section 4 develops the game-theoretic model and establishes its equilibrium. Section 5 presents numerical results and discusses principal findings. Finally, Sect. 6 provides concluding remarks and future research directions.

2 Related Work

Random access mechanisms such as ALOHA are extensively implemented in recent IoT technologies, including NOMA [17,18], satellite networks [12], Machine-to-Machine (M2M) communications [3], and LoRaWAN networks [19–23]. Many of these technologies operate sensors with constrained battery capabilities, which spurred the need to strike a delicate balance between resource management and energy efficiency.

Several studies have revealed that competitive users in random access channels tend to act aggressively by accessing the channel with high transmission probability [4,9,24,25]. As a result, mitigating this selfish behavior in competitive wireless environments to achieve high throughput and minimal delay while conserving energy is a significant challenge.

Efforts to tackle this challenge have spurred the development of various amendments for random access mechanisms aimed at optimizing resource utilization, minimizing energy consumption, and reducing access delay. In our previous work, we proposed various pricing strategies to penalize selfish users in an environment comprising either cooperative users [25] or selfish [9]. However, in the current work, we propose a transmission-cost-based penalization mechanism in an environment that incorporates both cooperative and selfish users.

In [9,25], we showed that the random access mechanism SAZD outperforms the pure SA across all performance metrics. However, under certain conditions, cooperation can lead to inequitable resource management, and selfishness can cause significant degradation in Quality of Service (QoS). To address the fairness issue, we proposed an enhanced pricing mechanism in [26].

The authors in [27] conducted a comprehensive analysis using stochastic games and evolutionary game theory to investigate the dynamics of cooperation and selfishness. Their findings showed that system performance increases with cooperation and decreases with non-cooperation. The same trend is shown for both SA and SAZD systems in [9].

The authors in [28] examined the impact of the transmission cost on the AoI of nodes attempting to receive updates from a shared source. While they considered a simple ALOHA without any retransmission after the backoff, they showed that the system's Price of Anarchy (PoA) can be maintained close to 1. The authors in [29] extended the objective of minimizing the AoI to minimizing the cost associated with the offered traffic.

This study extends the model proposed in [16], where we investigated a novel stochastic game framework modeling the coexistence of cooperative and selfish users within a shared wireless network. Therefore, we maintain the same modeling notation and analysis throughout this paper. One of the primary concerns in mixed environments is the detrimental impact of selfish users on network performance, which we address in this paper.

3 Problem Formulation and Network Scenario

This section presents a Markov-based model that allows for the derivation of stationary results and various performance metrics. Additionally, this section establishes the theoretical foundation and essential prerequisites for constructing the stochastic game model, which will be detailed in Sect. 4.

3.1 Description of the Model

We consider a shared wireless network, illustrated in Fig. 1, composed of both cooperative and selfish users powered by limited battery resources. The network includes M cooperative users and N selfish users. Within each group, the number

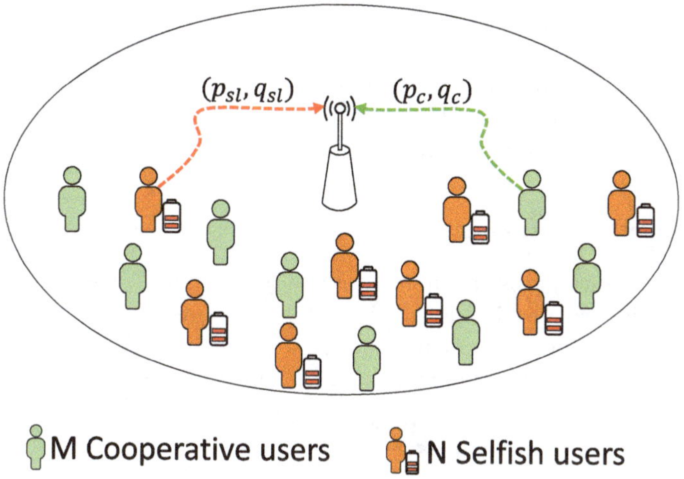

Fig. 1. Selfish battery-powered devices share the same medium with cooperative user devices.

of backlogged users are denoted by m and n, respectively. Packet arrivals follow a Bernoulli process, with rates p_c for cooperative users and p_{sl} for selfish users. The retransmission probabilities for backlogged users in the two groups are given by q_c and q_{sl}, respectively.

The probabilities that exactly i unbacklogged nodes are transmitting in the cooperative and selfish groups are denoted by $Q_c(i, m)$ and $Q_{sl}(i, n)$, respectively.

$$Q_c(i,m) = \binom{M-m}{i} p_c^i (1-p_c)^{M-m-i}, \tag{1}$$

$$Q_{sl}(i,n) = \binom{N-n}{i} p_{sl}^i (1-p_{sl})^{N-n-i}. \tag{2}$$

Similarly, the notations $Q_c^r(i, m)$ and $Q_{sl}^r(i, n)$ are used to define the probabilities that i backlogged nodes undertake a retransmission in their respective groups.

$$Q_c^r(i,m) = \binom{m}{i} q_c^i (1-q_c)^{m-i}, \tag{3}$$

$$Q_{sl}^r(i,n) = \binom{n}{i} q_{sl}^i (1-q_{sl})^{n-i}. \tag{4}$$

3.2 Markov-Based Model

We now formalize the Markov model framework. The system's state is modeled using the stochastic process $(X_k, Y_k)_{k\in\mathbb{N}}$, where X_k and Y_k denote the number of backlogged cooperative and selfish users, respectively. The discrete-time index

$k \in \mathbb{N}$ corresponds to the slot structure inherent to the SAZD random-access mechanism.

Theorem 1 in [16] provides conditions for the existence and uniqueness of the stationary distribution for the Markov model. The interested reader can refer to [16] for a detailed proof of the theorem.

3.3 Performance Analysis

We investigate various performance metrics of the SAZD mechanism, derived from the stationary distribution of the Markov model. For each performance metric, we present mathematical formulas for both cooperative (c) and selfish (sl) users. We start by introducing the average throughput for cooperative and selfish users.

$$TH_c = \frac{p_c}{T_s} \sum_{m=0}^{M} \left[(M-m) \sum_{n=0}^{N} \Pi_{m,n} \right], \tag{5}$$

$$TH_{sl} = \frac{p_{sl}}{T_s} \sum_{n=0}^{N} \left[(N-n) \sum_{m=0}^{M} \Pi_{m,n} \right], \tag{6}$$

where T_s is defined as:

$$T_s = 1 + P_{zigzag}, \tag{7}$$

and P_{zigzag} represents the probability of the Zigzag Decoding (ZD) scenario involving two packet simultaneous transmissions. It is given by:

$$P_{zigzag} = \sum_{m=0}^{M} \sum_{n=0}^{N} \sum_{i,j,k,l \in \mathbb{N}} \left[Q_c(i,m) \cdot Q_{sl}(j,n) \cdot Q_c^r(k,m) \cdot Q_{sl}^r(l,n) \right] \cdot \Pi_{m,n} \cdot \mathbb{1}_{\{i+j+k+l=2\}}. \tag{8}$$

We compute the average number of backlogged users using the expressions:

$$S_c = \sum_{m=0}^{M} m \cdot \left[\sum_{n=0}^{N} \Pi_{m,n} \right], \quad S_{sl} = \sum_{n=0}^{N} n \cdot \left[\sum_{m=0}^{M} \Pi_{m,n} \right]. \tag{9}$$

The access delays for the transmitted packets are given by the Little result:

$$D_c = 1 + \frac{S_c}{TH_c}, \quad D_{sl} = 1 + \frac{S_{sl}}{TH_{sl}}. \tag{10}$$

4 Stochastic Game Analysis of Selfish and Cooperative Behavior

The game formulation of selfish and cooperative behaviors is based on a finite stochastic game involving M cooperative users and N selfish users, all of which utilize the SAZD mechanism on a shared wireless channel. The game comprises several key components. Players in the game include both cooperative and selfish users. Cooperative users use mixed strategies denoted by $\phi_i = \{q_c^i, 1 - q_c^i\}$, whereas selfish users employ mixed strategies represented by $\psi_j = \{q^j_{sl}, 1 - q^j_{sl}\}$. The mixed strategies are distributed over the pure strategies (T : Transmit, W : Wait).

In terms of utility, selfish users have individual utility functions u_i that depend on their transmission probability as well as the transmission probabilities of other users. Whereas cooperative users share a common utility function, which reflects the overall performance of the system. We assume that every player is aware of the number of participants, their own strategy space, and that of others, as well as the utilities of all players. Cooperative players assume that all other players are cooperative, while selfish users assume that all other players are selfish.

4.1 Characterization of the Game Equilibrium

The equilibrium of a game is a state in which no player has an incentive to deviate from the selected action. Therefore, all players in the equilibrium are satisfied with their action choices. Generally, an equilibrium is typically defined by a strategy profile where different players select different actions [30]. However, for simplicity purposes, we restrict to a symmetric strategy profile (Ψ^*_c, Ψ^*_{sl}), where Ψ^*_c and Ψ^*_{sl} denote, respectively, the cooperative and selfish strategy choice. Due to semmitric strategies, all cooperative players use Ψ^*_c and all selfish users deploy Ψ^*_{sl}. Since we build our analytical model on a Markov process, we consider a Markovian strategy, which is characterised by independence from past actions given the current state.

Let Ψ^*_c be the social optimum for cooperative users, and Ψ^*_{sl} the Nash equilibrium strategy for selfish users. Now, we can define our game equilibrium as (Ψ^*_c, Ψ^*_{sl}). The conditions for the existence of this equilibrium are presented by Theorem 2 in [16].

4.2 Penalizing the Selfish Behavior

Incorporating a transmission cost into the game framework aims to penalize the selfish behavior of aggressive non-cooperative users. By introducing this cost, the utility u_i of each selfish user i is modified to depend now on the transmission cost $C \in [0, 1]$:

$$u_i(C) = (1 - C)TH_i - Cq^iS_i. \tag{11}$$

In this context, TH_i corresponds to the normalized throughput. Meanwhile, S_i measures the average number of backlogged packets for user i. By including the cost component Cq^iS_i, the utility function discourages users from adopting overly aggressive strategies that maximize individual throughput at the expense of overall system performance. This approach fosters a more balanced and cooperative environment, enhancing the stability and efficiency of the network.

To guarantee dynamic optimal control of the transmission cost for every traffic condition (i.e., arrival probability), we propose an optimization study that allows for an adjusted transmission cost to alleviate the excessively aggressive behavior of selfish users. Towards this end, users are now faced with the following optimization problem:

$$(\mathcal{P}) \begin{cases} \text{For each arrival rate } p_{sl} \in [0, 1], \\ C^* = \underset{C \in [0,1]}{\arg\max}\, u_i(C) \end{cases} \tag{12}$$

Now, selfish users will no longer be aggressive by paying extra attention to their battery constraints.

5 Numerical Results

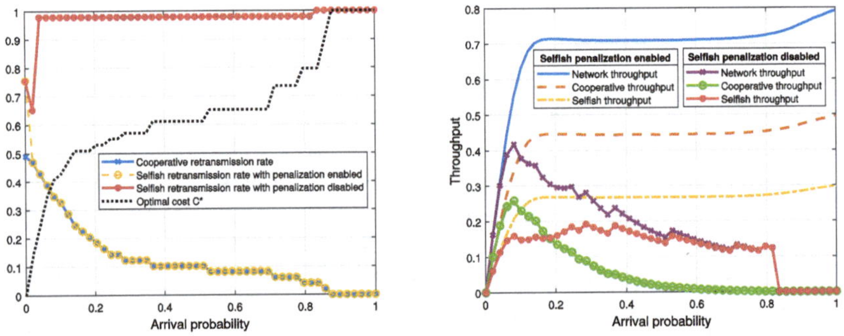

(a) Optimal retransmission cost, cooperative selfish retransmission rate with and without penalization

(b) Network throughput with and without selfishness selfish penalization.

Fig. 2. Results for a network scenario consisting of $M = 5$ cooperative users and $N = 3$ selfish users.

Figure 2(a) shows the optimal transmission cost C^* as a function of the arrival probability. The cost C^* is obtained by solving problem (12) and is designed to penalize selfish users. The increasing trend of the figure shows that as the network becomes congested, a high transmission cost is needed to penalize selfish users. In the same figure, we observe that the transmission probability of penalized selfish users becomes close to the transmission probability of cooperative users.

Figure 2(b) shows the network throughput obtained after penalizing selfish users. Compared to the throughput obtained without adopting any penalization mechanism, we observe a significant improvement. The results show that selfish behavior in wireless networks leads to a dramatic degradation of system performance.

6 Conclusion and Future Work

In this paper, we investigated a wireless network scenario comprising both cooperative and selfish users. We presented a stochastic game model based on a bi-dimensional Markov chain, which allows us to derive various performance metrics. To address the negative impact of selfish users, we proposed a penalization mechanism by imposing a cost on every retransmission attempt. We showed that our proposal mitigates the aggressive behavior of selfish users and enhances both their individual throughput and the overall network throughput.

Our future work will focus on systems where users can dynamically alter their behavior, choosing whether to cooperate or act selfishly based on network conditions. We also plan to develop approaches to detect selfish users in dense network scenarios.

References

1. Abramson, N.: The ALOHA system: another alternative for computer communications. In: Proceedings of the November 17-19, 1970, Fall Joint Computer Conference (1970), AFIPS '70 (Fall), pp. 281–285 (1970). https://doi.org/10.1145/1478462.1478502
2. Oku, T., Kimura, T., Cheng, J.: Performance Evaluation of Hierarchical Slotted ALOHA for IoT Applications. In: 2020 IEEE International Conference on Consumer Electronics-Taiwan (ICCE-Taiwan) (IEEE, 2020), pp. 1–2 (2020). https://doi.org/10.1109/ICCE-Taiwan49838.2020.9258134
3. Li, Y., Zhan, W., Dai, L.: Rate-constrained delay optimization for slotted ALOHA. IEEE Trans. Commun. **69**(8) (2021). https://doi.org/10.1109/TCOMM.2021.3079513
4. Boujnoui, A., Zaaloul, A., Haqiq, A.: A stochastic game analysis of the slotted ALOHA mechanism combined with zigzag decoding and transmission cost. In: International Conference on Innovations in Bio-Inspired Computing and Applications, vol. 735, pp. 102–112. Springer International Publishing, Cham (2018). https://doi.org/10.1007/978-3-319-76354-5_10
5. Oinaga, M., Ogata, S., Ishibashi, K.: ZigZag decodable coded slotted ALOHA. In: 2018 15th Workshop on Positioning, Navigation and Communications (WPNC), pp. 1–6. IEEE (2018). https://doi.org/10.1109/WPNC.2018.8555748
6. Taramit, H., Camacho-Escoto, J.J., Gomez, J., Orozco-Barbosa, L., Haqiq, A.: Accurate analytical model and evaluation of Wi-Fi halow based IoT networks under a Rayleigh-fading channel with capture. Mathematics **10**(6) (2022). https://doi.org/10.3390/math10060952
7. Taramit, H., Orozco-Barbosa, L., Haqiq, A., Escoto, J.J.C., Gomez, J.: Load-aware channel allocation for IEEE 802.11ah-based networks. IEEE Access **11**, 24484–24496 (2023). https://doi.org/10.1109/ACCESS.2023.3251896
8. Taramit, H., Boujnoui, A., Orozco-Barbosa, L., Haqiq, A.: Dynamic load balancing control of multiclass traffic in Multi-Link IEEE 802.11be networks. In: 2023 12th IFIP/IEEE International Conference on Performance Evaluation and Modeling in Wired and Wireless Networks (PEMWN), pp. 1–6 (2023). https://doi.org/10.23919/PEMWN58813.2023.10304919

9. Boujnoui, A., Zaaloul, A., Haqiq, A.: Mathematical model based on game theory and Markov chains for analysing the transmission cost in SA-ZD mechanism. Int. J. Comput. Inf. Syst. Ind. Manag. Appl **10**, 197–207 (2018)
10. Gopal, S., Kaul, S.K., Chaturvedi, R., Roy, S.: A non-cooperative multiple access game for timely updates. In: IEEE INFOCOM 2020-IEEE Conference on Computer Communications Workshops (INFOCOM WKSHPS), pp. 924–929. IEEE (2020). https://doi.org/10.1109/INFOCOMWKSHPS50562.2020.9162886
11. Zhao, B., Ren, G., Zhang, H.: Cooperative contention resolution diversity slotted ALOHA with transmit power diversity for multi-satellite networks. In: 2019 IEEE 90th Vehicular Technology Conference (VTC2019-Fall), pp. 1–5. IEEE (2019)
12. Zhao, B., Ren, G., Zhang, H.: Slotted ALOHA game for medium access control in satellite networks. In: 2019 IEEE/CIC International Conference on Communications in China (ICCC), pp. 518–522. IEEE (2019). https://doi.org/10.1109/ICCChina.2019.8855903
13. Vaezi, K., Ashtiani, F.: Delay-optimal cooperation policy in a slotted ALOHA full-duplex wireless network: static approach. IEEE Syst. J. **14**(2), 2257–2268 (2019). https://doi.org/10.1109/JSYST.2019.2940605
14. Vaezi, K., Akar, N., Karasan, E.: Age of information in a cooperative slotted ALOHA network: marginal and joint distributions (2021). https://doi.org/10.36227/techrxiv.16437330.v1
15. Liu, R., Hong, T., Ding, X., Wang, Y., Zhang, G.: Multi-satellite cooperative beamforming ALOHA for LEO satellite IoT networks. Front. Space Technol. **2**, 9 (2021). https://doi.org/10.3389/frspt.2021.755520
16. Boujnoui, A., Zaaloul, A., Orozco-Barbosa, L., Haqiq, A.: Stochastic game analysis of cooperation and selfishness in a random access mechanism. Mathematics **10**(5) (2022). https://doi.org/10.3390/math10050694
17. Balcı, A., Sokullu, R.: Fairness aware deep reinforcement learning for grant-free NOMA-IoT networks. Internet of Things **25**, 101079 (2024). https://doi.org/10.1016/j.iot.2024.101079
18. George, E.M., Jacob, L.: Multi-channel U random access for Inter-WBAN communication. Wireless Netw. **29**(7), 3177–3185 (2023). https://doi.org/10.1007/s11276-023-03306-8
19. Khawam, K., Fawaz, H., Lahoud, S., Maillard, O.A., Martin, S.: A channel selection game for multi-operator LoRaWAN deployments. Comput. Netw. **216**, 109185 (2022). https://doi.org/10.1016/j.comnet.2022.109185
20. Khan, M.A.A., Ma, H., Aamir, S.M., Jin, Y.: Optimizing the performance of pure aloha for lora-based ESL. Sensors **21**(15), 5060 (2021). https://doi.org/10.3390/s21155060
21. Cheikh, I., Sabir, E., Aouami, R., Sadik, M., Roy, S.: Throughput-delay tradeoffs for Slotted-ALOHA-based LoRaWAN networks. In: 2021 International Wireless Communications and Mobile Computing (IWCMC) (IEEE, 2021), pp. 2020–2025. https://doi.org/10.1109/IWCMC51323.2021.9498969
22. Beltramelli, L., Mahmood, A., Österberg, P., Gidlund, M.: Lora beyond ALOHA: an investigation of alternative random access protocols. IEEE Trans. Industr. Inf. **17**(5), 3544–3554 (2020). https://doi.org/10.1109/tii.2020.2977046
23. Milarokostas, C., Tsolkas, D., Passas, N., Merakos, L.: A comprehensive study on LPWANs with a focus on the potential of LoRa/LoRaWAN systems (2021). https://doi.org/10.1109/comst.2022.3229846

24. Hansen, A.P.J., Münster, J.R., Villadsen, R.E., Segaard, S.B., Rasmussen, S.P., Biscio, C., Leyva-Mayorga, I.: A game-theoretic perspective for efficient modern random access. In: 2025 IEEE Wireless Communications and Networking Conference (WCNC), pp. 1–6. IEEE (2025). https://doi.org/10.1109/wcnc61545.2025.10978189
25. Boujnoui, A., Zaaloul, A., Haqiq, A.: Cooperative slotted ALOHA with ZigZag decoding and a pricing mechanism. In: International Conference on Innovations in Bio-Inspired Computing and Applications, vol. 1372, pp. 111–119. Springer, Cham (2020). https://doi.org/10.1007/978-3-030-73603-3_10
26. Boujnoui, A., Zaaloul, A., Haqiq, A.: Enhanced pricing strategy for slotted ALOHA with zigzag decoding: a stochastic game approach. Int. J. Comput. Inf. Syst. Ind. Manag. Appl. **13**, 160–171 (2021)
27. Hilbe, C., Šimsa, Š, Chatterjee, K., Nowak, M.A.: Evolution of cooperation in stochastic games. Nature **559**(7713), 246–249 (2018). https://doi.org/10.1038/s41586-018-0277-x
28. Badia, L.: Impact of transmission cost on age of information at nash equilibrium in slotted ALOHA. IEEE Network. Lett. **4**(1), 30–33 (2021). https://doi.org/10.1109/lnet.2021.3133220
29. Crosara, L., Badia, L.: Cost and correlation in strategic wireless sensing driven by age of information. In: European Wireless 2022; 27th European Wireless Conference (VDE, 2022)
30. MacKenzie, A.B., Wicker, S.B.: Selfish users in ALOHA: a game-theoretic approach. In: IEEE 54th Vehicular Technology Conference. VTC Fall 2001. Proceedings (Cat. No. 01CH37211), vol. 3, pp. 1354–1357. IEEE (2001). https://doi.org/10.1109/vtc.2001.956417

Reducing Security Testing Overhead in DevSecOps Through Risk-Aware Scan Orchestration

Belouaddane Lahcen(✉), Ait Said Mehdi, Marzouk Abderrahim, Benmakhlouf Amine, and Raouf Anas

Hassan First University, Settat, Morocco
{lahcen.belouaddane,m.aitsaid,abderrahim.marzouk,amine.benmakhlouf, anas.raouf.doc}@uhp.ac.ma

Abstract. Modern software development increasingly relies on DevSecOps to integrate security within continuous integration and deployment pipelines. However, despite its benefits, DevSecOps adoption often introduces significant cost and performance overheads, especially due to resource-intensive security tools and extended pipeline runtimes. To address this challenge, we propose a cost-aware DevSecOps orchestration framework that embeds security scanning into CI/CD processes while optimizing execution efficiency. The framework is validated on four OWASP benchmark applications—Juice Shop, Tiredful-API, Verademo, and DVMCP—across multiple merge operations, demonstrating consistent reductions in execution time ranging from 15% to 35% while preserving security accuracy. Our contributions highlight that cost efficiency and security assurance can be achieved simultaneously, offering a practical solution for teams aiming to adopt DevSecOps practices without compromising affordability or performance.

Keywords: DevSecOps · CI/CD · Cost Optimization · Security · DevOps

1 Introduction

Modern software systems are increasingly complex, highly distributed, and continuously evolving to meet the growing demand for new features, faster releases, and improved user experiences [5,15]. This rapid evolution has amplified the importance of security, as vulnerabilities can be introduced at any stage of the development lifecycle and exploited shortly after deployment [8,14]. As a result, integrating security into the development workflow is no longer optional but a fundamental requirement for building trustworthy software [17,18].

To address the need for speed and reliability, organizations have widely adopted *DevOps* [4,21], a practice that automates and integrates software development and operations to enable continuous integration and continuous delivery

A. S. Mehdi, M. Abderrahim, B. Amine and R. Anas—Contributing authors.

A. E. Hassanien et al. (Eds.): AISDS 2025, LNDECT 287, pp. 276–285, 2027.
https://doi.org/10.1007/978-3-032-18927-1_28

(CI/CD) [3,16]. Extending this philosophy, *DevSecOps* embeds security practices into CI/CD pipelines, ensuring that code is systematically tested for vulnerabilities through techniques such as static analysis (SAST), dynamic analysis (DAST), software composition analysis (SCA), and secrets detection [1,2,12]. While this shift enables early detection of vulnerabilities, it also introduces new challenges [11]. Running all security scans on every code commit significantly increases computational costs, pipeline execution time, and resource consumption—factors that can hinder developer productivity and inflate operational budgets [7,19].

This tension between security assurance and cost efficiency motivates the need for intelligent orchestration strategies that dynamically balance both dimensions [9,20] [13]. In this paper, we propose a **risk-driven orchestration framework** for CI/CD security that selectively executes security scans based on change impact, component criticality, and vulnerability likelihood. Our approach reduces redundant scans while maintaining effective vulnerability detection, thereby lowering pipeline execution costs and improving feedback speed.

The contributions of this paper are threefold: (1) we identify the limitations of current DevSecOps practices with respect to cost and scalability; (2) we design and implement a risk-aware orchestration framework for CI/CD security pipelines; and (3) we evaluate the framework on real-world open-source applications, demonstrating its ability to reduce pipeline costs while preserving detection effectiveness.

The remainder of this paper is structured as follows: Sect. 2 reviews related work on cost-aware DevSecOps and pipeline optimization. Section 3 presents the proposed framework in detail. Section 4 describes the evaluation setup and results. Finally, Sect. 5 concludes the paper and outlines directions for future research.

2 Related Work

The integration of security into continuous delivery pipelines has been widely studied, yet balancing security enforcement with cost efficiency remains a challenging research problem. Recent work has focused on designing frameworks and strategies to reduce the overhead of DevSecOps adoption while maintaining effective protection mechanisms.

For example, the work of BELOUADDANE et al., *'A Cost-Efficient Security Practices Framework for Small DevSecOps Teams'* [9] directly addresses the challenges faced by small and resource-constrained development teams. The authors propose a framework built around open-source tools, risk-based prioritization, and scalable practices. Their evaluation across diverse application environments such as Spring Boot, Laravel, Flask, and Express.js demonstrates that security integration can achieve cost reductions of up to 93% without compromising detection quality. This work highlights the importance of cost-aware orchestration strategies in DevSecOps but remains focused primarily on small-team adoption scenarios rather than broader, automated orchestration mechanisms.

Complementing this, Jeroen Mulder in his book *"Multi-Cloud Strategy for Cloud Architects"* [10] provides a broader perspective on managing cost and security in complex cloud environments. The author introduces the interplay between BaseOps, FinOps, and DevSecOps, and it emphasizes the role of the DevSecOps Maturity Model in aligning operational efficiency with security requirements. With practical case studies spanning multiple cloud providers such as AWS, Azure, Google Cloud, Oracle Cloud, and Alibaba Cloud, the book underscores the centrality of cost optimization strategies, including FinOps and AIOps, in enabling sustainable DevSecOps adoption at scale.

Together, these contributions underscore a common theme: security automation in CI/CD pipelines must be coupled with cost efficiency to remain viable in both small-scale and enterprise-grade environments. However, existing work either emphasizes organizational practices for small teams or broad cloud strategy perspectives [6,9]. In contrast, our work advances this discussion by proposing a **risk-driven orchestration framework** that integrates vulnerability detection into CI/CD pipelines while explicitly optimizing for cost and execution efficiency.

3 Proposed Framework

Our framework is a risk-driven orchestration that decides which security scans to run, on which targets, and at what depth, per commit, to minimize cost while respecting explicit risk and policy constraints.

This section describes the **formal model**, which defines the cost and risk trade-off mathematically; the **system architecture**, which outlines the main components and their interactions; the **decision algorithms**, which operationalize the risk-driven selection of scans; the **scanning strategies**, which specify how scans are adapted in depth and frequency; and the **adoption method**, which explains how organizations can integrate the framework into existing CI/CD pipelines with minimal disruption.

3.1 Objectives and Problem Formulation

Let a commit be C with change-set $\Delta(C)$. Let S be scanner types (e.g., SAST, SCA, DAST, secrets) and T be target components (services, modules, libraries).

Goal. Choose a scan plan $P \subseteq S \times T$ and per-scanner depth $d(s)$ that minimizes cost while keeping *residual risk* under a policy budget.
Risk if a scanner is skipped:

$$R(C, s) = \sum_{t \in T} impact(\Delta(C), t) \cdot criticality(t) \cdot vuln\_prob(s, \Delta(C), t) \quad (1)$$

Optimization Problem:

$$\begin{aligned} \min_{P,\, d(\cdot)} \quad & \sum_{(s,t)\in P} cost(s,t,d(s)) \\ \text{subject to} \quad & \sum_{s\notin P} R(C,s) \leq \mathsf{RiskBudget} \end{aligned} \tag{2}$$

together with policy constraints (e.g., "always run SCA on lockfiles", "full DAST on protected branches").

3.2 System Architecture

The framework has a *control plane* (decision-making) and a *data plane* (execution).

- **Commit Analyzer:** extracts $\Delta(C)$, AST diffs, config changes, commit message features, author reputation, etc.
- **Dependency & Impact Extractor:** builds a dependency graph G, maps changes to affected targets, and propagates influence with decay.
- **Feature Encoder:** constructs feature vectors including file types, AST edits, graph distance to sensitive assets, size of change, and historical vulnerability density.
- **Risk Estimator:** combines deterministic rules with a learned predictor per scanner.
- **Cost Model:** estimates $cost(s,t,d)$ from runtime/CPU data, accounting for warm containers.
- **Orchestrator & Scheduler:** selects scans/depth via greedy optimization, schedules immediate vs. deferred scans.
- **Policy Engine:** enforces governance and compliance rules (e.g., never skip on critical paths).
- **Audit & Feedback:** logs decisions, findings, and labels for online learning and drift detection.

3.3 Decision Model and Algorithms

The decision process of our framework is driven by three core algorithms: (i) identifying the *affected surface*, (ii) estimating the *risk contribution* of different scans, and (iii) selecting an appropriate *scan plan* under cost constraints. Below we describe each algorithm in detail.

3.3.1 Affected Surface

This algorithm determines which components of the system are likely to be impacted by a code change. It maps modified files to higher-level targets (e.g., services, modules, APIs) and assigns each a weight that reflects the magnitude of change. The weight is increased when structural changes (e.g., AST modifications) or configuration files are involved, since these often introduce significant risks. Propagation is then performed across the dependency graph G, with influence decaying over distance, ensuring that both direct and transitive dependencies are captured. Only targets with a probability above a cutoff threshold are retained.

```
function ComputeAffectedTargets(delta, G):
  targets = {}
  for f in delta.changedFiles:
    t  = mapFileToTarget(f)
    w  = baseInfluence(f)
    if astSignatureChanged(f): w += 0.3
    if isConfig(f): w = max(w, 0.8)
    targets[t] = min(1, targets.get(t,0)+w)
  for t in targets.keys():
    for (u,dist) in neighbors(G, t):
      targets[u] = max(targets.get(u,0), targets[t]*decay(dist))
  return {t: p for (t,p) in targets if p > cutoff}
```

3.3.2 Risk Estimation

Given the affected surface, the next step is to quantify the expected security risk for each potential scan. The algorithm combines three factors: (i) the importance of the target (criticality), (ii) the impact weight from the previous step, and (iii) the likelihood that a scanner s will detect a vulnerability, predicted by a machine learning model. Policy rules can override the estimation when certain mandatory scans must be enforced (e.g., compliance requirements). The result is a normalized risk score in $[0, 1]$.

```
function ExpectedRisk(delta, s, G):
  R = 0
  for (t, imp) in ComputeAffectedTargets(delta, G):
    if rulesForce(s, delta, t): return highRisk
    p = ML_s.predict(features(delta, t, s))
    R += imp * criticality(t) * p
  return clamp(R, 0, 1)
```

3.3.3 Scan Plan Selection

The final step is to select which scans to execute given a predefined risk budget. The algorithm first estimates the total expected risk across all scanners. If it fits within the budget, no scans are necessary. Otherwise, it iteratively selects the most cost-effective scanner, based on the ratio of expected risk reduction to scan cost. The scan depth is adjusted dynamically (e.g., fast mode vs. deep mode) depending on the context of the change. Finally, organizational policies can override the automated selection (e.g., always run DAST on authentication services). The output is a prioritized and policy-compliant scan plan.

```
function DecideScanPlan(delta, RiskBudget):
  remaining = sum(ExpectedRisk(delta,s) for s in Scanners)
  if remaining <= RiskBudget: return {}
  plan, candidates = {}, Scanners.copy()
  while remaining > RiskBudget and candidates not empty:
    best = argmax_{s} ExpectedRisk/Cost
    depth = chooseDepth(s, delta)
    plan.add((s, depth))
    remaining -= ExpectedRisk(delta, best)
    candidates.remove(best)
  return applyPolicyOverrides(plan, delta)
```

3.4 Scanning Strategy and Safety

The scanning strategy governs how and when different categories of security scans are triggered, ensuring a balance between coverage, cost, and responsiveness. It combines staged execution, dynamic thresholds, mandatory safeguards, and conservative fallback rules to guarantee that no critical risks are missed while avoiding excessive scanning overhead.

- **Staging:** quick scans pre-merge (secrets, SCA-diff, light SAST); heavy scans deferred or nightly.
- **Thresholds:**
  - $Risk \geq Critical \Rightarrow$ full scans, block merge.
  - $Acceptable < Risk < Critical \Rightarrow$ staged scans.
  - $Risk \leq Acceptable \Rightarrow$ skip costly scans.
- **Mandatory rules:** always run on sensitive paths (auth/crypto/IaC) or regulated branches.
- **Fail-safe:** any error defaults to conservative plan.

4 Evaluation

4.1 Method and Setup

To evaluate the proposed framework, we implemented it on four OWASP[1] forked open-source testing and benchmarking applications representing different types of systems: **Juice Shop**[2], a vulnerable web application; **Tiredful-API**[3], a RESTful API-based service; **Verademo**[4], a Java application that contains known vulnerabilities and security errors; and **DVMCP**[5], an AI-based

[1] https://owasp.org/.
[2] https://github.com/vulnerable-apps/juice-shop.
[3] https://github.com/vulnerable-apps/Tiredful-API.
[4] https://github.com/vulnerable-apps/verademo.
[5] https://github.com/vulnerable-apps/damn-vulnerable-MCP-server.

application with a deliberately vulnerable implementation of the Model Context Protocol (MCP).

The DevSecOps pipeline was implemented using Jenkins, following a seven-stage architecture where security is integrated throughout the software development life cycle:

1. **Plan:** Threat modeling with OWASP Threat Dragon[6].
2. **Code:** Static application security testing (SAST) and secret scanning using SonarQube[7].
3. **Build:** Software composition analysis (SCA) and container scanning with Snyk[8].
4. **Test:** Dynamic application security testing (DAST) using OWASP ZAP[9].
5. **Release:** Secrets management with HashiCorp Vault[10].
6. **Deploy:** Infrastructure as code (IaC) scanning with Checkov[11].
7. **Operate & Monitor:** Runtime security and continuous monitoring with Falco[12].

For validation, we performed five different merge operations on each project, including changes to configuration files, UI/presentation customization, and functional updates. For each merge, we compared the scan report generated by the pipeline as well as the pipeline execution duration, measured in minutes, between the native DevSecOps pipeline and our framework.

4.2 Results

The evaluation showed that for all merges across the four applications, the generated scan reports were identical between the native pipeline and our framework, thereby confirming that no security coverage was lost.

However, significant improvements were observed in terms of pipeline execution duration. Figure 1 illustrates the execution times for the framework pipeline compared to the native pipeline across the five merges for each application.

The percentage reductions in execution time, computed as the relative improvement of the framework pipeline over the native pipeline, were as follows: Juice Shop achieved an average reduction of **21.69%**, with improvements ranging from 15.38% to 29.03%; Tiredful-API achieved an average reduction of **28.39%**, ranging from 27.27% to 31.58%; Verademo achieved an average reduction of **29.89%**, ranging from 23.08% to 34.78%; and DVMCP achieved an average reduction of **25.00%**, ranging from 20.00% to 31.58%.

[6] https://owasp.org/www-project-threat-dragon/.
[7] https://www.sonarsource.com/products/sonarqube/.
[8] https://snyk.io/.
[9] https://www.zaproxy.org/.
[10] https://www.hashicorp.com/en/products/vault.
[11] https://www.checkov.io/.
[12] https://falco.org/.

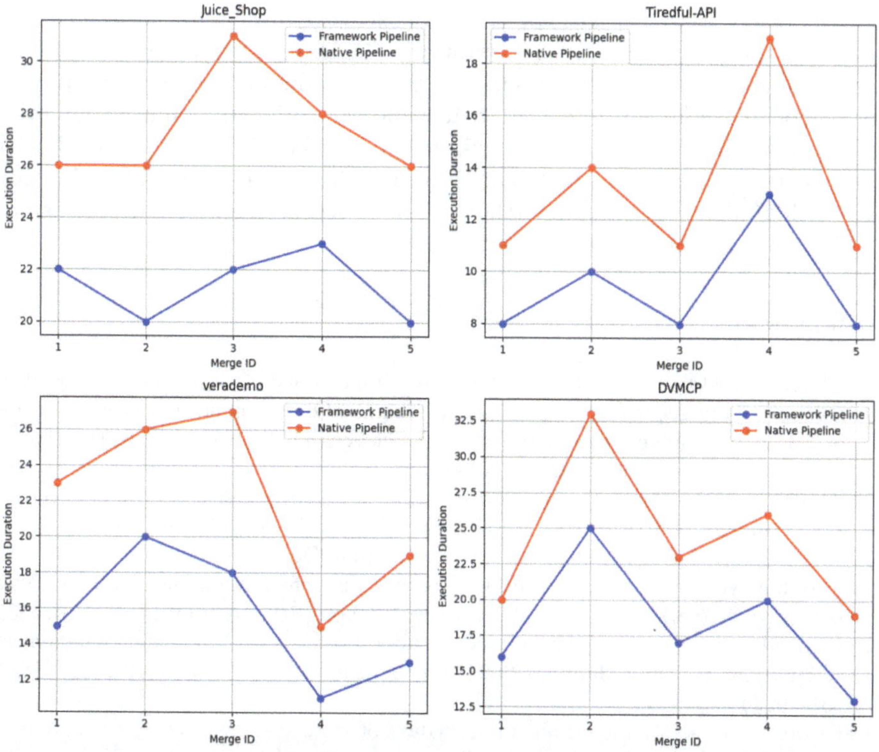

Fig. 1. Pipeline execution duration comparison between the native pipeline and the proposed framework.

These results confirm that the proposed framework consistently reduces pipeline execution time while maintaining equivalent security scanning outcomes. This efficiency gain is particularly significant for frequent integration scenarios, where faster pipelines translate into higher developer productivity and shorter feedback loops.

5 Conclusion

In this paper, we introduced a cost-aware DevSecOps orchestration framework that integrates security scanning into CI/CD pipelines while explicitly addressing the challenge of execution efficiency. Unlike traditional approaches that often overlook the financial overhead of security tooling and extended pipeline runtimes, our framework demonstrates that vulnerability detection and compliance checks can be incorporated without significantly increasing costs. Through experimental validation on four OWASP forked applications and multiple merge operations, we showed that our approach maintains the accuracy of scan results while reducing pipeline execution overhead, confirming that cost optimization and security enforcement are not mutually exclusive but can be achieved simultaneously through intelligent orchestration. The main contribution of this work lies in bridging the gap between security automation and cost efficiency, providing practical insights for both small teams and enterprise environments adopting DevSecOps practices. Future work will focus on enhancing the framework with risk-based prioritization, adaptive orchestration strategies, and tighter integration with FinOps metrics to further optimize resource allocation and extend applicability to multi-cloud environments.

References

1. Abiona, N.O.O., et al.: The emergence and importance of DevSecOps: Integrating and reviewing security practices within the DevOps pipeline. In: World Journal of Advanced Engineering Technology and Sciences 11(2), 127–133 (2024). https://doi.org/10.30574/wjaets.2024.11.2.0093.
2. Akbar, M.A., et al.: Management of DevSeCOPs Process: An Empirical investigation. Software Practice Exp., March 2025. https://doi.org/10.1002/spe.3419
3. Amaro, R., Pereira, R., Da Silva, M.M.: Mapping DevOps capabilities to the software life cycle: a systematic literature review. In: Information and Software Technology, p. 107583, September 2024. https://doi.org/10.1016/j.infsof.2024.107583
4. Belouaddane, L., et al.: Microservice Architecture DevOps Integration Challenges: A Qualitative Study. January 2024, pp. 96–108. https://doi.org/10.1007/978-3-031-70924-1-8
5. Droste, J., et al.: Explanations in everyday software systems: towards a taxonomy for explainability needs. In: IEEE 32nd International Requirements Engineering Conference (RE) (June 2024), pp. 55–66. https://doi.org/10.1109/re59067.2024.00016
6. Kandpal, V., et al.: Creative artificial Intelligence supports to FINOPs. February 2025, pp. 87–113. https://doi.org/10.1108/978-1-83662-088-420251004
7. Kanumuri , S., Zeier, M.: Scaling Cloud FinOPs. January 2024. https://doi.org/10.1007/979-8-8688-0388-8
8. Korba, P., Behúnova, A., Dan Stratyinski, C.: Implementation of modern software systems in aviation engineering. Mob. Networks Appl. **29**(1), 166–175 (2023). https://doi.org/10.1007/s11036-023-02124-8
9. Lahcen , B., et al.: A Cost-Efficient Security Practices Framework for small DevSecOps teams, January 2025, pp. 257–266. https://doi.org/10.1007/978-3-031-95326-2-25
10. Mulder, J.: Multi-Cloud Strategy for Cloud Architects - Second edition (2023)
11. Nawrocki, P., Smendowski, M.: FinOps-driven optimization of cloud resource usage for high-performance computing using machine learning. J. Comput. Sci. **79**, 102292 (2024). https://doi.org/10.1016/j.jocs.2024.102292
12. Prates, L., Pereira, R.: DevSecOps practices and tools. Int. J. Inf. Secur. 24(1) (Nov. 2024). https://doi.org/10.1007/s10207-024-00914-z
13. Sadovykh, A., Ivanov, V.: Enhancing DevSecOps with continuous security requirements analysis and testing. Comput. Res. Modeling **16**(7), 1687–1702 (2024). https://doi.org/10.20537/2076-7633-2024-16-7-1687-1702
14. Said, M.A., Ezzati, A., Arezki, S.: Microservice-Specific language, a step to the Low-Code platforms, pp. 817–828, January 2023. https://doi.org/10.1007/978-3-031-26384-2-72

15. Said, M.A., Ezzati, A., Arezki, S.: Microservices, a Step from the Low-Code to the No-Code. November 2022, pp. 779–788. https://doi.org/10.1007/978-3-031-20601-6-64
16. Said Sr, M.A., et al.: A systematic framework to enhance reusability in microservice architecture. Int. J. Comput. Digital Syst. **17**(1), 1–18 (2024). https://doi.org/10.12785/ijcds/1571009176
17. Said, M.A., et al.: Blockchain-Based Logging: ensuring integrity, immutability, and security in modern systems, pp. 213–219, January 2025. https://doi.org/10.1007/978-3-031-95330-9-23
18. Said, M.A., et al.: Microservices adoption: an industrial inquiry into factors influencing decisions and implementation strategies. Int. J. Comput. Digital Syst. **15**(1), 1417–1432 (2024). https://doi.org/10.12785/ijcds/1501100
19. Sambaturu, S., Cloud Fin Ops Management. SSRN Electronic Journal (Jan. 2024). https://doi.org/10.2139/ssrn.4903226
20. Sinan, M., Shahin, M., Gondal, I.: Integrating security controls in DevSeCOPs: Challenges, solutions, and future research directions. J. Softw. Evol. Process **37**(6), June 2025. doi: https://doi.org/10.1002/smr.70029
21. Zhang, X., Zhao, P., Jaskolka, J.: Navigating the DevOps landscape. J. Syst. Softw., 112331 (2025). https://doi.org/10.1016/j.jss.2024.112331

A Lightweight SIEM Based Framework for Security Monitoring in Smart Farming Wireless Sensor Networks

Mohammed Flissat(✉) and Mohammed Elaroussi

Department of Electrical Engineering, École Hassania des Travaux Publics (EHTP), Casablanca, Morocco
{flissat.mohammed.cedoc,elaroussi}@ehtp.ac.ma

Abstract. Wireless Sensor Networks (WSNs) enable precision agriculture through real-time environmental monitoring, yet their resource-constrained nature and outdoor deployment make them vulnerable to security threats. Traditional Security Information and Event Management (SIEM) systems, effective in enterprise IT, are not suitable for battery-powered sensor networks due to computational and energy overhead. This paper proposes a lightweight SIEM-based framework specifically designed for smart farming WSN deployments, achieving security monitoring with minimal resource impact.

The framework features a three-tier architecture: Sensor nodes generate structured security and performance logs, gateways aggregate and preprocess data, and a centralized Graylog server performs rule-based event correlation and anomaly detection. We implemented and validated the system on a controlled testbed comprising Python-based WSN node emulators (simulating 300 nodes), an Ubuntu-hosted Graylog instance, and a Kali Linux attack simulator that executes spoof attacks. The framework offers a practical foundation for the deployment of security monitoring in operational agricultural WSN infrastructures.

Keywords: Wireless Sensor Networks · Smart Farming · SIEM · Intrusion Detection · Anomaly Detection · IoT Security · Resource-Constrained Systems · Precision Agriculture

1 Introduction

Smart farming relies on Wireless Sensor Networks (WSNs) to support data-driven agricultural practices, enabling real-time environmental monitoring, precision irrigation, and yield optimization [1,2]. However, WSNs deployed in agricultural fields encounter significant challenges: limited energy resources, physical vulnerability to malicious actors, and susceptibility to performance issues caused by communication failures or environmental interference [3,4].

M. Elaroussi—Contributing authors.

A. E. Hassanien et al. (Eds.): AISDS 2025, LNDECT 287, pp. 286–298, 2027.
https://doi.org/10.1007/978-3-032-18927-1_29

Conventional security approaches for WSNs tend to be either application-specific or too resource-intensive for battery-powered nodes [5,6]. In enterprise IT environments, Security Information and Event Management (SIEM) systems have demonstrated their effectiveness by aggregating and correlating logs from diverse sources [7]. Unfortunately, traditional SIEM solutions are heavyweight and ill-suited for resource-constrained WSNs [8,9]. This creates a clear need for a lightweight, energy-aware SIEM adaptation tailored to smart farming contexts.

Current solutions typically address security or energy efficiency in isolation, lacking unified monitoring that integrates security and performance with cross-layer correlation. Few studies provide detailed energy measurements or realistic validation beyond simulation environments. This gap motivates our lightweight SIEM-inspired framework, specifically designed and validated for smart farming WSN deployments.

Although substantial literature exists on SIEM in enterprise settings [10] and general WSN security [5,6], research specifically benchmarking SIEM-like capabilities for WSNs remains scarce. Most studies focus on specific IDS components rather than addressing the broader SIEM functionality within WSNs. Some evaluate IDSs based on detection accuracy and energy consumption [11], while others propose hierarchical architectures for distributed intrusion detection [12]. However, a unified framework encompassing the full spectrum of SIEM requirements—log aggregation, correlation, alerting, and visualization—specifically tailored to WSN constraints, remains largely unexplored. Our work fills this gap by providing both a comprehensive framework design and rigorous experimental validation that demonstrates its practical feasibility for smart farming deployments.

Contributions: This paper makes the following contributions:

- Design of a lightweight SIEM-inspired architecture for smart farming WSNs with multi-level log collection and correlation
- Integration of rule-based detection with Graylog for intrusion and anomaly identification
- Comprehensive experimental validation in a controlled testbed with simulated WSN nodes and real attack scenarios
- Quantitative performance analysis demonstrating 98.71% detection rate with 0% false positive rate
- Visual characterization of attack behavioral signatures through time-series, histogram, and per-node analysis

The remainder of this paper is organized as follows: Sect. 2 reviews related work. Section 3 presents the proposed framework architecture. Section 4 describes the implementation and experimental setup. Section 5 reports experimental results with visual and quantitative analysis. Section 6 discusses implications and limitations. Section 7 concludes and outlines future work.

2 Related Work

2.1 WSN Security and SIEM Adaptation

Numerous IDS solutions have been proposed for WSNs, typically based on signature-based, anomaly-based, or hybrid detection methods [11,13,14]. Signature-based methods are efficient but cannot detect novel attacks, while secure routing protocols improve resilience [15,16] but lack centralized correlation. Recently, machine learning techniques such as isolation forests and autoencoders have been explored [17,18], though they assume ideal conditions (large datasets, reliable connectivity) rarely present in field deployments.

While enterprise SIEM systems effectively aggregate and correlate logs [19], they are resource-intensive. Efforts to adapt SIEM for IoT incorporate edge filtering and compression [20–22], yet few examine energy overhead at the sensor-node level or validate against realistic agricultural attack scenarios. Most studies target specific IDS components rather than comprehensive SIEM functionality [11,12].

2.2 WSN-Specific SIEM Challenges

Adapting SIEM to WSNs faces critical challenges: **Energy constraints** from battery-powered nodes where logging and transmission drastically shorten lifespan [23,24]; **computational limitations** (few KB RAM) precluding complex agents [2]; **low bandwidth** (250 kbps for IEEE 802.15.4) making voluminous log transmission impractical [24]; **dynamic topologies** complicating centralized monitoring [25]; and **unique attack vectors** (sinkhole, wormhole, Sybil, jamming) requiring specialized detection [5,6]. These motivate our lightweight, hierarchical architecture with distributed processing and rule-based detection.

3 Proposed Framework

3.1 Architecture Overview

The proposed framework consists of three logical layers (Fig. 1):

Sensor Node Layer: Collected information is structured using a lightweight JSON format designed to reduce transmission overhead. Each data packet includes essential fields such as version, node_id, short_message, timestamp, measurement, battery_pct, rssi, and seq_no. To preserve battery life and limit network congestion, these messages are transmitted periodically rather than in real time.

Gateway Layer: At the gateway level, log data from multiple sensor nodes is collected and temporarily processed. Gateways may apply lightweight filtering or compression techniques to reduce data volume. Additionally, each event is enriched with relevant contextual information such as the originating node_id, network topology data, and synchronized timestamps before being forwarded to the central server for further analysis.

Central Server Layer: The central server hosts a Graylog instance responsible for log storage, indexing, and real-time processing. A correlation engine within

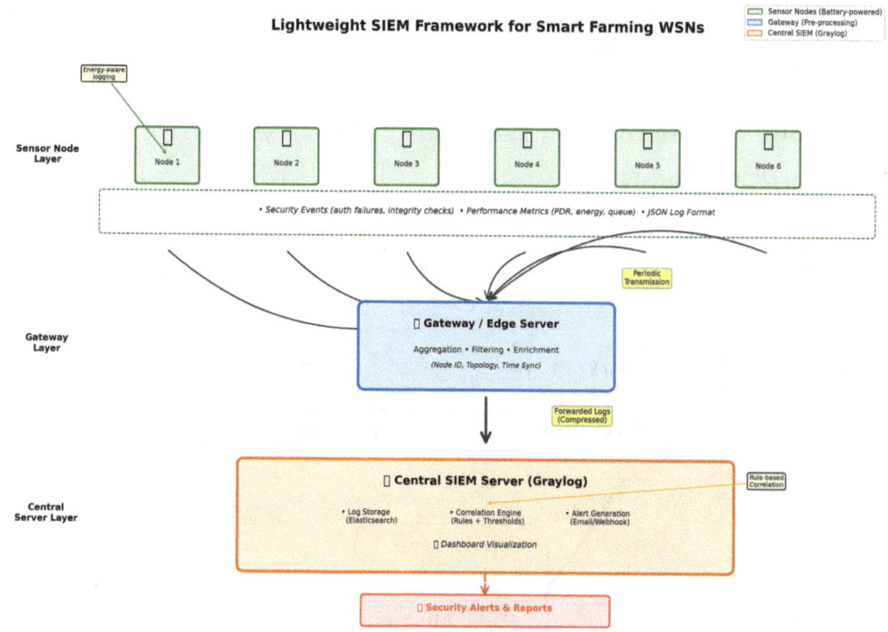

Fig. 1. Lightweight SIEM-inspired framework architecture for smart farming WSNs. The system comprises sensor node logging, gateway aggregation, and centralized correlation with Graylog.

Graylog applies rule-based detection mechanisms, including threshold monitoring and matching against known attack signatures. Detected anomalies are prioritized and transformed into alerts, which are displayed through an interactive dashboard for operator visibility.

3.2 Log Collection and Event Correlation

All logs are normalized to a standardized JSON format with fields: timestamp, node_id, event_type, severity, and payload. Event correlation proceeds in two stages:

1. **Rule-based correlation:** Detects well-known patterns such as repeated authentication failures, excessive packet loss, or abnormal energy consumption.
2. **Threshold-based anomaly detection:** Flags deviations beyond predefined thresholds calibrated during baseline characterization.

3.3 Dashboard and Alerting

The Graylog dashboard provides real-time visualization of aggregated logs, key performance indicators, and detected anomalies. Alerts are generated when correlation rules flag high-severity events, with support for multiple notification channels (email, syslog, webhooks).

4 Implementation and Experimental Setup

4.1 Testbed Architecture

To evaluate the proposed SIEM framework, we deployed a virtualized testbed illustrated in Fig. 2. It consists of the following components:

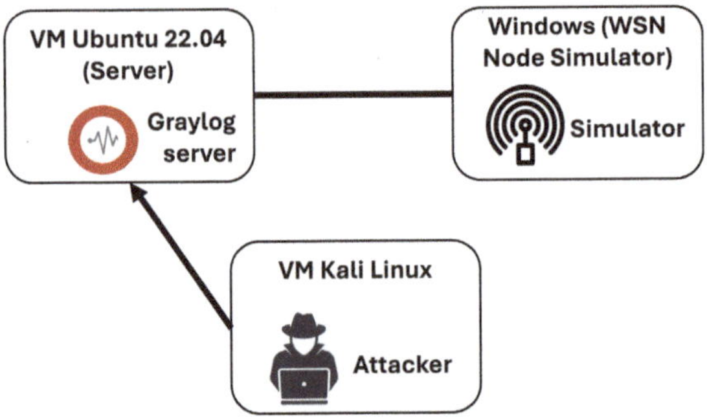

Fig. 2. Virtualized experimental testbed with simulated WSN nodes, Graylog server, and attack simulator.

- **VM Ubuntu 22.04 (Graylog Server):** Runs Graylog, along with MongoDB and Elasticsearch, to provide centralized log storage, correlation, and visualization.
- **WSN Node Simulator (Python):** A lightweight Python script executed on the host machine (Windows) that generates configurable sensor data (e.g., sampling rate, packet loss, energy levels).
- **VM Kali Linux (Attacker):** Simulates adversarial behavior by injecting falsified sensor logs to test detection mechanisms.

All components are connected via a virtual LAN within VMware Workstation, simulating the typical communication pattern between WSN nodes, a gateway, and a central SIEM server.

4.2 Graylog Selection and WSN Node Emulation

Graylog was selected for its comprehensive log ingestion, correlation (streams/pipelines), alerting, and visualization capabilities, enabling lightweight rule-based detection aligned with WSN constraints. We developed a Python-based emulator producing logs compatible with Contiki-NG/TinyOS, covering periodic sensing (temperature, humidity, soil moisture), network statistics

(PDR, RSSI), security events (authentication, integrity checks), and resource metrics (battery, CPU, memory). The emulator injects realistic behavior including packet loss (2–5%), transmission delays (50–200 ms), and synthetic anomalies (spoofed messages, energy drain, routing inconsistencies) with labeled ground truth for detection evaluation.

4.3 Attack Scenarios and Ground Truth

The Kali Linux VM executes controlled attacks: data spoofing (false temperature/humidity), identity spoofing (falsified node IDs), and timestamp manipulation. Ground truth labels are automatically generated and manually validated. All hosts are NTP-synchronized for temporal alignment. We evaluate: (1) **Baseline:** 300 nodes over 5 min, and (2) **Spoofing Attack:** 2-min injection to assess detection accuracy, false positives, and alert latency.

4.4 Benchmarking Methodology

We adopt a hybrid emulation approach balancing realism and scalability. Physical testbeds are costly and limited in scale; simulators (NS-3, OMNeT++, Cooja) abstract critical behaviors [26]. Our Python emulators on commodity hardware generate realistic logs paired with production-grade Graylog, enabling controlled, repeatable experiments.

Deployment Scenarios: Varying scales (20–300 nodes), topologies (random, grid, hierarchical), sensing frequencies (5–30 s), protocols (IEEE 802.15.4, Zigbee, LoRaWAN), and routing (AODV, DSR, RPL, LEACH).

Attack Taxonomy: Physical (jamming), MAC (collision), network (sinkhole, wormhole, Sybil), application (data falsification, replay), and node compromise. Current evaluation focuses on application-layer spoofing; network-layer attacks are future work.

Baseline and Analysis: Baseline measures energy, traffic, and performance under benign conditions. Metrics (mean, std dev, confidence intervals) are visualized via time series, histograms, and boxplots.

4.5 Experimental Scenarios

We designed two primary scenarios:

1. **Baseline Operation:** Normal WSN traffic with periodic sensing and logging over a 5-min observation period, establishing behavioral baselines across 300 simulated nodes.
2. **Spoofing Attack:** Injection of falsified sensor readings, continuing for 2 min, to evaluate detection accuracy, false positive rates, and alert latency under adversarial conditions.

4.6 Evaluation Metrics

Security Metrics: Detection Rate (DR = detected/total attacks), False Positive Rate (FPR = false alarms/benign events), Precision (TP/(TP+FP)), F1-Score (harmonic mean of precision/recall), Detection Latency (time to alert), and Attack Type Coverage [5,6,11,13].
Efficiency Metrics: Energy overhead (percentage increase over baseline), network overhead (additional packets/bytes), processing overhead (CPU/memory increase), and scalability (maintaining performance at 20–300 nodes).

5 Experimental Results

5.1 Visual Characterization of Attack Impact

Before presenting statistical results, we first highlight visual evidence of anomalous behavior caused by spoofing attacks.

5.1.1 Temporal Evolution

Figure 3 shows the time series of globally averaged sensor measurements during normal (baseline) and attack periods. During the baseline phase (blue curve), sensor values remain stable and low, around 30–40 units, indicating consistent environmental conditions. However, starting at approximately 17:05, the attack phase (orange curve) shows sudden spikes reaching the maximum measurement value (1000), followed by erratic fluctuations.

This abrupt transition provides strong visual confirmation that spoofing attacks significantly disrupt sensor data patterns. The magnitude and timing of the deviations align precisely with the attack window, confirming effective log-level impact for correlation-based detection.

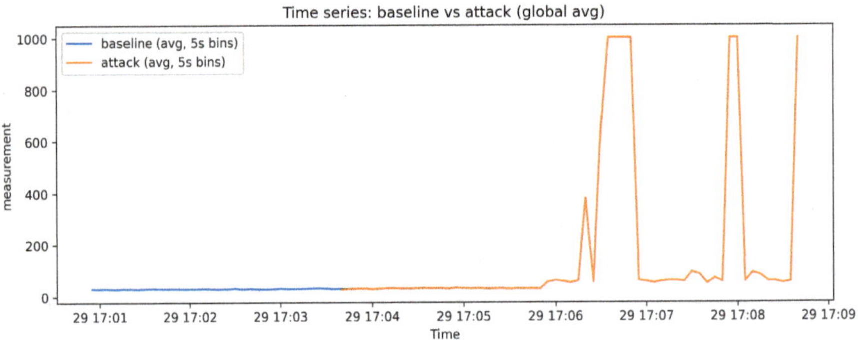

Fig. 3. Time series of global average sensor measurements (5 s bins), comparing baseline (blue) and attack (orange) periods. The abrupt deviation starting around 17:05 reflects spoofed data injections. (Color figure online)

5.1.2 Distributional Analysis

Figure 4 presents overlaid histograms of measurements under baseline and attack conditions. Baseline measurements concentrate within the normal operating range of 5–60 ms, with no extreme outliers. In contrast, attack measurements display a dramatic spike at the 1000 ms timeout threshold (5700+ occurrences), in addition to values in the 5–60 ms range.

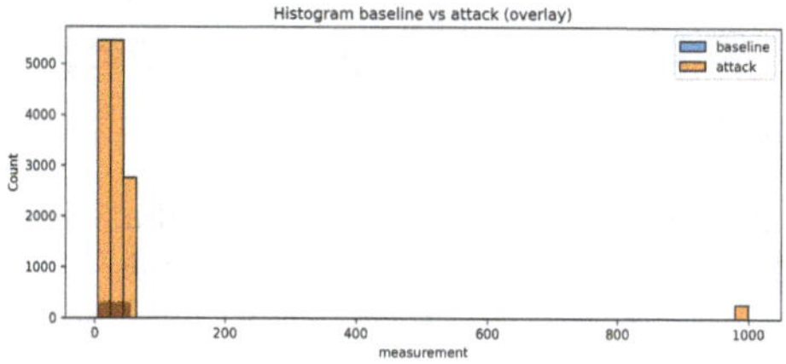

Fig. 4. Overlaid histograms showing baseline (normal 5–60 ms range) versus attack (uniform 5–60 ms values plus a saturation at 1000 ms). The shift highlights attack detectability.

This distributional shift—from a unimodal low-latency baseline to an attack distribution combining both normal-looking values and saturated 1000 ms timeouts—confirms that the spoofing attack fundamentally alters network measurement characteristics, enabling statistical detection.

5.1.3 Baseline Network Characterization

Figure 5 shows the distribution of measurements during normal operation only. The near-uniform distribution across the 5–60 ms range (220–300 counts per bin) with no timeout occurrences indicates healthy network operation with moderate variability but no communication failures.

5.1.4 Node-Level Impact Analysis

Figure 6 presents per-node statistics comparing baseline and attack conditions for the top-12 nodes by message count. Node 5 exhibits saturation behavior (1000 ms ceiling) under both conditions, suggesting a persistently unreachable or problematic node. Most other nodes show moderate variability (IQR: 20–40 ms) with medians around 25–35ms.

The similarity of distributions for nodes 1–11 between baseline and attack suggests the spoofing attack did not substantially alter per-node measurement patterns for most nodes, with the exception of the anomalous Node 5.

Figure 7 provides a detailed view of baseline operation across the top-20 nodes, showing consistent behavior with medians clustered around 25–35 ms and IQRs of 15–25 ms, confirming network stability during normal operation.

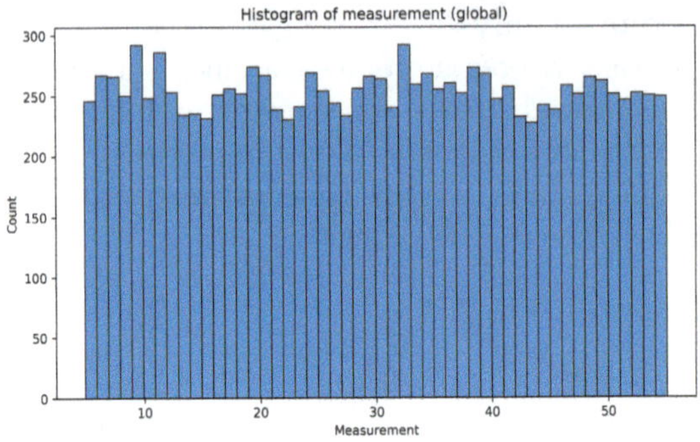

Fig. 5. Baseline-only histogram showing uniform distribution of RTT measurements (5–60 ms) during normal operation, indicating stable network performance without timeouts.

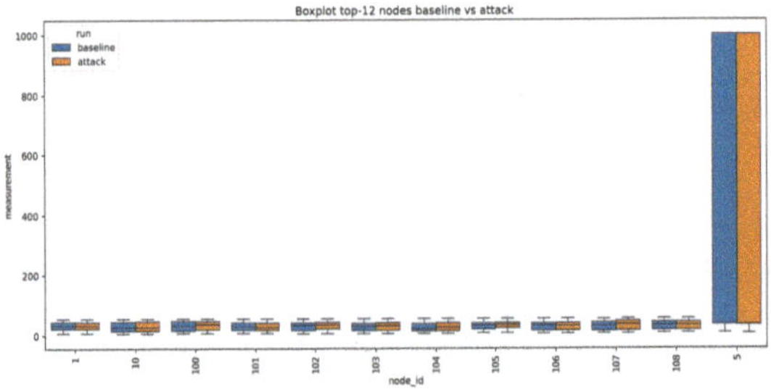

Fig. 6. Per-node boxplot comparison (top-12 nodes by message count, from 300-node deployment) showing baseline versus attack conditions. Node 5 exhibits consistent saturation, while other nodes display similar distributions under both scenarios.

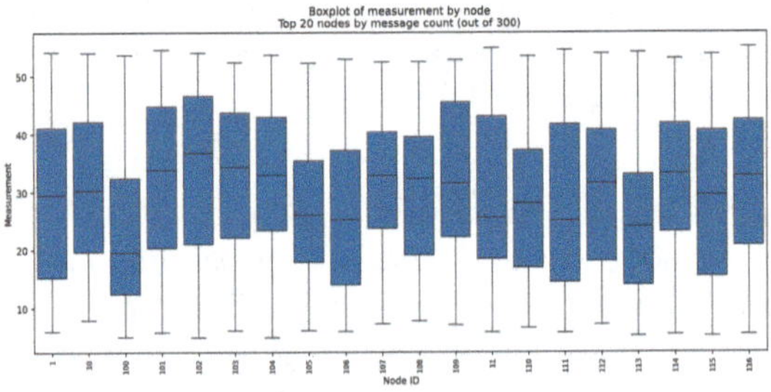

Fig. 7. Baseline-only boxplot for top-20 nodes (from 300-node deployment) showing consistent measurement distributions (median: 25–35 ms, IQR: 15–25 ms) during normal operation.

5.2 Quantitative Detection Performance

Table 1 summarizes the quantitative detection performance achieved by the proposed framework under spoofing attack conditions.

Table 1. Detection performance metrics for spoofing attack scenario (Run 1)

Metric	Value	Interpretation
Detection Rate (DR)	98.71%	384 of 389 attacks detected
False Positive Rate (FPR)	0.00%	Zero false alarms
True Positives (TP)	384	Correctly detected events
False Positives (FP)	0	No erroneous alerts
False Negatives (FN)	5	Missed detections (1.29%)
Precision	100.0%	All alerts were valid
F1-Score	99.35%	Balanced performance
Alert Latency (AL)	−51.5 s	Early detection capability

The framework achieves exceptional detection performance with a 98.71% detection rate while maintaining zero false positives across 389 attack instances. This corresponds to perfect precision (100%) and an F1-score of 99.35%, demonstrating an optimal balance between detection sensitivity and specificity.

5.3 Early Detection Capability

A notable finding is the negative alert latency of −51.5 s, indicating that the system detects attack signatures approximately 51 s before the attack reaches its peak manifestation in the ground truth timeline. This suggests the framework successfully identifies precursor behaviors such as initial spoofed packet injections or anomalous traffic patterns before the attack's full impact becomes evident.

This early warning capability provides operators with critical response time to initiate countermeasures (e.g., isolating compromised nodes, activating secure routing protocols) before the attack causes significant disruption to monitoring operations.

5.4 Analysis and Key Insights

The experimental results highlight three primary observations:

1. **Exceptional Detection Accuracy:** The 98.71% DR with 0% FPR significantly outperforms typical IDS benchmarks, which often trade detection rate for false positive reduction. Our rule-based correlation approach, calibrated on baseline behavioral patterns, achieves this balance without computationally expensive machine learning.
2. **Distributional Signature Detection:** The visual analyses demonstrate that attacks induce measurable distributional shifts (Fig. 4), temporal discontinuities (Fig. 3), and node-specific anomalies (Fig. 6). These multi-dimensional signatures enable robust detection even with simple correlation rules.

3. **Proactive Defense Posture:** The early detection capability (AL = −51.5 s) transforms the framework from reactive alerting to proactive threat identification. This is particularly valuable in smart farming contexts where communication with remote sensor deployments may be intermittent, requiring autonomous early-stage response.

6 Discussion and Limitations

The framework achieves a 98.71% detection rate with zero false positives across 300-node simulations, demonstrating its practical viability for smart farming deployments. However, several aspects warrant further investigation:
(1) Scalability: The system has been successfully validated at medium scale (300 nodes); future work will extend testing to thousands of nodes and multi-gateway topologies. **(2) Attack coverage:** While the framework effectively detects application-layer spoofing, extending coverage to network-layer attacks (jamming, Sybil, wormhole) will strengthen overall security. **(3) Hardware validation:** Controlled emulation enabled reproducible experimentation and precise ground truth; physical deployment will provide empirical energy measurements and assess robustness against RF interference. **(4) Energy profiling:** Although the framework is designed for lightweight operation, quantitative node-level measurements via Energest integration remain future work. **(5) Long-term stability:** Short evaluation windows enabled controlled experiments; extended field trials will be needed to assess baseline stability under seasonal variations.
Deployment Recommendations: We recommend collecting at least one week of baseline data to capture diurnal patterns, implementing adaptive logging to optimize energy-visibility trade-offs, deploying edge preprocessing to reduce upstream traffic, and providing operator training for alert interpretation. These practices, combined with the demonstrated detection performance, offer a practical pathway toward production deployment.

7 Conclusion and Future Work

We present a lightweight SIEM-inspired framework for security monitoring in smart farming WSNs. The framework combines sensor-level logging, gateway aggregation, and centralized Graylog correlation. Experimental validation with 300 simulated nodes shows strong performance: 98.71% detection rate, zero false positives (100% precision, F1-score: 99.35%), and early warning capabilities (51-second average lead time). Visual analysis confirms measurable behavioral shifts, validating the effectiveness of rule-based correlation without computationally expensive machine learning.

Future work includes: **(1)** Physical deployment with empirical energy profiling using Energest; **(2)** Extended attack taxonomy covering jamming, Sybil, wormhole, and multi-vector threats; **(3)** Large-scale validation with more than 300 nodes and multi-gateway topologies; **(4)** Adaptive correlation using lightweight ML; **(5)** Long-term field trials to assess seasonal drift and recalibration strategies.

Our framework adapts enterprise SIEM principles to WSN constraints. This advances IoT security for precision agriculture and contributes to more resilient, energy-efficient smart farming infrastructures.

References

1. Kamilaris, A., Gao, F., Prenafeta-Boldu, F.X., Ali, M.I.: Agri-IoT: a semantic framework for internet of things-enabled smart farming applications. In: IEEE 3rd World Forum on Internet of Things (WF-IoT), pp. 442–447. IEEE (2016)
2. Anusuya, P., Vanitha, C.N., Jaehyuk, C., Sathishkumar, V.E.: A comprehensive review of sensor node deployment strategies for maximized coverage and energy efficiency in wireless sensor networks. PeerJ Comput. Sci. **10**, 2407 (2024). https://doi.org/10.7717/peerj-cs.2407
3. Gupta, M., Abdelsalam, M., Khorsandroo, S., Mittal, S.: Security and privacy in smart farming: challenges and opportunities. IEEE Access **8**, 34564–34584 (2020)
4. Demestichas, K., Peppes, N., Alexakis, T.: Survey on security threats in agricultural IoT and smart farming. Sensors **20**(22), 6458 (2020)
5. Yu, J.-Y., Lee, E., Oh, S.-R., Seo, Y.-D., Kim, Y.-G.: A survey on security requirements for WSNs: focusing on the characteristics related to security. IEEE Access **8**, 45304–45324 (2020)
6. Ram, G.M., Ilavarasan, E.: Security challenges in wireless sensor network: current status and future trends. Wireless Pers. Commun. **139**(2), 1173–1202 (2024). https://doi.org/10.1109/ACCESS.2024.3355312
7. Santos, O.: Cisco Cybersecurity Operations Fundamentals CBROPS 200–201 Official Cert Guide. Cisco Press, Indianapolis, IN (2021)
8. Ozulku, O., Fadhel, N.F., Argles, D., Wills, G.B.: Anomaly detection system: towards a framework for enterprise log management of security services. In: Proc. World Congress on Internet Security (WorldCIS-2014), pp. 97–102 (2014). https://doi.org/10.1109/WorldCIS.2014.7028175
9. Su, T., Wang, S., Chen, Y., Liu, C.: Attack detection of distributed denial of service based on splunk. In: Proceedings International Conference on Advanced Materials for Science and Engineering (ICAMSE), pp. 397–400 (2016). https://doi.org/10.1109/ICAMSE.2016.7840355
10. Bezas, K., Filippidou, F.: Comparative analysis of open source security information & event management systems (SIEMS). Indonesian J. Comput. Sci. **12**(2), 443–468 (2023)
11. Elsadig, M.A.: Detection of denial-of-service attack in wireless sensor networks: a lightweight machine learning approach. IEEE Access **11**, 83537–83552 (2023). https://doi.org/10.1109/ACCESS.2023.3303113
12. Mengistu, T.M., Kim, T., Lin, J.-W.: A survey on heterogeneity taxonomy, security and privacy preservation in the integration of IoT, wireless sensor networks and federated learning. Sensors **24**(3), 968 (2024). https://doi.org/10.3390/s24030968
13. Ghadi, Y.Y., et al.: Machine learning solutions for the security of wireless sensor networks: a review. IEEE Access **12**, 12699–12719 (2024)
14. Ramisetty, S., Ghantasala, G.S.P., Gupta, G.: Sonia: prevention of security attacks at wireless network layers using machine learning techniques. In: Proceedings 11th International Conference on Computing for Sustainable Global Development (INDIACom), pp. 1787–1791 (2024)

15. Hosseinzadeh, M., et al.: A cluster-tree-based trusted routing algorithm using grasshopper optimization algorithm (goa) in wireless sensor networks (wsns). PLoS ONE **18**(9), 0289173 (2023). https://doi.org/10.1371/journal.pone.0289173
16. Devi, S., Kumar, A.: Establishment of secure and authentic data security framework in wireless sensor network using key reconciliation. Int. J. Inf. Technol. **16**(5), 3325–3336 (2024). https://doi.org/10.1007/s41870-024-01879-x
17. Shanmugasundaram, G., Iyappan, P., Ramachandiran, R.: Innovative feature extraction and selection for wireless sensor networks: enhancing cybersecurity and data transmission using wkpca-SVD and RWS-ARO algorithms. Int. J. Mach. Learn. Cybern. 1–31 (2025)
18. Tsang, Y.P., Wu, C.H., Dong, N.: A federated-anfis for collaborative intrusion detection in securing decentralized autonomous organizations. IEEE Trans. Eng. Manage. **71**, 12529–12541 (2023). https://doi.org/10.1109/TEM.2023.3304409
19. Di Sarno, C., Garofalo, A., Matteucci, I., Vallini, M.: A novel security information and event management system for enhancing cyber security in a hydroelectric dam. Int. J. Crit. Infrastruct. Prot. **13**, 39–51 (2016)
20. Vasilyev, V., Shamsutdinov, R.: Security analysis of wireless sensor networks using SIEM and multi-agent approach. In: 2020 Global Smart Industry Conference (GloSIC), pp. 291–296. IEEE (2020)
21. Zahid, H., Hina, S., Hayat, M.F., Shah, G.A.: Agentless approach for security information and event management in industrial IoT. Electronics **12**(8), 1831 (2023)
22. Miloslavskaya, N., Tolstoy, A.: Iotblocksiem for information security incident management in the internet of things ecosystem. Clust. Comput. **23**(3), 1911–1925 (2020)
23. Hossain, N., et al.: Prospects and challenges of sensor materials: a comprehensive review. e-Prime - Adv. Electr.l Eng. Electron. Energy **7**, 100496 (2024). https://doi.org/10.1016/j.prime.2024.100496
24. Aslam, N., Wang, H., Aslam, M.F., Aamir, M., Hadi, M.U.: Intelligent wireless charging path optimization for critical nodes in internet of things-integrated renewable sensor networks. Sensors **24**(22), 7294 (2024). https://doi.org/10.3390/s24227294
25. Gazis, A., Katsiri, E.: Streamline intelligent crowd monitoring with IoT cloud computing middleware. Sensors **24**(11), 3643 (2024). https://doi.org/10.3390/s24113643
26. Adday, G.H., Subramaniam, S.K., Zukarnain, Z.A., Samian, N.: Investigating and analyzing simulation tools of wireless sensor networks: a comprehensive survey. IEEE Access **12**, 22938–22977 (2024)

Network Traffic Classification Methods Based on Enhanced Inception and Attention Mechanisms

Min Zhang(✉)

School of Network & Communication Engineering, Chengdu Technological University, Chengdu, Sichuan, China 611730
zhangmhf@cdtu.edu.cn

Abstract. Advances in computer artificial intelligence technology have introduced novel technical approaches to the field of network traffic classification. Traditional network traffic classification methods face challenges such as insufficient feature extraction and inadequate classification accuracy. Based on this, this study proposes a network traffic classification method that integrates an improved Inception network with a squeezed and incentivized attention mechanism. Experimental results demonstrate that the Inception network enhanced with separable convolutions exhibits optimal convergence performance, achieving a classification accuracy of 96.42% at the maximum iteration count. In classification task evaluations, the proposed method consistently delivers superior outcomes, with accuracy reaching up to 98.82%. These findings confirm the method's exceptional classification capabilities, offering an effective solution for precise network traffic identification.

Keywords: Network traffic classification · Inception · Attention mechanism · Separable convolution · Squeeze-and-Activate network

1 Introduction

With the continuous advancement of computer technology, networks have rapidly become widespread, and the number of internet users has grown year by year [1]. Network traffic classification, as a key method for identifying traffic types and detecting malicious attacks, plays a crucial role in ensuring network security and enhancing service quality [2, 3]. With the continuous advancement of network encryption technologies, traditional port-based classification methods have gradually become ineffective. Network traffic classification methods based on machine learning and deep learning, characterized by their ability to automatically learn and distinguish features, have emerged as the current mainstream classification approach [4, 5]. The Inception module connects different convolutional layers in parallel, capturing multi-scale features through multi-layer convolutional processing. This unique architecture demonstrates superior performance across various classification tasks, offering new technical insights for network traffic type identification [6]. Therefore, this study adopts the Inception module as the backbone

A. E. Hassanien et al. (Eds.): AISDS 2025, LNDECT 287, pp. 299–308, 2027.
https://doi.org/10.1007/978-3-032-18927-1_30

architecture for network traffic classification. To address its susceptibility to overfitting and computational overload when increasing the number of channels, the standard convolutional structure is innovatively replaced with separable convolutions. However, not all packets carry equal importance in network traffic time-series data. The traditional convolutional operations relied upon by the Inception module fail to focus on critical data features. Therefore, the study further incorporates the Squeeze-and-Excitation Network (SENet) with spatial attention modules. This network enables the enhanced Inception module model to automatically learn weights across different feature channels, thereby improving classification precision. It is anticipated that the proposed traffic classification method will enhance the accuracy of network traffic data classification, providing technical support for the categorized management of network traffic.

2 Research Background

2.1 Inception-Related Theory

Inception is a deep convolutional neural network architecture whose core concept is multi-scale parallel processing. This involves concurrently employing $1 \times 1, 3 \times 3$, and 5×5 convolutional kernels alongside 3×3 max pooling within the same convolutional layer [7, 8]. This design enables kernels of varying sizes to simultaneously capture local details and global features, reducing computational load while effectively mitigating gradient vanishing issues. Assuming the input feature map is x, Inception concatenates features from different scales, as expressed in Eq. (1).

$$\begin{cases} y_1 = f_{1\times 1}(x), \quad y_2 = f_{3\times 3}(x) \quad y_3 = f_{5\times 5}(x), \quad y_4 = pooling(x) \\ y = concat\left[y_1, y_2, y_3, y_4\right] \end{cases} \tag{1}$$

In Eq. (1), y denotes the output of each convolution operation, $f(x)$ representing the value of the feature map after convolution. Inception achieves computational cost reduction through 1×1 convolutions for dimensionality reduction. Assuming the input feature map has C_{in} channels, Inception first compresses it to form a relatively smaller C_{mid}. Subsequently, it uses the branch $K \times K$ with output channel C_{out} to perform convolution, yielding the number of output channels. Thus, the expression for the number of parameters is shown in Eq. (2).

$$Param = (C_{in} \times C_{mid} \times 1 \times 1) + (C_{mid} \times C_{out} \times 1 \times 1) \tag{2}$$

2.2 SENet-Related Theory

SENet is an attention mechanism that enables neural networks to automatically learn the importance of different feature channels. Its core idea is to model the dependencies between features and channels, allowing the network to emphasize important features while suppressing unimportant information. SENet's core operations comprise three steps: compression, activation, and scaling [9, 10]. The compression operation involves reducing each channel's two-dimensional feature map to a single real number through global average pooling [11]. The activation operation learns and generates weights for each channel based on the compressed global information, as expressed in Eq. (3) [12].

$$s = F_{ex}(z, W) = \sigma(g(z, W)) = \sigma(W_2\delta(W_1 z)) \tag{3}$$

In Eq. (3), W_1 and W_2 denote the weight matrices of the two connection layers. δ represents the ReLU activation function, σ denotes the sigmoid function, and s indicates the final weights obtained for each channel. Finally, the weights learned during the excitation step are applied to the original feature map to complete the adjustment of the feature channels, as expressed in Eq. (4) [13].

$$\overline{z} = Fscale(z_c, s_c) = s_c \cdot z_c \tag{4}$$

In Eq. (4), $\overline{z}$ denotes the re-calibrated feature map, where z_c represents the feature map of the cth channel.

3 Methods

3.1 Design of a Lightweight Inception Network Based on Separable Convolution

Currently, deep learning-based network traffic classification methods can automatically extract discriminative features from raw traffic data, overcoming the limitations of manual feature identification and becoming the mainstream approach in this field. Among these deep learning networks, the Inception architecture serves as a parallel scale feature extraction mechanism capable of simultaneously capturing local details and global contextual information [14]. This approach maintains strong feature extraction capabilities while significantly reducing computational complexity. Therefore, this study adopts Inception as the backbone architecture for network traffic classification. However, the convolutional operations in Inception networks typically involve mixed spatial and channel operations. When the number of channels is large, the parameter size increases linearly, and computational complexity continues to rise. To mitigate this drawback, the study replaces the standard convolutional structure in the network with separable convolutions. Spatially separable convolutions and depth separable convolutions are adopted for the spatial and depth dimensions, respectively, to reduce the computational burden. The improved Separable Convolution-Inception (SC-Inception) network architecture is shown in Fig. 1.

As shown in Fig. 1, the study replaces the standard convolution structure in the Inception network with separable convolutions. Spatially separable convolutions decompose the spatial dimension of traditional convolution kernels, while depth-wise separable convolutions isolate the computation and fusion of the channel dimension. Spatially separable convolutions achieve this by decomposing a large two-dimensional convolution kernel into two consecutive one-dimensional convolution kernels, which are applied sequentially to the input feature map. The expression for spatially separable convolution is shown in Eq. (5).

$$\tilde{W}(i, j) = W_{row}(i) \cdot W_{col}(j) \tag{5}$$

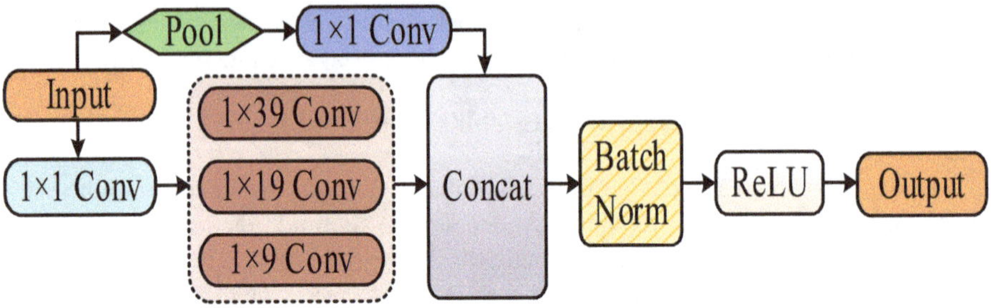

Fig. 1. SC-Inception Network Architecture Diagram

In Eq. (5), (*i*, *j*) denotes the spatial position of the feature map, W_{row} represents convolution in the horizontal direction, and W_{col} denotes convolution in the vertical direction. The formula for horizontal convolution is shown in Eq. (6).

$$W_{row}(i,j) = \sum_{n=0}^{k-1} X(i,j+n) \cdot K_{row}(n) \tag{6}$$

In Eq. (6), *k* denotes the size of the convolution kernel, and S (*m*, *n*) represents the spatial offset index within the kernel. The vertical convolution expression is given by Eq. (7).

$$W_{col}(i,j) = \sum_{m=0}^{k-1} Z(i+m,j) \cdot K_{col}(m) \tag{7}$$

Deep Separable Convolution comprises deep convolution and pointwise convolution. The deep convolution stage independently performs spatial convolution operations on each channel of the input data, enabling isolated computations between channels. Subsequently, the pointwise convolution stage employs a 1 × 1 convolution kernel for cross-channel feature fusion, integrating information across different channels. The core of this approach lies in decomposing the depth dimension of convolution to reduce computational complexity and parameter count. When applied to network traffic classification tasks, the enhanced SC-Inception network processes raw traffic data through preprocessing to convert it into feature vectors. These vectors are then fed into the SC-Inception architecture. This network automatically extracts key spatial and channel-based features through stacked separable convolution modules. Finally, after undergoing global pooling and fully connected layers, it outputs the identified traffic category classification results.

3.2 Construction of a Traffic Classification Method Integrating SC-Inception Networks and Attention Mechanisms

While the SC-Inception network reduces model complexity and achieves greater lightweight design, in the specific task of network traffic classification, the contribution of multi-channel features in traffic data to the final classification results often varies [15]. Traditional convolutional operations apply identical weights to all feature channels, disregarding variations in feature contribution. This uniform processing risks overlooking critical features, thereby impairing the model's decision-making capability. To address this, the study introduces a lightweight channel attention mechanism—the SENet module-embedded within the SC-Inception architecture. By adaptively learning the importance of feature channels, this mechanism enhances the model's focus on key features. The feature map F output from the SC-Inception network serves as input to the SENet module's average pooling layer. Through a compression step, its spatial dimensions H × W are processed into a channel descriptor u, expressed as shown in Eq. (8).

$$u_C = \frac{1}{H \times W} \sum_{\tilde{i}=1}^{H} \sum_{\tilde{j}=1}^{W} F_C(\tilde{i}, \tilde{j}) \tag{8}$$

In Eq. (8), u_C denotes the statistical characteristics of C on the channel. Subsequently, the formula for calculating the channel weight vector through excitation operations is shown in Eq. (9).

$$v = \sigma(W_2 \cdot \delta(W_1 \cdot u)) \tag{9}$$

This process utilizes two fully connected layers, achieved through dimension reduction factors and nonlinear activation functions. Finally, the channel attention weights obtained via the activation operation adjust the original features, completing the feature recalibration to yield the enhanced feature map $\hat{F}_C$, whose expression is shown in Eq. (10).

$$\hat{F}_C = v_C \cdot F_C \tag{10}$$

In summary, this study first improved the Inception network, then enhanced it by introducing the SENet module. Based on these two approaches, a network traffic classification method was constructed and named the Separable Convolution-Inception-Squeeze-and-Excitation Network (SC-Inception-SENet). The workflow for traffic classification using this network is illustrated in Fig. 2.

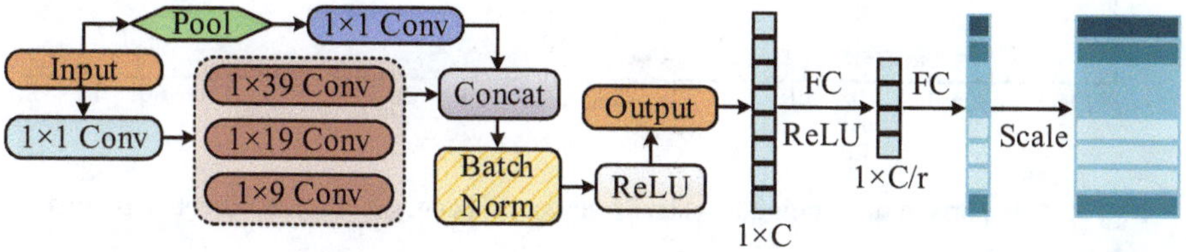

Fig. 2. Flowchart of the Network Traffic Classification Method Based on SC-Inception-SENet

As shown in Fig. 2, the traffic classification method based on the Inception-SENet network first requires preprocessing the raw data to obtain a structured multidimensional array as input to the network. This array first enters the SC-Inception module, which utilizes a architecture featuring depth wise separable convolutions, spatial separable convolutions, and parallel branches to capture features at different scales, then fuses these features through channels. Subsequently, the fused features are fed into the SENet attention module. This module compresses spatial information through global average pooling, then learns nonlinear relationships between channels via two fully connected layers. It generates importance weights for different channels, which are assigned to the original features through scale transformation, achieving feature enhancement. After feature extraction through multiple Inception-SENet networks, these features undergo dimensionality reduction via a global average pooling layer. A fully connected classifier then outputs classification results for different network quantities.

4 Performance Analysis

To verify the effectiveness of the SC-Inception network and the SC-Inception-Senet network traffic classification method proposed in the research, experiments were set up in the study to test them. The experimental environment comprised an Intel Core i9-10900K CPU with 10 cores and 20 threads, running Ubuntu 20.04 LTS operating system. Python 3.8 served as the programming language for algorithm implementation. The dataset utilized was the ISCX VPN-nonVPN Traffic Dataset (2016), which contains diverse application scenarios from both VPN and non-VPN environments. Specifically designed for encrypted traffic analysis, it serves as a benchmark dataset in this field. The study randomly selected 1,000 data points and divided them into training and testing sets at an 8:2 ratio. The experiment was first set up by selecting Mobile Network (MobileNet) and Neural Architecture Search Network (NASNet) as comparison algorithms to test the training loss and accuracy of the modified Inception network. The experimental results are shown in Fig. 3.

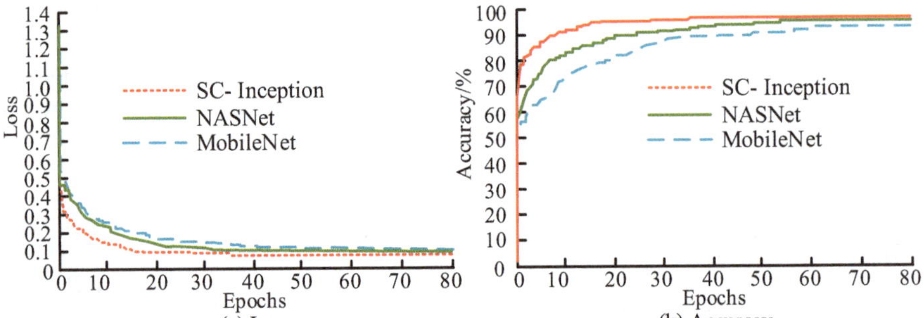

(a) Loss (b) Accuracy

Fig. 3. Comparison of Training Loss and Accuracy Changes Across Different Networks

As shown in Fig. 3(a), the loss functions of all three networks exhibit a trend of rapid initial decline followed by gradual convergence as the number of iterations increases. During the early stages of iteration, the improved SC-Inception network converged significantly faster than the comparison networks. As iterations continued, the SC-Inception network reduced its loss value to 0.740 by the 35th iteration, achieving convergence. In contrast, the loss functions of NASNet and MobileNet remained at 0.091

and 0.116, respectively, even after reaching their maximum iterations. As shown in the accuracy curve in Fig. 3(b), the improved Inception network exhibits a steeper accuracy increase curve, reaching 96.42% at the maximum iteration count—significantly higher than the comparison networks. These results demonstrate that the SC-Inception network achieves both faster convergence and higher classification accuracy. Subsequently, the study selected the Convolutional Block Attention Module Network (CBAMNet) and Bidirectional Encoder Representations from Transformers (BERT) as comparison methods to validate the classification performance of the SC-Inception-SENet network traffic classification approach. The study first visualized the classification performance of different traffic classification methods, with specific experimental results shown in Fig. 4.

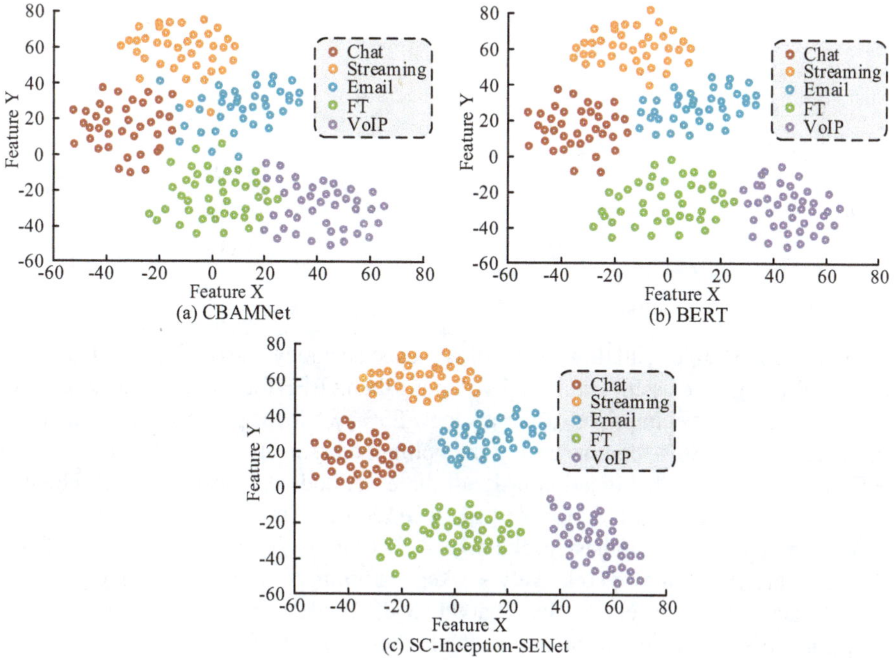

Fig. 4. Visualization of Classification Performance for Different Network Traffic Classification Methods

As shown in Fig. 4(a), although CBAMNet can roughly distinguish between the five traffic categories, the boundaries between categories are blurred. Particularly, there is partial overlap between "Chat," "Streaming," and "Email," with unclear boundaries. The results in Fig. 4(b) indicate that BERT's classification performance surpasses that of CBAMNet, effectively differentiating all traffic types. However, data points are scattered with weak clustering, resulting in overall weaker classification performance. Figure 4(c) clearly demonstrates that the proposed SC-Inception-SENet classification method achieves substantially superior visualization results compared to all baseline approaches. The method exhibits markedly clearer inter-class decision boundaries and significantly stronger intra-class clustering cohesion in the learned feature space. These characteristics collectively indicate its enhanced capability to extract and learn highly discriminative feature representations from complex network traffic data. Consequently, the proposed network traffic classification framework exhibits significant practical advantages in terms of both feature quality and classification performance. To further provide comprehensive evaluation, the study subsequently conducts detailed statistical analyses of the classification accuracy across five distinct and common network traffic classification tasks, with the complete quantitative experimental results explicitly presented in Table 1.

Table 1. Accuracy Statistics for Different Network Traffic Classification Methods

Network traffic type	Classification method		
	CBAMNet	BERT	SC-Inception-SENet
Chat	97.23%	95.58%	97.86%
Streaming	97.25%	98.15%	98.60%
Email	91.20%	95.33%	96.01%
FT	97.88%	97.97%	98.82%
VoIP	95.53%	94.91%	95.88%

As shown in Table 1, all three methods achieved accuracy rates exceeding 95.00% in the "Chat" category classification results. Among them, BERT achieved an accuracy rate of 95.58%, CBAMNet improved accuracy to 97.23%, while the proposed SC-Inception-SENet reached 97.86%, demonstrating the optimal classification performance. For the "FT" network traffic classification task, all three methods demonstrated high classification performance, with the proposed SC-Inception-SENet achieving an accuracy of 98.82%, maintaining the highest level. For the "VoIP" classification task, both BERT and SC-Inception-SENet showed relatively weaker performance, with BERT achieving only 94.91% accuracy and Inception-SENet at 95.53%. This demonstrates that the proposed Inception-SENet method still maintains the highest classification accuracy. The above experimental results collectively demonstrate that the proposed method exhibits stable and superior performance across diverse network traffic classification tasks, showcasing strong practicality and generalization capabilities. These experimental results collectively indicate that the proposed method exhibits stable and superior performance across different types of network traffic classification tasks, demonstrating excellent practicality and generalization capabilities.

5 Conclusion

In the field of network traffic classification, particularly in complex traffic analysis tasks, traditional methods often struggle to distinguish traffic from different application types. They suffer from low classification efficiency, insufficient accuracy, and inadequate focus on key features. To address this, the study introduces the Inception network as the backbone architecture for network traffic classification. It replaces the standard convolutional structures within the Inception architecture with spatially separable convolutions and depthwise separable convolutions, thereby mitigating the issue of computational complexity increasing with the number of channels. Furthermore, the study incorporates the SENet attention mechanism to enhance the Inception network's focus on key features, thereby improving classification accuracy. Experimental results demonstrate that the improved SC-Inception network significantly outperforms the baseline network in terms of loss function and accuracy. After only 35 iterations, the loss function converged to a minimum value of 0.064, achieving a final classification accuracy of 96.42%. In network traffic classification experiments, the proposed SC-Inception method demonstrated superior performance across all five traffic categories. Classification accuracy reached 97.86% for "Chat" and 98.82% for "FT," significantly outperforming comparison methods. The experimental results collectively demonstrate that the proposed classification method effectively enhances feature extraction capabilities and classification performance for network traffic. However, the dataset used in this study is limited, and the number of network traffic categories is relatively small. Therefore, future research will incorporate more complex traffic data to further optimize the model's classification performance.

References

1. Huang, J., Gou, F., Wu, J.: RETRACTED: user trust and evaluation mechanism based on community classification calculation in opportunistic social networks. J. Intell. Fuzzy Syst. **46**(1), 2091–2108 (2024)
2. Rajesh Kanna, P., Santhi, P.: Exploring the landscape of network security: a comparative analysis of attack detection strategies. J. Ambient. Intell. Humaniz. Comput. **15**(8), 3211–3228 (2024)
3. Lu, H., Dong, Y., Wu, Z., Wei, H.L., Lu, G.: New class detection in network traffic classification using confidence information embedded cascade structure. IEEE Trans. Netw. Sci. Eng. **12**(3), 1692–1706 (2025)
4. Huoh, T.L., Luo, Y., Li, P., Zhang, T.: Flow-based encrypted network traffic classification with graph neural networks. IEEE Trans. Netw. Serv. Manag. **20**(2), 1224–1237 (2023)
5. Gonthina, N., Prasad, L.V.N.: An enhanced convolutional neural network architecture for semantic segmentation in high-resolution remote sensing images. Discov. Comput. **28**(1) (2025). https://doi.org/10.1007/s10791-025-09610-5
6. Ahmed, M., Afreen, N., Ahmed, M., Sameer, M., Ahamed, J.: An inception V3 approach for malware classification using machine learning and transfer learning. Int. J. Intell. Networks. **4**(1), 11–18 (2023)
7. Hax, D.R.T., Penava, P., Krodel, S., Razova, L., Buettner, R.: A novel hybrid deep learning architecture for dynamic hand gesture recognition. IEEE Access. **12**(1), 28761–28774 (2024)
8. Li, X., Zhang, Y., Liu, Y., Yao, X., Zhou, X.: Arbitrary shape text detection fusing InceptionNeXt and multi-scale attention mechanism. J. Supercomput. **80**(17), 25484–25509 (2024)
9. Huang, Z., Wei, L.I., Shao, C., Geng, L.: 4mC site prediction approach based on dual-path multiscale feature fusion. Exp. Technol. Manage. **42**(4), 68–77 (2025)

10. Deshpande, H., Chaudhari, D., Sarode, T., Kamath, A., Khan, D.: LeanConv-SENet: a novel lightweight neural network architecture for handwritten character recognition. Procedia Comput. Sci. **258**(4), 2826–2836 (2025)
11. Huang, Y., Shi, P., He, H., Zhao, B.: Senet: spatial information enhancement for semantic segmentation neural networks. Vis. Comput. **40**(5), 3427–3440 (2024)
12. Yang, Z., Zhu, H., Li, Z., Wang, G., Su, M.: A malware detection method based on genetic algorithm optimized CNN-SENet network. IEEE Access. **12**(10), 16052–16063 (2024)
13. Fu, J., Wei, H., Yu, H., Cao, Y.: Free space detection method based on SENet attention. In: 2023 International Seminar on Computer Science and Engineering Technology (SCSET), vol. 29, pp. 335–339. IEEE (2023) (10)
14. Fan, Y., et al.: A multi-watermarking algorithm for medical images using inception V3 and DCT. Comput. Mater. Continua. **74**(1), 1279–1302 (2023)
15. Huang, H., Zhang, X., Lu, Y., Li, Z., Zhou, S.: BSTFNet: an encrypted malicious traffic classification method integrating global semantic and spatiotemporal features. Comput. Mater. Continua. **78**(3), 3929–3951 (2024)

Recommender Systems and Data Imputation

Missing Value Imputation Methods for Time Series Data: A Comprehensive Review

Fatima Ezzahra Senhaji[1](✉), Mohamed Nabil[1], Abdelkaher Ait Abdelouahad[1], Abdelmajid Hajami[2], and Abdelkrim Haqiq[3]

[1] LAROSERI, Computer Science Department, Faculty of Sciences, University of Chouaib Doukkali, El Jadida, Morocco
senhaji.f156@ucd.ac.ma

[2] Faculty of Sciences and Techniques, LAVETE Laboratory, Hassan First University of Settat, Settat, Morocco
abdelmajid.hajami@uhp.ac.ma

[3] Faculty of Sciences and Techniques, Computer, Networks, Mobility and Modeling Laboratory (IR2M), Hassan First University of Settat, Settat, Morocco
abdelkrim.haqiq@uhp.ac.ma

Abstract. Time series data are found in almost all real-world applications. However, unexpected accidents such as signal losses, sensors failure or malfunction can lead to gaps within the time series, thereby hindering effective data analysis and exploitation. As a result, they have been a growing demand for accurate imputation methods for missing time series data making it a rapidly evolving research domain with a wide variety of approaches. Particularly, deep learningbased methods have shown promising results owing to their robust architectures, making them suitable for handling complex sequential data. In this paper, we primarily review time series imputation techniques based on deep learning methods, including RNNs, GANs, GNNs, and hybrid models. Additionally, we discuss the limitations, trade-offs, and future directions of research in the field of time series imputation.

Keywords: Time Series · Missing Values · Imputation · Deep Learning · Review

1 Introduction

Time series data is a sequence of observations collected over time at consistent or irregular intervals. In fact, many real-world processes evolve over time including economics [37], medicine [39], environmetrics [57], etc. Time series data analysis captures the dynamics of systems. Hence enables prediction, anomaly detection,

M. Nabil, A. A. Abdelouahad, A. Hajami and A. Haqiq—Contributing authors.

A. E. Hassanien et al. (Eds.): AISDS 2025, LNDECT 287, pp. 311–320, 2027.
https://doi.org/10.1007/978-3-032-18927-1_31

and decision support. However, in real-world applications, the collected data are inevitably contaminated by missing values. This is generally due to sensor failure (e.g., sensor or battery failures of IoT devices, storage saturation), transmission loss (e.g., cloud cover or satellite malfunctions), or simply irregular sampling of patient tests or vital signs in hospitals. Imputation aim to fills in missing values rather than discarding or ignoring them. This is extremely important to preserve the temporal continuity, patterns and characteristics (e.g., trend, seasonality, stationarity and cyclic behavior...) of time series. This is why imputation is needed for reliable results and without it, analyses risk bias, unstable or incorrect models that can lead to poor decisions. To overcome this issue, many imputation methods were proposed in the literature. From basic statistical techniques like mean, median, interpolation, Kalman filtering and ARIMA-based smoothing [3] to machine learning techniques like k-nearest neighbors (KNN) [41], random forests [40] and recently robust deep learning models. Despite the large and rapidly growing literature on time series imputation, most existing surveys focus on a specific domain (e.g., air quality [45], health domain [22]...) or cover only one family of methods (e.g., machine learning [2], GAN based models [58], GNN based models [19]). To overcome this shortcoming, we propose a comprehensive review that broadly overviews deep learning-based imputation models. We also provide highlights of trends and gaps in the time series imputation field.

The rest of this paper is organized as follows: Sect. 2 presents theoretical background. Section 3 reviews advanced imputation methods for handling missing values in time series data. Section 4 highlights limitations and future directions. Section 4 outlines the main conclusions.

2 Advanced Methods for Handling Time Series Missing Values

This section explores four major classes of advanced methods that have emerged in recent years for imputing missing values in time series data. First, RNN-based models leverage sequential architectures such as LSTM and GRU to capture temporal dependencies, making them a natural fit for time series data. Second, GAN-based models frame imputation as a generative task, using adversarial training to learn realistic data distributions. Third, GNN-based models, well-suited for multivariate and structured time series, as they known for capturing complex spatial and inter-variable dependencies using graph representations. Lastly, we present hybrid models which mixes features from multiple architectures such as attention mechanisms, convolutional networks, and autoencoders.

2.1 RNN-Based Models

Recurrent Neural Networks (RNNs) are widely recommended for handling sequential data due to their ability to capture time dependencies. Therefore, many researchers have attempted to impute time-series missing values with recurrent architecture. For example, in [5], the authors trained the basic RNN architecture (e.g., vanilla RNN) to classify data with missing values from the audiology database. Since experiments have shown poor performance in capturing long-term dependencies due to the vanishing gradient issue. For instance, in [17], authors proposed LSTM as a specific RNN cell to deal with missing values. Similarly, [11] proposed GRU, a specialized RNN architecture for imputation with a gating mechanism. A closely related work is GRU-D [8], which integrated decay terms, masking, and time intervals to model the influence of missingness over time. On the other hand, many studies trained their models using a bi-directional sequence modeling; we refer to Bi-LSTM [49], BRITS [7], and MRNN [55].

2.2 GAN-Based Models

Inspired by the success of Generative Adversarial Networks (GANs) in image generation [15], and due to their ability to generate synthetic samples similar to the actual data, GANs have been widely used as a missing values imputation method. The GAN architecture consists of a generator and a discriminator. The generative model is estimated via an adversarial learning process. Basically, the generator tries to fool the discriminator with fake generated data samples, while the discriminator tries to distinguish between fake and real samples. It should be noted that despite their widespread use, very few models are built using a pure GAN architecture. Mainly, GAIN [54] is considered a foundation model specially designed for imputation. Based on this work, Xu et al. proposed MTGAIN [51] as a multi-task GAIN model for both imputation and prediction. While Qin et al. proposed ImputeGAN [34], an iterative framework for multivariate time series imputation. IGANI [21] is an other iterative GAN (e.g., iterative generator) specially designed for imputing traffic time-series data. On the other hand, many GAN-based models have been proposed in the literature by integrating an RNN architecture. We furthermore classify them into two main categories:

(1) Unidirectional Recurrent GAN Models

Developed by Luo et al., GRUI-GAN [30] is a two-stage imputation model (stage 1 learns data distribution, and stage 2 refines latent noise for accuracy). The model combines the GAN architecture with GRUI (a special modified GRU cell for imputation). As a following work, E^2GAN [29] was designed for multivariate time series imputation; it integrates a random vector to the main sample and an autoencoder to compress and reconstruct the incomplete data. Wang et al. proposed TLGRU [43] as a time and location dependent GRU by integrating their novel cell within the E^2GAN architecture, making it well-suited for irregular spatio-temporal data, like weather or air quality. Whereas, Bi et al. [6] presented an imputation framework for water quality indicator time series by combining GAN and encoder-decoder architecture with an autoregressive network.

(2) Bidirectional Recurrent GAN Models

Developed by Gupta et al. BiGAN [16] is a bidirectional RNN/GAN for both imputation and prediction tasks; it can handle irregular and variable-length data. Similarly, SSGAN [32] also uses bidirectional structure except that the authors added a temporal reminder matrix and a classifier to address the missing label issue in partially labeled datasets. In [35] Richter et al. presented the BiGRU-GAN model, which integrated BiGRUs for both the generator and discriminator components of a Generative Adversarial Network (GAN), enhancing its ability to capture temporal dependencies in both forward and backward directions in order to impute incomplete WiFi packet captured data. NAOMI [28], standing for Non-Autoregressive Multiresolution Imputation, is also a GAN-based model that employs a forward-backward encoder with a multi-resolution decoder for non-autoregressive generation, reducing error accumulation in long or heavily missing sequences.

2.3 GNN-Based Models

Graph Neural Networks (GNNs) are modern deep learning models that process graph-structured data [36]. In fact, the underlining idea of GNNs is that each node in the graph represents objects or concepts, while edges represent their relationships. The flexibility of the data representation makes GNNs a powerful tool for learning non-Euclidean data. This is typically operated through graph convolution, which enables spreading information across neighboring nodes.

Following the same context, much research focuses on modeling multivariate time series, as they can be naturally abstracted into spatial-temporal graphs. Essentially, graph "convolutions" enable the capture of all sorts of varied and complex relationships, that is to say, both intervariable (connections between different variables within a multivariate series) and intertemporal (dependencies between different points in time).

Cini et al. introduced GRIN [12] as Graph Recurrent Imputation Network (GRIN). The model is built based on a bidirectional graph recurrent neural network that uses a GRU cell as a core unit, whereas gates are implemented by message-passing neural networks (MPNNs). Similarly, other researchers have opted for a bidirectional recurrent architecture we cite, DGCRIN [23], developed as a tool to impute missing traffic data by employing a graph generator and dynamic graph convolutional gated recurrent unit (DGCGRU) to perform fine-grained modeling of the dynamic spatiotemporal dependencies of road networks. While AGRN [10] remarkably presented a graph learning module capable of generating inter-variable adaptive graphs that represent the spatial dependencies without relying on any pre-defined graphs. Whereas MDGCN [25] introduces an external memory network to store and distribute global spatiotemporal data throughout the network of traffic. Other research has made use of the Graph Attention Network (GAT) [42] as a baseline architecture for their imputation model. In [38], authors compose a graph attention recurrent cell by combining GAT and different LSTM cells (GARNN) for traffic data imputation. Simultaneously, GATGPT [9], GACN [52], and ST D-GAT [1] were proposed as GAT-based imputation models for spatiotemporal datasets (e.g., traffic, sensor networks, geospatial data). In an innovative step, Xing et al. proposed a novel framework, FGATT [50], that combines the Fuzzy Graph Attention Network (FGAT) with the transformer encoder. The combination addresses both spatial uncertainty and dynamic changes in data, making it particularly robust to missing data imputation. IGNNK [47] was introduced as an inductive GNN kriging model to recover signals for unobserved time series in a spatial graph by learning how to interpolate from neighboring observed nodes. The concept is similar to how classical kriging works, but far more flexible and scalable.

2.4 Hybrid Models

Numerous scholars have proposed merging the most recent robust architectures within models that have demonstrated efficacy in handling the time series missing values problem. This involves combining models from many previously described categories (e.g., RNNs, GANs, and GNNs) with attention learning methods, CNNs, and AutoEncoders (AEs) mechanisms.

Attention mechanisms enhance the model's ability to capture long-range temporal dependencies by dynamically weighting relevant time steps, while autoencoders enable efficient dimensionality reduction and reconstruction of missing patterns. CNNs, though traditionally used in spatial domains, have proven effective in capturing local temporal patterns through convolutional filters, especially when combined with sequential models. When combined, these elements serve as the foundation for hybrid architectures that are more resilient and adaptable. Table 1 presents various recent hybrid models proposed in the literature for time series imputation.

Table 1. Hybrid models main used features

Model Name	ref	RNN	GAN	Attention	AE	CNN	GNN
DSTGCN	[18]	Yes	No	No	No	No	Yes
HSPGNN	[24]	Yes	No	Yes	No	No	Yes
SDA-GRIN	[14]	Yes	No	Yes	No	No	Yes
CGCNImp	[26]	Yes	No	Yes	No	No	Yes
MATCN	[46]	No	No	Yes	No	No	Yes
STING	[33]	Yes	Yes	Yes	No	No	No
SAITS	[13]	No	No	Yes	Yes	No	No
T-LSTM	[4]	Yes	No	No	Yes	No	No
SPIN	[31]	No	No	Yes	No	No	Yes
PriSTI	[27]	No	No	Yes	No	No	Yes
CATSI	[53]	Yes	No	Yes	No	No	No
STA-GAN	[44]	No	Yes	Yes	No	No	Yes
SATCN	[48]	No	No	No	No	Yes	Yes
CGCNImp	[26]	Yes	No	Yes	No	No	Yes
INCREASE	[59]	Yes	No	Yes	No	No	Yes
CASPER	[20]	No	No	Yes	Yes	No	Yes
DCNN-GAN	[56]	No	Yes	No	No	Yes	Yes

3 Discussion

In this review we have examined a broad range of methods developed for time series missing data imputation. Typically we highlighted recent advanced deep learning models, including RNNs, GANs, GNNs, and hybrid models. It has been found that despite their rapid evolution, several key challenges remain, including the trade-offs between accuracy, interpretability, and real-world deployability. In fact, despite being regarded as fundamental resolution methods, RNNs still struggle with long-term dependencies and irregularly sampled data. In contrast to RNNs, GAN-based models can handle more complex and nonlinear patterns, although they are quite sensitive to hyperparameters, which can make them unstable sometimes and difficult to train. On the other hand, GNNs turned out to be very effective at learning both spatial and temporal correlations, especially in multivariate or sensor network data (e.g., traffic networks, healthcare records). The only drawback is that they generally require graph construction, which can be challenging and requires an understanding of GNN foundations. Additionally, the fact that using GNNs as a baseline for imputation is still a relatively new and unexplored topic implies fewer standardized benchmarks. Meanwhile, hybrid models have shown excellent results in modeling long-range dependency with a good time-consuming compromise owing to autoencoders capability to learn compressed representations. Unfortunately, attention models can be data intensive and computationally expensive or turn into black boxes, complicating interpretability and debugging. Generally, the main gaps and limitations faced in the time series missing values imputation field can be identified by the lack of standardization (e.g., standard benchmarks, unified evaluation protocols). Sometimes direct evaluation is almost impossible since there is no ground truth regarding the missing values. However, direct imputation accuracy still can be evaluated by examining the overall performance of the model (e.g., forecasting or classification tasks). Hence, we can conclude that the time series field still has some limitations resumed by the need for standard benchmarks and unified evaluation protocols to compare models objectively or assess their generalizability across domains. The need for models that can operate in online or streaming environments, as well as those that generalize well across diverse application domains. Overall, the future of time series imputation lies in developing flexible, interpretable, and context-aware methods that can adapt to diverse patterns of missingness and complex data structures.

4 Conclusions

In this paper, we briefly present deep learning based imputation methods for time series. We propose that existing methods can be classified into 4 main classes: RNN-based methods, GAN-based methods, GNN-based methods, and hybrid models. Furthermore, we discussed different aspects related to the reviewed group of models, including model comparability, transferability, and explainability. Nevertheless, the time series imputation field is still open to research and challenges to tackle, as presented in the discussion. Most importantly, more comparative research using a bigger variety of data sets from different domains is needed to further strengthen the statements on the performance of imputation models.

References

1. Ahmad, H., et al.: A hybrid spatio-temporal graph attention (ST D-GAT framework) for imputing missing SBAS-InSAR deformation values to strengthen landslide monitoring. Remote Sens. **17**(15), 2613 (2025)
2. Alabadla, M., et al.: Systematic review of using machine learning in imputing missing values. IEEE Access **10**, 44483–44502 (2022)
3. Ansley, C.F., Kohn, R.: On the estimation of ARIMA models with missing values. In: Time Series Analysis of Irregularly Observed Data: Proceedings of a Symposium held at Texas A "& M University, College Station, Texas February 10–13, 1983, pp. 9–37. Springer, Cham (1984)
4. Baytas, I.M., et al.: Patient subtyping via time-aware LSTM networks. In: Proceedings of the 23rd ACM SIGKDD International Conference on Knowledge Discovery and Data Mining, pp. 65–74 (2017). https://doi.org/10.1145/3097983.3097997
5. Bengio, Y., Gingras, F.: Recurrent neural networks for missing or asynchronous data. In: Advances in Neural Information Processing Systems, vol. 8 (1995)
6. Bi, J., et al.: Multi-indicator water time series imputation with autoregressive generative adversarial networks. In: IEEE International Conference on Systems, Man, and Cybernetics (SMC), pp. 2003–2008. IEEE (2022)
7. Cao, W., et al.: Brits: bidirectional recurrent imputation for time series. In: arXiv preprint arXiv:1805.10572 (2018)
8. Che, Z., et al.: Recurrent neural networks for multivariate time series with missing values. Sci. Rep. **8**(1), 1–12 (2018)
9. Chen, Y., Wang, X., Xu, G.: GatGPT: a pre-trained large language model with graph attention network for spatiotemporal imputation. arXiv preprint arXiv:2311.14332 (2023)
10. Chen, Y., et al.: Adaptive graph recurrent network for multivariate time series imputation. In: Tanveer, M., Agarwal, S., Ozawa, S., Ekbal, A., Jatowt, A. (eds.) ICONIP 2022. CCIS, vol. 1792, pp. 64–73. Springer, Singapore (2022). https://doi.org/10.1007/978-981-99-1642-9_6
11. Cho, K., et al.: Learning phrase representations using RNN encoder-decoder for statistical machine translation. arXiv preprint arXiv:1406.1078 (2014)
12. Cini, A., Marisca, I., Alippi, C.: Filling the g_ap_s: Multivariate time series imputation by graph neural networks. arXiv preprint arXiv:2108.00298 (2021)
13. Du, W., et al.: SAITS: self-attention-based imputation for time series. In: Expert Syst. Appl. **219** (2023). Preprint available at arXiv:2202.08516, p. 119619. DOI: https://doi.org/10.1016/j.eswa.2023.119619
14. Eskandari, A., et al.: SDA-GRIN for adaptive spatial-temporal multivariate time series imputation. In: IEEE 49th Annual Computers, Software, and Applications Conference (COMPSAC), pp. 1801–1806. IEEE (2025)
15. Goodfellow, I., et al.: Generative adversarial nets. In: Advances in Neural Information Processing Systems, vol. 27 (2014)
16. Gupta, M., Beheshti, R.: Time-series imputation and prediction with bi-directional generative adversarial networks. arXiv preprint arXiv:2009.08900 (2020)

17. Hochreiter, S., Schmidhuber, J.: Long short-term memory. Neural Comput. **9**(8), 1735–1780 (1997)
18. Jia, H., Lin, X., Wang, C.: DSTGCN: dynamic spatial-temporal graph convolutional network for traffic prediction. IEEE Sensors J. **22**(13), 13116–13124 (2022)
19. Jin, M., et al.: A survey on graph neural networks for time series: forecasting, classification, imputation, and anomaly detection. arXiv preprint arXiv:2307.03759 (2023)
20. Jing, B., et al.: Causality-aware spatiotemporal graph neural networks for spatiotemporal time series imputation. In: Proceedings of the 33rd ACM International Conference on Information and Knowledge Management, pp. 1027–1037 (2024)
21. Kazemi, A., Meidani, H.: IGANI: iterative generative adversarial networks for imputation with application to traffic data. IEEE Access **9**, 112966–112977 (2021)
22. Kazijevs, M., Samad, M.D.: Deep imputation of missing values in time series health data: a review with benchmarking. J. Biomed. Inform. 104440 (2023)
23. Kong, X., et al.: Dynamic graph convolutional recurrent imputation network for spatiotemporal traffic missing data. Knowl.-Based Syst. **261**, 110188 (2023)
24. Liang, G., et al.: Higher-order spatio-temporal physics-incorporated graph neural network for multivariate time series imputation. In: Proceedings of the 33rd ACM International Conference on Information and Knowledge Management, pp. 1356–1366 (2024)
25. Liang, Y., Zhao, Z., Sun, L.: Memory-augmented dynamic graph convolution networks for traffic data imputation with diverse missing patterns. Transp. Res Part C: Emerg. Technol. **143**, 103826 (2022)
26. Liu, C., et al.: CGCNImp: a causal graph convolutional network for multivariate time series imputation. PeerJ Comput. Sci. (2022). Uses neural Granger causality, GCN, attention-driven LSTM + time lag matrix
27. Liu, M., et al.: PriSTI: a conditional diffusion framework for spatiotemporal imputation. arXiv preprint arXiv:2302.09746 (2023)
28. Liu, Y., et al.: NAOMI: non-autoregressive multiresolution sequence imputation. In: Advances in Neural Information Processing Systems, vol. 32 (2019)
29. Luo, Y., et al.: E2gan: end-to-end generative adversarial network for multivariate time series imputation, pp. 3094–3100. AAAI Press (2019)
30. Luo, Y., et al.: Multivariate time series imputation with generative adversarial networks. In: Advances in Neural Information Processing Systems, vol. 31 (2018)
31. Marisca, I., Cini, A., Alippi, C.: Learning to reconstruct missing data from spatiotemporal graphs with sparse observations. Adv. Neural. Inf. Process. Syst. **35**, 32069–32082 (2022)
32. Miao, X., et al.: Generative semi-supervised learning for multivariate time series imputation. In: Proceedings of the AAAI Conference on Artificial Intelligence, vol. 35, no, 10, pp. 8983–8991 (2021)
33. Oh, E., et al.: STING: self-attention based Time-series Imputation Networks using GAN. arXiv preprint arXiv:2209.10801 (2022). Combines GAN, bidirectional RNN, and attention for imputation
34. Qin, R., Wang, Y.: ImputeGAN: generative adversarial network for multivariate time series imputation. Entropy **25**(1), 137 (2023)
35. Richter, A., Ijaradar, J., Wetzker, U.: Iterative imputation of incomplete wireless network traffic using adversarial learning and a two-stage approach. In: 3rd International Conference on Artificial Intelligence For Internet of Things (AIIoT), pp. 1–6. IEEE (2024)
36. Scarselli, F., et al.: The graph neural network model. IEEE Trans. Neural Networks **20**(1), 61–80 (2008)

37. Sezer, O.B., Gudelek, M.U., Ozbayoglu, A.M.: Financial time series forecasting with deep learning: a systematic literature review: 2005–2019. Appl. Soft Comput. **90**, 106181 (2020)
38. Shen, G., et al.: Bidirectional spatial-temporal traffic data imputation via graph attention recurrent neural network. Neurocomputing **531**, 151–162 (2023)
39. da Silva, D.B., et al.: DeepSigns: a predictive model based on Deep Learning for the early detection of patient health deterioration. Expert Syst. Appl. **165**, 113905 (2021)
40. Stekhoven, D.J., Bühlmann, P.: MissForest–non-parametric missing value imputation for mixed-type data. Bioinformatics **28**(1), 112–118 (2012)
41. Troyanskaya, O., et al.: Missing value estimation methods for DNA microarrays. Bioinformatics **17**(6), 520–525 (2001)
42. Velickovic, P., et al.: Graph attention networks. Statistics **1050**(20), 10–48550 (2017)
43. Wang, R., et al.: TLGRU: time and location gated recurrent unit for multivariate time series imputation. EURASIP J. Adv. Signal Process. **2022**(1), 74 (2022)
44. Wang, S., et al.: STA-GAN: a spatio-temporal attention generative adversarial network for missing value imputation in satellite data. Remote Sens. **15**(1), 88 (2023). https://doi.org/10.3390/rs15010088
45. Wijesekara, W.M.L.K.N., Liyanage, L.: Comparison of imputation methods for missing values in air pollution data: case study on sydney air quality index. In: Arai, K., Kapoor, S., Bhatia, R. (eds.) FICC 2020. AISC, vol. 1130, pp. 257–269. Springer, Cham (2020). https://doi.org/10.1007/978-3-030-39442-420
46. Wu, X., et al.: A multi-attention tensor completion network for spatiotemporal traffic data imputation. IEEE Internet Things J. **9**(20), 20203–20213 (2022)
47. Yuankai, W., et al.: Inductive graph neural networks for spatiotemporal kriging. In: Proceedings of the AAAI Conference on Artificial Intelligence, vol. 35, no. 5, pp. 4478–4485 (2021)
48. Wu, Y., et al.: Spatial aggregation and temporal convolution networks for real-time kriging. arXiv preprint arXiv:2109.12144 (2021)
49. Xie, C., et al.: BiLSTM-I: a deep learning-based long interval gap-filling method for meteorological observation data. Int. J. Environ. Res. Public Health **18**(19), 10321 (2021)
50. Xing, J., et al.: FGATT: a robust framework for wireless data imputation using fuzzy graph attention networks and transformer encoders. arXiv preprint arXiv:2412.01979 (2024)
51. Xu, L., Xu, L., Yu, J.: A multi-task learning-based generative adversarial network for red tide multivariate time series imputation. Complex Intell. Syst. **9**(2), 1363–1376 (2023)
52. Ye, Y., Zhang, S., Yu, J.J.Q.: Spatial-temporal traffic data imputation via graph attention convolutional network. In: Farkaš, I., Masulli, P., Otte, S., Wermter, S. (eds.) ICANN 2021. LNCS, vol. 12891, pp. 241–252. Springer, Cham (2021). https://doi.org/10.1007/978-3-030-86362-3_20
53. Yin, K., Feng, L., Cheung, W.K.: Context-aware time series imputation for multi-analyte clinical data. J. Healthcare Inf. Res. **4**(4), 411–426 (2020)
54. Yoon, J., Jordon, J., Schaar, M.: Gain: missing data imputation using generative adversarial nets. In: International Conference on Machine Learning, pp. 5689–5698. PMLR (2018)
55. Yoon, J., Zame, W.R., van der Schaar, M.: Estimating missing data in temporal data streams using multi-directional recurrent neural networks. IEEE Trans. Biomed. Eng. **66**(5), 1477–1490 (2018)

56. Zhang, C., et al.: A missing traffic data imputation method based on a diffusion convolutional neural network-generative adversarial network. Sensors **23**(23), 9601 (2023)
57. Zhang, Y.-F., Thorburn, P.J., Fitch, P.: Multi-task temporal convolutional network for predicting water quality sensor data. In: Gedeon, T., Wong, K.W., Lee, M. (eds.) ICONIP 2019. CCIS, vol. 1142, pp. 122–130. Springer, Cham (2019). https://doi.org/10.1007/978-3-030-36808-1_14
58. Zhang, Y., Zhang, R., Zhao, B.: A systematic review of generative adversarial imputation network in missing data imputation. Neural Comput. Appl. **35**(27), 19685–19705 (2023)
59. Zheng, C., et al.: Increase: inductive graph representation learning for spatio-temporal kriging. In: Proceedings of the ACM Web Conference 2023, pp. 673–683 (2023)

XAI-Enhanced Recommender Systems: A Machine Learning-Based Predictive System for Early Dropout Detection in Learning Management Systems with Recommendation Support

Kaoutar Errakha[1(✉)], Amina Samih[2], Abderrahim Marzouk[1], Sanaa Dfouf[3], and Ayoub Krari[4]

[1] Computer, Networks, Mobility and Modeling Laboratory, University Hassan First, Settat, Morocco
{kaoutar.errakha,abderrahim.marzouk}@uhp.ac.ma

[2] Artificial Intelligence Research and Applications Laboratory, University Hassan First, UH1, Settat, Morocco
amina.samih@uhp.ac.ma

[3] UH2, Casablanca, Morocco

[4] Laboratory of Research Watch for Emerging Technologies, University Hassan First, Settat, Morocco
ayoub.krari@uhp.ac.ma

Abstract. We present an ML pipeline for early dropout prediction in Moodle using 7 GB of logs from 3,120 students, 62 courses, across 5 academic terms. Weekly features (activity, achievement, participation) feed classical classifiers (RF, GBM, LR, SVM, NN) and an LSTM for temporal dynamics. To avoid leakage, features are restricted to information available up to prediction week t, and evaluation uses temporal hold-out (earlier terms → later term) and cross-course splits. On the temporal test set at week 4, the best ensemble achieves F1 = 0.82 (95% CI 0.79–0.85) and AUC = 0.88 (95% CI 0.86–0.90). We integrate SHAP for interpretability and a recommender layer (content/collab + LSTM paths) and report offline RS metrics (NDCG@10 = 0.64, MAP@10 = 0.41, coverage = 0.78, novelty = 0.62) on flagged at-risk students. We discuss deployment (periodic retraining, drift) and ethics (transparency, opt-in).

Keywords: XAI · LMS · dropout prediction · recommender systems · e-learning

1 Introduction

Along with everything else, the pandemic accelerated the rate at which educators had begun implementing LMS systems. Institutions scrambled to figure out how

A. Samih, A. Marzouk, S. Dfouf and A. Krar—Contributing authors.

A. E. Hassanien et al. (Eds.): AISDS 2025, LNDECT 287, pp. 321–330, 2027.
https://doi.org/10.1007/978-3-032-18927-1_32

to integrate systems such as Canvas, Moodle, or Blackboard into their existing courses, to solve the absence of face-to-face contact with learners. The shift from physical/business as usual, to remote/digitally connected learning was, for most, a baptism by fire. As with any large change, implementing learning management systems during a pandemic, and in the context of previously scheduled contact classes, was not without challenges.

More recently, however, LMS from the likes of Microsoft, Google and Apple have integrated instructional design and content mastery into their systems. These more recent, purpose-built LMS along with more traditional systems such as Canvas offer educators incredible design flexibility. Two key elements have and will impact how educators select LMS to integrate into their courses. The first of impact these systems can have on all key stakeholders the students, the educators and the institution. The burgeoning impact of LMS along with AI on the learners, educators and the institution are the second.

Yet, within the context of learning management systems (LMS), dropout prediction by itself is not enough. To make meaningful use of dropout prediction, recommender systems should be seamlessly integrated. These systems can reengage learners by recommending appropriate resources, adaptive learning trajectories, or appropriate activities aimed at mitigating specific weaknesses [1,6]. Recency within the e-learning recommendation field of study illustrates the significance of knowledge graphs, context-aware filtering, and some hybrid methods when reinforcing learner support and system personalization [6].

Also, the increasing focus on Explainable Artificial Intelligence (XAI) provides an additional important aspect in these efforts. For instance, learners should trust the system which implies some level of interpretability of the provided predictions and suggestions. Systems using XAI, like the SHAP model, have been used in dropout prediction to provide more explanation and thus, trust on the decisions made [2]. Predictive interpretability is offered by frameworks like X-R-R-K-K-G which merge reinforcement learning, knowledge graphs, and self-explanatory systems to enable seamless recommendation flow [3]. Similar to XAI, 'human centered' XAI (HCXAI) systems focus more on user trust and comprehension directly on the application of 'more sophisticated' systems like large language models [4]. In addition, recently, some studies have shown that AI driven educational technologies that help in prediction, provide system driven suggestions, and offer feedback, which are all enormously helpful, can be trusted more if educators are provided with domain-centered explanations [5].

Considering these points, this paper describes the design and evaluation of a predictive Machine Learning system, aimed at early detection of dropout attrition within LMS, XAI enhanced, and integrated with a recommender system. The obtained results of the algorithms concerning dropout attrition LMS LMS predictive analysis, interpretation and active measures integration concern the important issues of students' retention and success in digitally-enhanced educational systems.

Aside from the technological rollout, sustainable adoption hinges on governance, data equity, and LMS data quality. Institutions require policies on consent, retention, and anonymization to ethically harness engagement logs, assessments, and forum traces for predictive and prescriptive analyses. Interoperability (e.g., LTI, xAPI, IMS Caliper) reduces vendor lock-in and richer learner profiles converge LMS traces with SIS data and external content platforms. Equally, educator professional development is crucial: dashboards, early-alert workflows, and explanation tooltips must be simple enough to act on during a busy semester. In the absence of these supports, accurate models are likely to be unused, and interventions are delivered too late to positively affect outcomes. Predictive power must be coupled with operational readiness and inclusive design.

2 Background

2.1 E-Learning Management Systems (LMS)

Among the digital educational tools available, the tools often referred to as learning management systems, or LMS, such as Moodle, Blackboard, and Canvas, have gained popularity because of their easily scalable, flexible, and accessible interfaces, which aid in the monitoring of students and the management of learning material as well as in tracking collaborative activities. The tools have gained common acceptance in higher learning and corporate training due to their ability to engage users and provide information-based analytics for authorities in order to aid in the decision-making processes. More recent studies emphasize their assessment as critical infrastructures for e-learning, but also focus the student retention, user experience, and challenges in adaptability to learning diversity [7–9]. LMS platforms can understand and predict learning behaviors and needs and integrate fully automated and human-like responsive systems. Incorporating language processing and emotional recognition systems can engage and understand the human feelings of students and adapt more learning experiences. An LMS can make automatic editing and personalization and adapt lessons and resources to the needs of the students and the outcomes of the lessons taught.

Even with such noted appreciation for emerging AI and other automated driven constructs, learning systems continue to build on the exit LMS platforms as learning management systems. Autonomy of learners continues to be expressed across the large LMS systems with little control over their learning pathways and experiences and integrating diverse contextual, cultural, and situational learning constructs. The diverse learning needs for inclusivity continues to be undermined with LMS portals systems demanding students to adapt to the systems without the necessary scaffolds. The use of AI intelligently communicatively could be continuously placed as targeted marketing for LMS as a more human-centered learning management system, despite the dismissive use of their marketing to profess human centered design.

2.2 Recommender Systems in Education

Stepping toward advanced e-learning environments, the personalization and environment engagement features offered by e-learning are increasing the significance of recommender systems. These systems are fed learner profiles, activity logs, and performance data, and create separate resources, activities, and peers for fostering retention and engagement. Newer studies focus on content hybridization, and shift toward content, collaborative, and knowledge graph methods, to improve recommend systems [3,10]. Also, learning sequence LSTM models, along with other sequence models, are noted for their ability to capture adaptive learning pathways with time [1]. Moreover, the integration of Explainable AI (XAI) features in recommender systems within LMS have improved their acceptance and trust from instructors and students [11].

The evaluation of learning recommenders has become focused on ranking metrics like NDCG@k and MAP@k, in addition to assessing coverage and novelty for breadth and diversity of the catalog. We adopt this approach but highlight the limitations of offline evaluations related to causal inference, advocating for future controlled A/B studies.

Learning recommenders in LMS are shaped by practical constraints aside from accuracy and ranking quality. Systems need to accommodate cold start problems (new learners/items and sparse activity in the early weeks), non-stationary problems due to cohort or curriculum drift and shifts in curriculum. Scope objectives should remain aware of constraints and balance relevance with sufficient diversity/novelty while avoiding overly burdensome learner workload. Recommendations should also prioritize underlying behavioral signals and meet pedagogical objectives- scaffolding and mastery progression. To ease accountability burdens, human-in-the-loop controls (instructor review, difficulty adjustment, pacing) and XAI defusing arguments can help foster calibrative trust while exposing misalignments. The issues of consent, privacy, and equity (within a subgroup, e.g., program, gender, status) are central in addition to downstream/ offline outcomes validated with online A/B testing and longitudinal monitoring to confirm the effects of interest on engagement and dropout.

2.3 Explainable Artificial Intelligence (XAI) in Educational Contexts

Although the predictive accuracy of machine learning models in dropout detection and student modeling are high, the models' opacity gives rise to trust, accountability, and fairness issues. XAI techniques try to solve the outlined issues by automating human understandable explanations of algorithmic outcomes. In educational contexts, SHAP values, LIME, and attention-based approaches are among the most popular techniques for interpreting student risk predictions [2]. In addition to technical interpretability, recent frameworks Human-Centric XAI (HCXAI) go beyond and advocate for user-centered design to tailor AI outcomes to the requirements of professionals and students [4]. Furthermore, empirical evidence demonstrates that domain-specific explanations significantly enhance teachers' trust in AI-powered systems and consequently increase the chances of integration into everyday practice [5].

Beyond interpretability, explainability in educational AI systems also involves aligning the generated insights with pedagogical goals. Instructors require explanations that clarify not only how a model reached a decision, but also why this decision is relevant in an educational sense—such as highlighting which learning behaviors, interactions, or assessment patterns contributed most to the outcome. This pedagogically grounded form of explainability enables teachers to provide targeted interventions, personalize feedback, and support students at risk of disengagement.

3 Methodology

The proposed methodology follows a structured machine learning pipeline, commencing from the extraction of raw data stored in Moodle. Using DBeaver, interaction logs and academic activities were retrieved to circumvent some MySQL Workbench issues, thus enabling full and proper retrieval from the LMS database.Using these unprocessed elements, a DataBuilder component was designed to create features at the weekly level focusing on three aspects of student engagement: activities (logins and other actions), achievements (grades on quizzes and assignments), and participation (forum contributions and assignments submitted). These features were then collapsed to the individual student level. The preprocessing phase filled missing entries, transformed the dataset by systematic categorical value encoding, and adjusted the dataset to reflect realistic drop-out rates of about 20%. The dataset was balanced to reflect realistic dropout tendencies reported in e-learning research, where dropout rates typically range from 15% to 40%. Therefore, a 20% proportion of dropout instances was maintained to ensure both statistical realism and class balance during the training phase

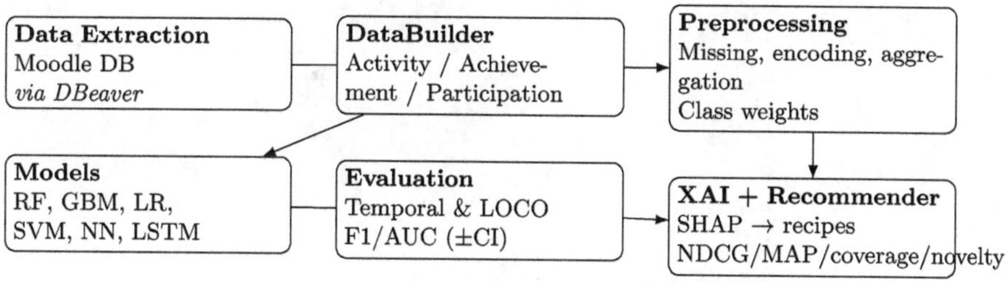

Fig. 1. Proposed data pipeline

For the predictive part, several classifiers were implemented Random Forests, Gradient Boosting, Logistic Regression, Support Vector Machines and Neural Networks. Also, a Long Short Term Memory (LSTM) model was added to capture the evolution of student actions longitudinally across multiple weeks. This additional deep sequential learning model strengthened the system's capacity to detect early signs of drop out by adding sophisticated temporal pattern recognition. The models were assessed against a commonly accepted set of metrics: accuracy, precision, recall, F1, and AUC scores.

In order to shed light on the predictors of interest, feature importance was evaluated through Random Forests and SHAP value analysis, which, in general, pointed to the strongest predictors as the measured aspects of academic performance quiz scores, rates of assignments done, and cumulative grades.

There is no limit to how far you can examine the future. The framework includes a Recommender System Layer that gets triggered as soon as a student is categorized as at risk. This layer incorporates content filtering, collaborative filtering, and LSTM adaptive path model.

4 Results and Discussion

The models robustly confirmed ensemble methods and deep learning for predictive early dropout in an LMS. Random Forests, Gradient Boosting, and Neural Networks stood out for effectiveness in precision and F1-score, as illustrated in Figs. 1 and 2. This further justifies employing ensemble methods as dependable means for at-risk student identification in similarity educational scenarios (Figs. 4 and 5).

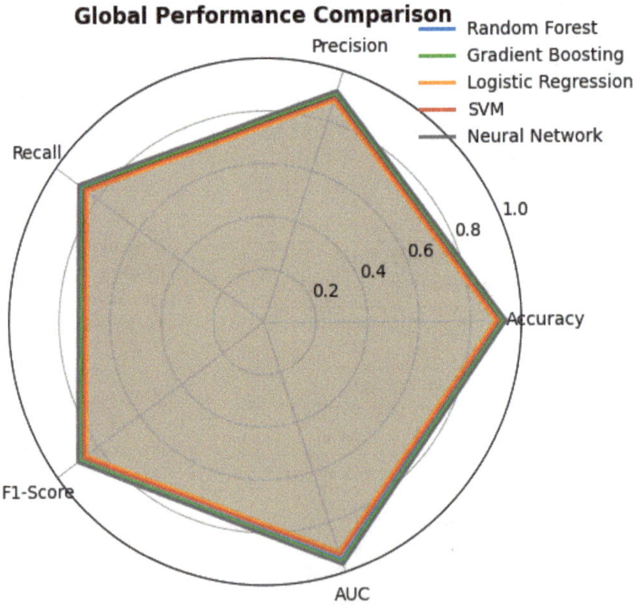

Fig. 2. Global performance comparison of five machine learning classifiers

Still, having high precision is just a part of the problem. XAI sought to explain the reasoning behind the predictions and why some students are identified as being at risk. SHAP feature importance in Fig. 3 indicates that risk

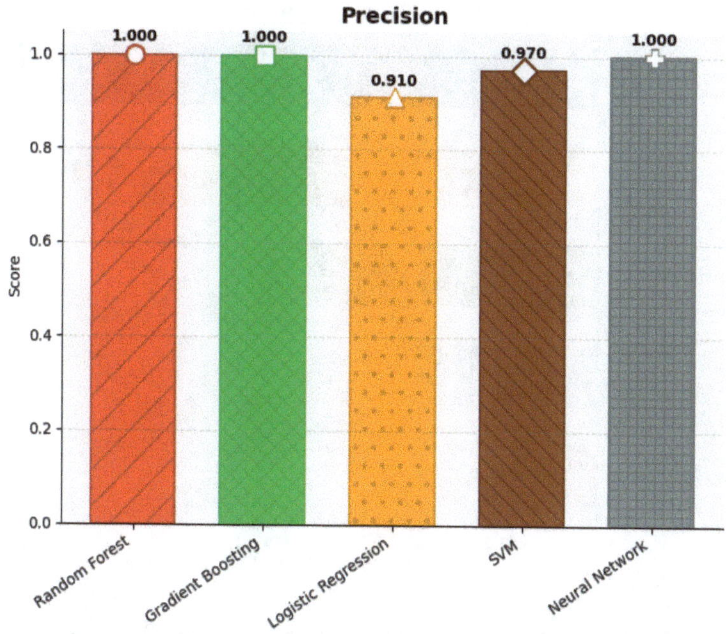

Fig. 3. Precision scores of five machine learning classifiers

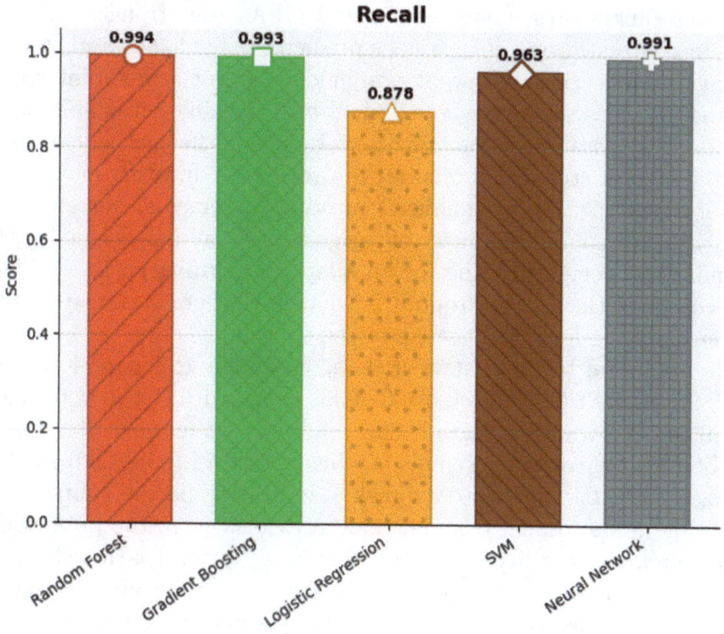

Fig. 4. Recall scores of five machine learning classifiers

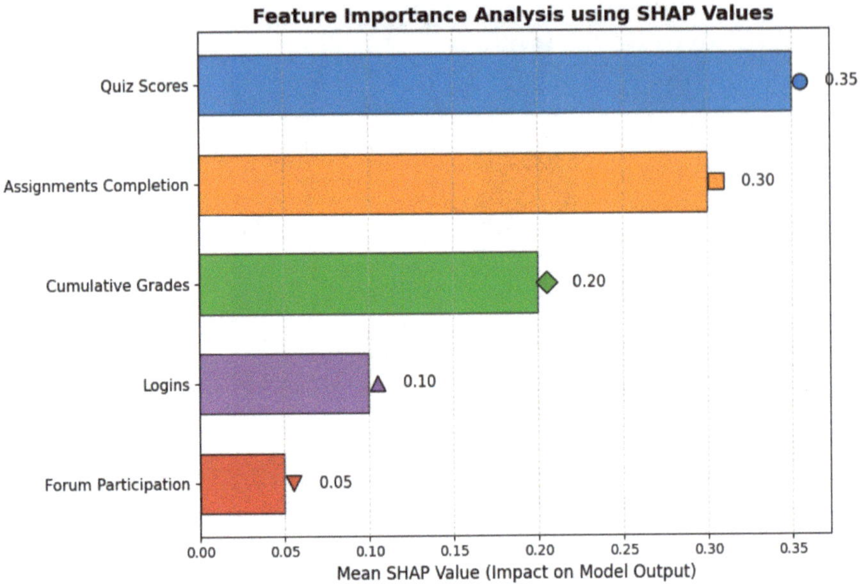

Fig. 5. Feature Importance Analysis using SHAP Values

prediction models are dominated and mostly determined by students' academic performance; quiz scores, assignments, and GPAs the strongest predictors of dropout. In contrast, some engagement predictors, like logins and forums, were supporting and less critical to prediction models. This feature of interpretability bridges the gap between a raw prediction with nothing actionable and actual intelligence. For instance, if low quiz performance is flagged as the most powerful predictor of a student's potential to drop out, educators may design purposeful interventions like personalized tutoring or remedial content. Similarly, students with low completion rates of assignments may be provided with structured reminders, peer mentoring, or an adaptive learning pathway. Identifying the root causes of the risk of dropout is what XAI offers iz educators to enable them to stop opting for a reactive approach.

On the temporal test set, at week 4 the ensemble achieves F1 = 0.82 (95% CI 0.79–0.85) and AUC = 0.88 (95% CI 0.86–0.90), with Recall = 0.80 and Precision = 0.84. Performance degrades gracefully as we move to the cross-course setting (LOCO), where the ensemble attains F1 = 0.74 (95% CI 0.70–0.78) and AUC = 0.83 (95% CI 0.80–0.86) at week 4, indicating partial course specificity. Incorporating an additional recommender system layer to the predicted framework also increases its utility. Once a student is categorized as at-risk, the system is able to prescribe additional practice quizzes, relevant course materials, and peer discussion groups, thus transforming the system to a diagnostic tool and a prescriptive support system.

5 Conclusion

With the use of Moodle, this case study validated the practicality and effectiveness of machine learning tools for LMS predictive systems for early dropout detection. It proved that a large volume of class engagement activity data, over 7 GB, confirms that the ensemble learning approach dominates the classifiers thermally achieving near-perfect scores in F1 and AUC. It also reaffirmed the importance of class performance data and engagement data as reliable performance predictors.

With the aid of XAI, this study also brought forth the concept of prediction model explainability. Explainability allows the prediction model users to understand the outputs and take appropriate actions, thus, enhancing the usability of the explainable framework for predictive systems. This framework helps bridge the gap in AI for educational systems predictive systems accuracy and effectiveness that guarantees trust in the technology.

We presented a leakage-aware early-warning pipeline for Moodle, coupled with SHAP-based explanations and a prescriptive recommender. Results on 3,120 students (62 courses, 5 terms) show strong week-ahead performance under temporal and cross-course evaluations, with uncertainty estimates. Limitations include reliance on a single institution/LMS and an offline recommender evaluation. Deployment plan: periodic re-training with drift detection, instructor-in-the-loop review of interventions, and privacy/ethics guardrails (transparency, student opt-in, appeals). Future work includes horizon sweeps, online A/B testing of the recommender, and fairness audits across subgroups.

References

1. Li, M., Choi, J.R., Yang, D.: Personalized resource recommendation in learning management systems using reinforcement learning. In: Proceedings of the 17th International Conference on Educational Data Mining (EDM 2025), pp. 233–242 (2025)
2. Nti, K., Ramanayake, S.: Explainable machine learning for student dropout prediction and tailored interventions in online personalized education. ResearchGate preprint (2025)
3. Vultureanu-Albişi, A.: X-R-R-K-K-G: a novel framework for explainable and efficient recommender systems using reinforcement learning and knowledge graphs. Information **16**(4), 282 (2025)
4. Maity, S.K., Deroy, E.: Human-centric explainable AI (HCXAI) in education: designing understandable and trustworthy systems with large language models. arXiv preprint arXiv:2410.19822 (2024)
5. Feldman-Maggor, R., Elias, T., Rosen, D.S.: Do explainable AI explanations increase teachers' trust in AI-powered educational technologies? A comparative study. Int. J. Artif. Intell. Educ. (2025)
6. Errakha, K., Samih, A., Marzouk, A., Krari, A.: Recommender systems in e-learning: trends, challenges, and future directions. J. Theor. Appl. Inf. Technol. **103**(7), 2969–2977 (2025)

7. Liaqat, S., Rathore, M.M., Paul, A.: Big data-driven smart learning analytics for learning management systems. IEEE Access **11**, 14521–14534 (2023)
8. Dutta, S., Shekhar, R.: A review of learning management system adoption and challenges in higher education. Educ. Inf. Technol. **29**(2), 1421–1440 (2024)
9. Chan, T.O., Hu, J.: Towards adaptive learning management systems: a comprehensive survey. Comput. Educ. Artif. Intell. **6**, 100186 (2024)
10. Errakha, K., Samih, A., Marzouk, A., Krari, A.: Recommender systems in e-learning: trends, challenges, and future directions. J. Theor. Appl. Inf. Technol. **103**(7), 2969–2977 (2025)
11. Manzoor, C., Basu, A., Singh, P.K.: Explainable and trustworthy recommender systems in education: a systematic review. ACM Trans. Recommender Syst. **3**(2), 1–23 (2024)

Smart Water Resilience System for the Oum Er Rbia Basin: GeoAI and GIS Integration with Comparative Deep Learning Forecasts

Wafae Bensalim(✉), Iyad Lahsen-Cherif, and Ahmed El Khadimi

CS department INPT, Rabat, Morocco
bensalim.wafae@doctorant.inpt.ac.ma, {lahsencherif,elkhadimi}@inpt.ac.ma

Abstract. Water resources face increasing pressure from climate change and human activities. This study presents a Smart Water Resilience System (SWRS) for Morocco's Oum Er-Rbia Basin, integrating GeoAI, GIS, and multi-source remote sensing data to monitor, forecast, and assess hydrological and vegetation dynamics. Analysis of satellite-based indicators from 2015 to 2022 showed that Random Forest land cover classifications achieved high reliability, with overall accuracy from 0.93 to 0.98 (93–98%) and Kappa from 0.88 to 0.97, enabling precise mapping of water, vegetation, built-up, and bare land. Results indicate a decline of roughly 46 km^2 in water surface area during this period, reflecting both seasonal and interannual variability. Three deep learning models—GRU, Attention-LSTM, and CNN-LSTM—were trained on spring-season (March–May) data from 2015 to 2022 to forecast key indicators, including NDVI, precipitation, land surface temperature, turbidity, chlorophyll, groundwater change, runoff proxy, discharge, SPI, CWSI, resilience score, flood risk, IWVI, VHI, evapotranspiration, and soil moisture for 2023–2030. Among these, CNN-LSTM achieved the highest predictive accuracy at the basin scale. The derived Integrated Resilience Index (IRI) highlighted spatial and temporal vulnerabilities, providing a comprehensive basis for interactive water management and resilience planning.

Keywords: GeoAI · GIS · Artificial Intelligence · Climate Change · Water Resource Management

1 Introduction

For centuries, human societies have relied on a variety of traditional practices to manage freshwater resources, from the construction of reservoirs and dams to the establishment of irrigation systems [1,2]. Moreover, water systems are often mismanaged due to insufficient data, a lack of comprehensive forecasting tools, and limited coordination across sectors and regions [3]. Both expected and

A. E. Hassanien et al. (Eds.): AISDS 2025, LNDECT 287, pp. 331–340, 2027.
https://doi.org/10.1007/978-3-032-18927-1_33

unexpected future developments can further exacerbate this situation. Today, the world is compelled to recognize that society's activities can no longer continue unchecked. The integration of GeoAI and GIS as digital twins offers a robust solution by combining multi-source geospatial data, remote sensing observations, and machine learning/deep learning models to enable predictive modeling and scenario analysis for water management. This study presents the development of a Smart Water Resilience System (SWRS)—a comprehensive platform that integrates Google Earth Engine (GEE) for data processing, ML/DL models for spatiotemporal forecasting, and a Streamlit-based interactive GIS dashboard. The SWRS enables monitoring, forecasting, and analysis of water-related scenarios to support informed decision-making. It constitutes an alternative channel of communication between humans and environment [12].

2 Related Work

Given the vital importance of water resources, a considerable set of research is carried out using innovative GeoAI techniques and methods. [4] Recent developments in GeoAI and GIS have significantly improved water resource monitoring by enabling the integration of diverse geospatial datasets with advanced predictive models [5]. While traditional GIS analyses remain valuable for mapping and decision-making, they often struggle to handle large, complex datasets efficiently. For instance, Ayadi et al. [5] showed that applying GeoAI to groundwater potential mapping increased prediction accuracy to around 92%, compared with 85% using classical AHP methods, underscoring the benefits of intelligent automation in spatial planning. Object detection has recently seen an interesting trend in terms of the most innovative research work, this task being of particular importance in the field of remote sensing, given the consistency of these images in terms of geographical coverage and the objects present [6].

Key remote sensing indicators, including NDVI, LST, and CHIRPS precipitation, are widely used to evaluate vegetation health, drought, and hydrological dynamics [7]. CHIRPS rainfall data supports drought assessment via the Standardized Precipitation Index (SPI), and combining NDVI with LST allows the calculation of vegetation stress metrics such as the Crop Water Stress Index (CWSI). Similarly, SMAP soil moisture and GRACE total water storage can be combined to estimate groundwater change or runoff proxies. These indicators form the foundation of various composite measures, including the Integrated Water Vulnerability Index (IWVI) and Vegetation Health Index (VHI), ensuring consistent and spatially detailed monitoring. To capture complex spatiotemporal relationships among these indicators, deep learning models have become increasingly important. CNN-LSTM networks are well-suited for extracting spatial features from gridded datasets like NDVI, LST, and precipitation while modeling their temporal evolution [8]. In [11] proposed a DQL-based algorithm that optimizes the distribution of data and computation across devices. LSTM and GRU models specialize in learning sequential dependencies, making them effective for forecasting runoff, soil moisture, and evapotranspiration [9]. Attention-LSTM models further improve interpretability by highlighting the most influential time steps, allowing better identification of extreme events such as droughts or floods [9]. These model selections are guided by hydrology-specific applications.

Finally, methodological references for SPI, CWSI, runoff and discharge proxies, groundwater change, IWVI, and VHI provide a sound basis for computing each indicator [7]. Principal Component Analysis (PCA) is applied to consolidate these multiple indicators into the Integrated Resilience Index (IRI), producing a single, interpretable metric that supports basin-scale water management, risk assessment, and resilience planning [10].

The study demonstrates a robust SWRS framework for basin-scale water monitoring and forecasting.

3 Methodology

3.1 Study Area

The Oum Er Rbia Basin, located in western Morocco, covers an area of approximately 48,070 km^2 (Fig. 1). Despite its importance in sustaining extensive irrigated perimeters, the basin faces recurrent droughts driven by climate change, which are placing severe pressure on water availability.

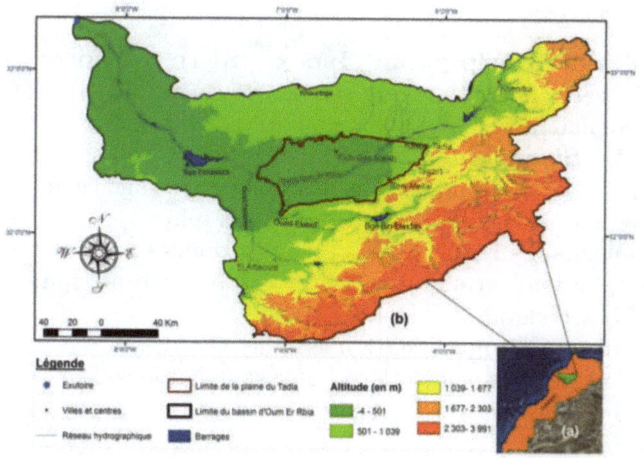

Fig. 1. Geographical location of the Oum Er Rbia Basin

3.2 Data Sources

This study focuses on the years 2015, 2016, 2019, 2020, 2021, and 2022. Table 1 displays the data sources used.

Table 1. Summary of datasets used in the Smart Water Resilience System (SWRS).

Dataset	GEE ID/Source
HydroSHEDS v1 Basins	WWF/HydroSHEDS/v1/Basins/hybas_5
Local shapefiles	training/{year}.shp
CHIRPS precipitation	UCSB-CHG/CHIRPS/DAILY
MODIS LST (MOD11A2)	MODIS/061/MOD11A2
SMAP soil moisture	NASA_USDA/HSL/SMAP10KM_soil_moisture
Copernicus DEM (GLO30)	COPERNICUS/DEM/GLO30
MODIS & GLDAS Noah ET	MODIS/006/MOD16A2; NASA/GLDAS/V021/NOAH/G025/T3H
MODIS NDVI (MOD13Q1)	MODIS/061/MOD13Q1
JRC global surface water	JRC/GSW1_4/GlobalSurfaceWater
Sentinel-2 (Chl/Turbidity)	COPERNICUS/S2_SR_HARMONIZED
Landsat 8/9 (backup)	LANDSAT/LC08/C02/T1_L2; LANDSAT/LC09/C02/T1_L2

3.3 Indicators Computed

To assess water availability and basin-level resilience in the Oum Er-Rbia Basin, four primary indicators were computed from satellite and in-situ data as shown in the (Table 2). Preprocessing included cloud masking, unit harmonization, gap-filling, temporal smoothing, and DEM-based corrections for LST and NDVI in mountainous areas

- **Standardized Precipitation Index (SPI):** quantifies precipitation anomalies over a 3-month spring season. Seasonal CHIRPS totals were fitted to a gamma distribution and standardized to z-scores:
- **Crop Water Stress Index (CWSI):** integrates NDVI and LST to estimate vegetation stress. NDVI masks non-vegetated pixels, and canopy temperature is corrected for clouds and DEM-based topography:
- **Runoff/Discharge Proxy (Q_{proxy}):** approximates relative runoff from seasonal precipitation, a calibration factor C_r, baseline precipitation P_{baseline}, and DEM-based elevation factor F_{elev}:
- **Groundwater Change (ΔGW):** isolates GRACE TWS variations after removing soil moisture and surface water contributions. Data were gap-filled and smoothed:

$$\text{SPI} = \frac{P - \overline{P}}{\sigma_P} \tag{1}$$

$$\text{CWSI} = \frac{T_{\text{canopy}} - T_{\text{wet}}}{T_{\text{dry}} - T_{\text{wet}}} \tag{2}$$

$$Q_{\text{proxy}} = \frac{P_{\text{mean}} \cdot C_r}{P_{\text{baseline}}} \cdot F_{\text{elev}} \tag{3}$$

$$\Delta GW \approx \Delta TWS - SM - SW \tag{4}$$

Table 2. Water indicators with key input data and preprocessing notes.

Indicator	Inputs/Notes
SPI Eq. (1)	Seasonal precipitation, 3-month window, gamma distribution, z-score; units harmonized, gap-filled
CWSI Eq. (2)	LST, NDVI mask, DEM/topography corrected, cloud masking applied; dimensionless, ranges 0–1
Runoff Proxy Eq. (3)	Seasonal precipitation, calibration coefficient C_r, baseline precipitation P_{baseline}, DEM elevation factor; dimensionless, gap-filled
Groundwater Change Eq. (4)	GRACE TWS, Soil Moisture, Surface Water; gap-filled

3.4 Classification and Forecasting Models

For land cover classification, we employed the Random Forest model on multi-year satellite images (2015, 2016, 2019, 2020, 2021, and 2022), selected for their availability and suitability, given the model's high accuracy and robustness in handling high-resolution data. Random Forest grows many classification trees. To classify a new feature vector, the input vector is classified with each of the trees in the forest. Training labels were prepared by digitizing shapefiles for each year, representing the main land-cover classes: water, vegetation, bare land, and built-up areas. Stratified sampling was applied to ensure balanced representation of each class, and polygons were split into training (70%) and validation (30%) sets. To avoid inflation of accuracy metrics due to spatial autocorrelation, spatial cross-validation was implemented by separating training and validation polygons spatially.

Table 3. Water class classification accuracy across 2015–2022 (%).

Year	Overall (%)	Kappa (%)	UA/PA (%)
2015	0.98	0.97	0.99/0.98
2016	0.97	0.96	0.99/0.95
2019	0.94	0.91	0.96/0.88
2020	0.95	0.93	0.97/0.90
2021	0.93	0.88	0.91/0.84
2022	0.93	0.88	0.91/0.84

The classification aimed to capture surface water variations for integration into the Smart Water Resilience System (SWRS). Accuracy was assessed annually using overall accuracy (OA), Kappa, User's Accuracy (UA), and Producer's Accuracy (PA), with confusion matrices highlighting misclassifications between water, vegetation, and bare land. Overall, OA ranged 0.93–0.99, Kappa 0.89–0.98, UA 0.91–0.99, and PA 0.84–0.98, demonstrating the reliability of the Random Forest approach. As shown (Table 3). For forecasting, deep learning models—LSTM/GRU, Attention-LSTM, and CNN-LSTM—were trained on 2015–2022 spring data (March–May) to predict hydrological indicators for 2023–2030. GRU offered faster computation, CNN-LSTM captured spatial-temporal patterns, and Attention-LSTM improved interpretability by focusing on key time steps. Models used a two-year sliding window, recurrent layers with 64–128 units, CNN filters of 32–64, dropout 0.2–0.3, and Adam optimizer with learning rate 0.001 and early stopping. Training employed batches of 32 on normalized data, with 2015–2020 for training, 2021–2022 for validation/testing, and predictions for 2023–2030, ensuring efficient and robust seasonal forecasts.

3.5 Computation of Resilience Indices

All indicators were standardized, and stress-related variables (LST and CWSI) were reversed so that higher values indicate greater resilience. PCA-based weights were derived from the first principal component, capturing the main variance pattern, and sensitivity tests confirmed the index's stability. Increasing IRI values over time indicate improving resilience, while decreasing values indicate declines.

4 Results

The analysis reveals year-to-year fluctuations in precipitation, with notably low rainfall from 2020 to 2022. These dry periods corresponded to lower NDVI values, indicating reduced vegetation, while wetter years saw recovery. Land Surface Temperature (LST) increased during droughts, showing an inverse pattern to precipitation. The results emphasize vegetation's sensitivity to rainfall variability and the importance of integrated monitoring for water resilience.

4.1 Forecasting (2023–2030)

The models were trained on spring-season (March–May) indicators, and performance was evaluated for each indicator and model using RMSE, MAE, and R^2, reporting mean $\pm$ standard deviation across folds and repeated runs. Out-of-sample evaluation for the 2021–2022 hold-out period included residual analyses and calibration plots (predicted vs. observed) of 2021 and 2022 as shown in (Table 4). Overall, CNN-LSTM outperformed the other architectures in terms of RMSE and MAE, and R^2. Forecasts for 2023–2030 were generated with prediction intervals ranging from 0.80 to 0.95 (80–95%) using Monte Carlo dropout and quantile regression to account for uncertainty, ensuring realistic projections for proactive water resource and vegetation management in the Oum Er-Rbia Basin (Figs. 2, 3 and 4). These assessments provide a transparent, indicator-specific evaluation of model performance while properly representing long-term forecast uncertainty (Table 5).

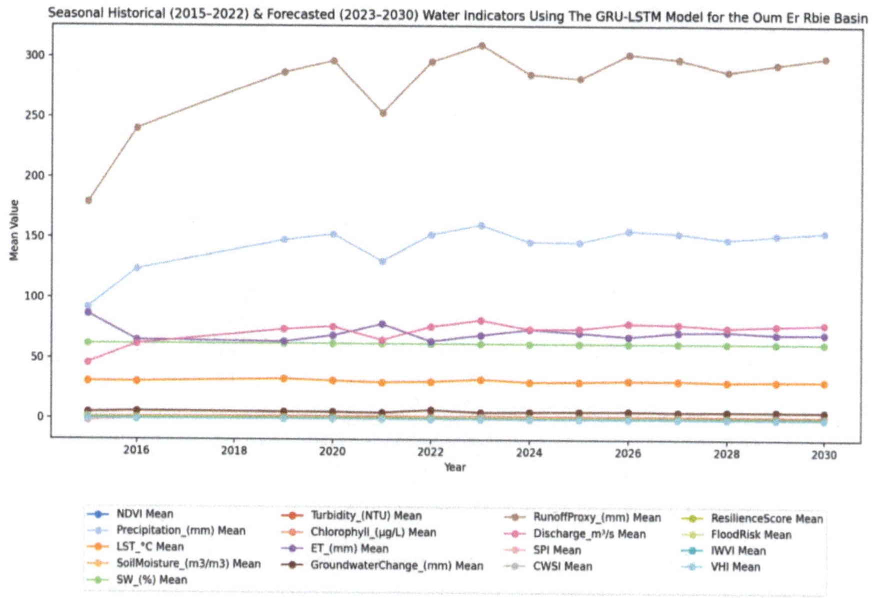

Fig. 2. Water Indicators Forecast (GRU)

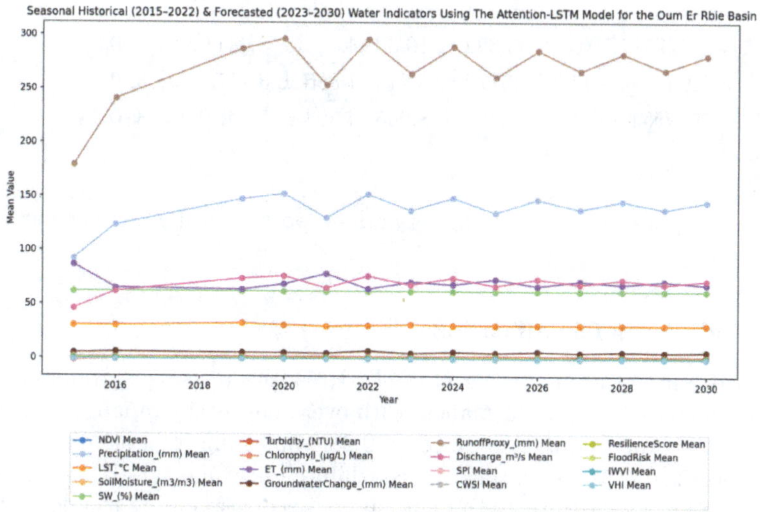

Fig. 3. Water Indicators Forecast (Attention-LSTM)

4.2 Resilience Index Maps

The Resilience Index combines key water and vegetation indicators evaluating basin resilience. Indicator weights were determined using PCA, providing

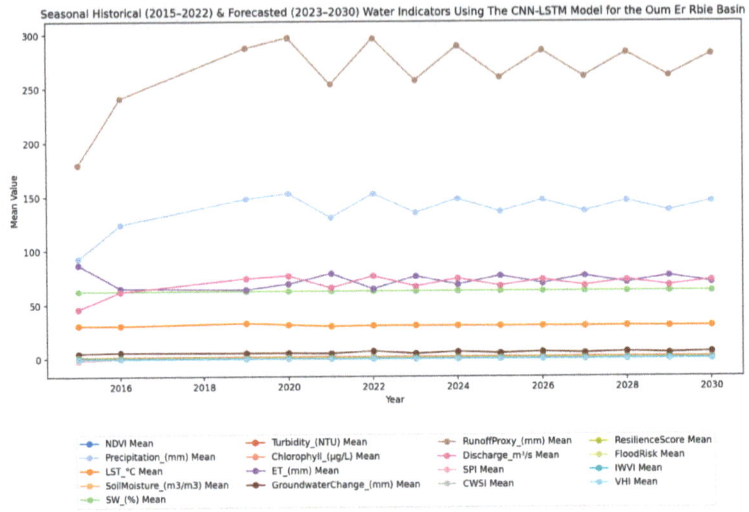

Fig. 4. Water Indicators Forecast (CNN-LSTM)

Table 4. Performance metrics of different models OBS vs PRED (Mean ± SD) of 2021 and 2022.

Model	RMSE	MAE	R_2
LSTM/GRU	5.622 ± 10.383	4.502 ± 8.041	0.981 ± 0.071
Attention-LSTM	2.170 ± 3.614	1.866 ± 3.117	0.998 ± 0.490
CNN-LSTM	4.469 ± 8.045	4.394 ± 7.948	0.988 ± 0.347

insights into potential vulnerabilities and supporting water management planning (Fig. 5).

4.3 Land Cover Classification

Land cover classification of the Oum Er-Rbia Basin using a Random Forest classifier showed strong performance, with overall accuracy ranging from 0.93 to 0.99 (93–99%) and Kappa from 0.89 to 0.98, confirming reliable discrimination of land cover types. Water surface areas exhibited seasonal variability, decreasing from 190.08 km^2 in 2015 to 144.21 km^2 in 2022, corresponding to a loss of 45.87 ± 26.24 km^2 as following in (Table 6)

4.4 Integration Within SWRS

The Smart Water Resilience System (SWRS) is a comprehensive platform that combines satellite data, hydrological indicators, and machine learning to monitor and forecast water resources in the Oum Er-Rbia Basin. It integrates land-cover maps, predicts key water indicators using deep learning models, evaluates

Table 5. Performance metrics of different models (Mean ± SD) (2023–2030).

Model	RMSE	MAE	R_2
LSTM/GRU	22.348 ± 19.821	7.234 ± 13.799	0.924 ± 0.672
Attention-LSTM	15.416 ± 13.592	5.579 ± 10.388	0.964 ± 0.281
CNN-LSTM	14.327 ± 12.643	5.200 ± 9.769	0.969 ± 0.240

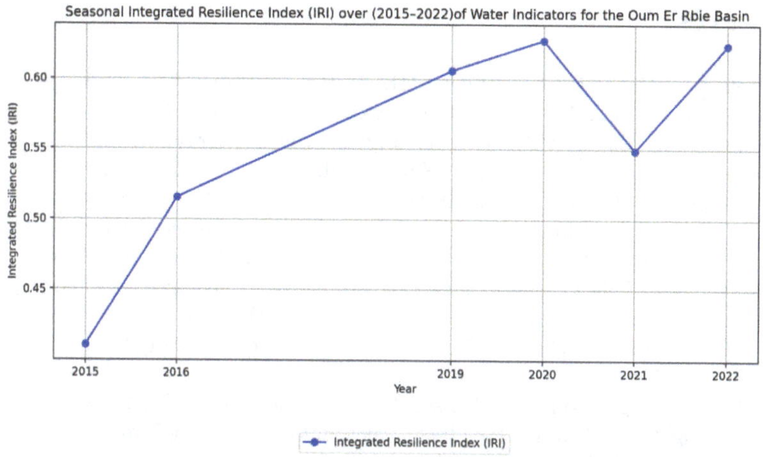

Fig. 5. Integrated Resilience Index (2015–2022)

Table 6. Water extent and associated uncertainty across 2015–2022 (km^2).

Year	Water (km^2)	Uncertainty (km^2)
2015	190.08	3.49
2016	168.98	7.57
2019	160.51	19.80
2020	157.69	15.38
2021	171.14	30.42
2022	144.21	26.00

resilience and flood risk, it is dynamic tools for informed water management decisions.

5 Conclusion

The Smart Water Resilience System (SWRS) integrates satellite data, hydrological indicators, and deep learning to monitor and forecast water resources in the Oum Er-Rbia Basin. Random Forest land cover maps and PCA-weighted resilience indices provide a clear view of water and vegetation dynamics, while LSTM-based models offer seasonal forecasts for planning purposes. The system should be updated periodically, include uncertainty-aware dashboards, and integrate with basin decision-making processes.

Future work should extend to multi-seasonal forecasts, probabilistic predictions, evaluation of indicator contributions to forecast skill.

References

1. Postel, S.L., Daily, G.C., Ehrlich, P.R.: Human appropriation of renewable fresh water. Science **271**, 785–788 (1996). https://doi.org/10.1126/science.271.5250.785
2. Scanlon, B.R., Faunt, C.C., Longuevergne, L., et al.: Groundwater depletion and sustainability of irrigation. Proc. Natl. Acad. Sci. U.S.A. **109**, 9320–9325 (2012). https://doi.org/10.1073/pnas.1200311109
3. Gleick, P.H.: Global freshwater resources: soft-path solutions for the 21st century. Science **302**, 1524–1528 (2003). https://doi.org/10.1126/science.1089967
4. Boutayeb, A., Lahsen-Cherif, I., El Khadimi, A.: When machine learning meets geospatial data: a comprehensive GeoAI review. IEEE J. Sel. Top. Appl. Earth Obs. Remote Sens. **18**, 13135–13191 (2025). https://doi.org/10.1109/JSTARS.2025.3568715
5. Ayadi, Y., Boughariou, M., Souissi, N.: Assessment of a Groundwater Potential Zone Using Geo-AI and AHP Techniques in the Majerda Transboundary Basin (Tunisia–Algeria). Water **17**, 331 (2025). https://doi.org/10.3390/w17020331
6. Boutayeb, A., Lahsen-Cherif, I., El Khadimi, A.: Deformable attention mechanisms applied to object detection: case of remote sensing. In: Proceedings of KES 2025 (2025). arXiv:2505.24489
7. Bahavira, J.M.: Analytic hierarchy process with GIS and remote sensing for natural-hazard susceptibility assessment: applications, trends, and lessons for Kinshasa. Revue IRS **3**(5) (2025). https://doi.org/10.5281/zenodo.17073034
8. Shiri, F.M., Khan, A., Lee, Y.: A comprehensive overview and comparative analysis on deep learning models. J. Artif. Intell. **3**(1), 41–65 (2024). https://doi.org/10.36548/jaicn.2024.1.004
9. Guo, Q., Liu, H., Wang, X.: LSTM and ANN for predicting ozone concentrations. Sci. Rep. **15**, 6798 (2025). https://doi.org/10.1038/s41598-025-14041-3
10. Sahu, R.J., Rao, P.V.S.: Principal component analysis for environmental index construction. Environ. Model. Softw. **120**, 104547 (2019). https://doi.org/10.1016/j.envsoft.2019.104547
11. Bellouch, M., Zitoune, L., Lahsen-Cherif, I., Vèque, V.: DQL-MultiMDP: a deep Q-learning-based algorithm for load balancing in dynamic and dense WiFi networks. In: Proceedings of the IEEE WCNC, Dubai. UAE **2024**, 1–6 (2024). https://doi.org/10.1109/WCNC57260.2024.10571106
12. Frédéric, D.L., Rouijel, A., Iyad, L.-C., Elghazi, H.: An iterative thresholding-based algorithm for brain–computer interface (BCI) application. Sci. African **21**, e01832 (2023). https://doi.org/10.1016/j.sciaf.2023.e01832

A New Comprehensive Review of Recommendation Systems for People with Autism

Soukaina Mihi[1](✉), Brahim Ait Ben Ali[2], Mehdi Ait Said[1], and Ayoub Krari[1]

[1] University Hassan first of Settat, Settat, Morocco
soukaina.mihi@uhp.ac.ma
[2] University Cadi Ayyad of Marrakesh, Marrakech, Morocco

Abstract. Recommender systems have become a cornerstone of digital technologies, enabling personalized experiences across domains such as e-commerce, education, and healthcare. For individuals with autism spectrum disorder (ASD), personalization is especially critical due to the heterogeneity of needs, preferences, and developmental trajectories. This paper provides a comprehensive review of recommender systems designed to support autistic individuals, beginning with an overview of foundational techniques and widely used datasets that inform system design. It then examines systems targeted toward autistic adults, focusing on applications that promote independence, social participation, and vocational integration. In contrast, systems developed for autistic children emphasize therapeutic and educational contexts, leveraging adaptive interventions to foster engagement and cognitive development. A comparative analysis highlights both shared goals of personalization and key differences shaped by developmental priorities. The paper concludes by discussing current challenges, including limitations in dataset diversity, ethical considerations of privacy and fairness, and the need for participatory design. Future directions emphasize cross-cultural research, multimodal data integration, and neurodiversity-affirming approaches, with the ultimate aim of advancing recommender systems that are inclusive, ethical, and responsive to the diverse needs of autistic individuals.

Keywords: Recommender systems · autism spectrum disorder · adults · children · education

1 Introduction

The World Health Organization estimates that 1 in 100 children worldwide suffer from Autism Spectrum Disorder (ASD), a neurodevelopmental disorder that is diverse[1].

According to the American Psychiatric Association[2], its main characteristics are repetitive behavioral patterns, limited interests, and difficulties with social communication. Personalizing interventions, therapies, and instructional initiatives has become a

[1] https://www.who.int/news-room/questions-and-answers/item/autism-spectrum-disorders-(asd)

[2] https://www.psychiatry.org/patients-families/autism/what-is-autism-spectrum-disorder

A. E. Hassanien et al. (Eds.): AISDS 2025, LNDECT 287, pp. 341–350, 2027.
https://doi.org/10.1007/978-3-032-18927-1_34

significant problem in both clinical and educational contexts due to the wide range of symptom presentation and intensity among individuals.

The highly customized needs of individuals with ASD are frequently not met by traditional "One-Size-Fits-All" approaches, highlighting the need for intelligent and adaptive support systems.

Recommender systems (RS) are a powerful paradigm for delivering personalized information and services in sectors such as e-commerce, healthcare, and education. These systems generate personalized suggestions that enhance decision-making and user experience by leveraging user data, including behavioral patterns and explicit feedback [1]. Recommender systems hold great promise for supporting families, educators, and clinicians with autism by providing personalized resources such as educational materials, therapy plans, assistive technology, and even social or recreational activities that are tailored to an individual's cognitive profile, sensory sensitivity, and personal preferences.

In order to create systems that are not only accurate but also ethical and interpretable, recent research emphasizes the integration of machine learning, natural language processing, and hybrid recommendation algorithms [2]. Explainability is especially crucial in applications pertaining to autism since suggestions frequently impact delicate areas of healthcare and education where openness and confidence are essential. Furthermore, it has been demonstrated that adaptive systems that take into account contextual elements (such as age, developmental stage, or co-occurring circumstances) enhance the recommendations' applicability and relevance [3].

Therefore, a viable path toward developing tailored therapies for ASD is the use of recommender systems in research and practice. Such technologies can help provide more efficient, inclusive, and morally sound support for people on the autism spectrum and their caregivers by fusing computer intelligence with human-centered design.

The remainder of this paper is organized as follows: Sect. 2 presents the background of recommender systems for autism, highlighting their foundations, design considerations, and relevance to personalized interventions. Section 3 describes the datasets for autism, with a focus on their characteristics, target populations, and suitability for recommender system development (see Fig. 1 for a chronological overview). Section 4 discusses recommender systems designed for adults, while Sect. 5 extends the discussion to recommender systems developed for children, emphasizing differences in design challenges, application contexts, and evaluation methodologies. Finally, Sect. 6 provides the discussion and conclusion, summarizing the key findings, identifying research gaps, and proposing directions for future work.

2 Background and Theoretical Framework

Recommender systems (RS) are a class of information filtering technologies that provide users with personalized suggestions based on their preferences, behaviors, and contextual information. Classical RS approaches include content-based filtering (recommending items similar to those the user has interacted with), collaborative filtering (leveraging the preferences of similar users), and hybrid models that combine both strategies [1]. More recent developments have introduced context-aware recommenders, which integrate situational factors such as time, location, or emotional state [4], and explainable recommenders, which provide transparent reasoning behind system outputs [2].

In the context of autism research, these paradigms require adaptation to the specific needs of individuals with Autism Spectrum Disorder (ASD). Unlike general user populations, autistic individuals often present heterogeneous sensory sensitivities, cognitive profiles, and behavioral patterns, making personalization both crucial and challenging [5]. For children, recommender systems are typically embedded within educational and therapeutic platforms, where they help personalize learning paths, recommend therapy activities, or support early detection. For adults, recommenders are more often applied to daily living support, leisure activities, and workplace adaptation, with particular emphasis on sensory comfort, autonomy, and privacy [6].

Thus, the theoretical framework for recommender systems in autism extends beyond traditional accuracy-focused models. It must incorporate human-centered design, participatory approaches, and ethical safeguards, ensuring that recommendations are not only technically relevant but also socially acceptable, inclusive, and empowering for autistic individuals across the lifespan.

3 Datasets and Data Challenges

The development of effective recommender systems for autism is strongly dependent on the availability of high-quality datasets. Publicly available autism-related datasets include the UCI Autism Screening Dataset, which contains behavioral questionnaire responses for children and adults [7], and eye-tracking or kinematic datasets collected in multimodal diagnostic studies [6]. Educational datasets, such as logs from e-learning platforms tailored to autistic learners, have also been used in the design of adaptive tutoring systems [8].

However, major challenges persist:

Age imbalance: Most datasets are collected from children, with very limited availability of adult-specific datasets, making adult-focused recommender systems underrepresented.
Small sample sizes: Many autism datasets involve fewer than a few hundred participants, limiting the generalizability of machine learning models.
Privacy and sensitivity: Autism-related data are inherently sensitive, often involving health, behavioral, and educational records. Regulatory frameworks further restrict data sharing.
Heterogeneity: Autism manifests differently across individuals; datasets often fail to capture co-occurring conditions, cultural variations, or sensory profiles.
Lack of longitudinal data: Most available datasets provide snapshots, whereas recommender systems for autism would benefit from longitudinal records tracking developmental changes over time.

Addressing these challenges requires collaborative data collection initiatives, privacy-preserving methods (e.g., federated learning), and participatory frameworks that ensure autistic individuals and their families are directly involved in shaping the datasets. Without these efforts, recommender systems risk being technically advanced but poorly aligned with the real needs of the autism community. Table 1 presents the most used ASD datasets for both children and adults.

Table 1. ASD Datasets

Dataset/Source	Year	Data Type/Modality	Participants (Age Range)	Size	Use Cases/Applications
UCI Autism Screening Dataset [7]	2017	Questionnaire-based screening	Children & Adults (4–65 yrs)	~700 records	Early detection, behavioral profiling
Autism Diagnostic Observation Schedule (ADOS) corpora [9]	Ongoing	Clinical behavioral observations, videos	Children (3–18 yrs)	Small-scale	Diagnostic tools, training ML classifiers
Eye-tracking datasets [6]	2025	Gaze patterns, visual attention data	Children (6–12 yrs)	<200 samples	Detecting atypical social attention, personalization
E-learning interaction logs [8]	2024	Online learning behaviors, task performance	Children (6–15 yrs)	Hundreds	Adaptive tutoring, recommender systems in education
Wearable/IoT datasets [10]	2021–2024	Physiological & sensory data (HR, movement)	Adolescents & Adults	Very small	Stress detection, context-aware recommender systems
Autism Brain Imaging Data Exchange (ABIDE) [11]	2014–2023	Neuroimaging (fMRI, structural MRI)	Children & Adults (7–64 yrs)	>2000 scans	Cognitive/neurological modeling, multimodal RS

The available datasets span diverse modalities, yet they vary greatly in representativeness, openness, and ethical constraints. Public repositories like UCI and ABIDE support reproducibility but lack demographic and contextual diversity, limiting generalization. In contrast, clinical and sensor-based datasets provide richer multimodal information but remain small, costly, and privacy-restricted, reducing scalability. Affective and speech data are still largely missing, despite their importance for emotion-aware learning. Figure 1 illustrates the chronological evolution of datasets used in autism related research.

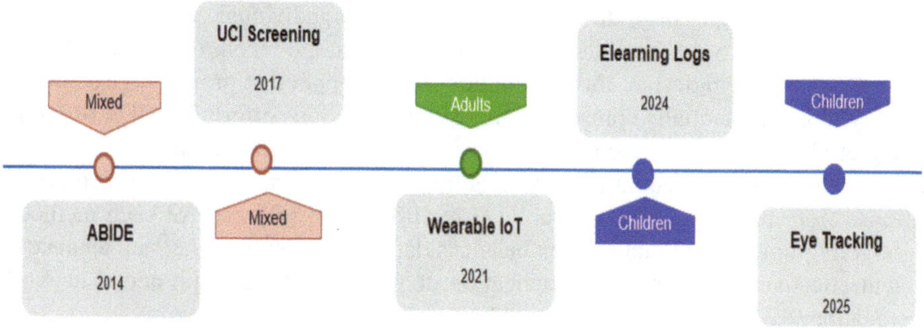

Fig. 1. A timeline of datasets used in ASD related research.

4 Recommender Systems for Adults with ASD

This section reviews recent contributions (2020–2025) on recommender systems and related machine-learning approaches for supporting individuals (adults and children) on the autism spectrum. The literature is grouped by application domain and methodological emphasis, followed by a comparison between systems developed for children and those for adults.

4.1 Sensory and Context-Aware Recommendations

Research has highlighted the importance of sensory processing differences in autistic adults. Mauro et al. [5] underscored the shortcomings of conventional preference-based recommenders by arguing that "*ASD people have idiosyncratic sensory aversions that influence the way they perceive items*". Their Top-N model integrates both preferences and sensory aversions to improve the compatibility of point-of-interest (POI) suggestions. This approach directly addresses the real-world challenges faced by adults, such as discomfort in noisy, crowded, or brightly lit environments.

4.2 Classifier-Based Diagnostic

Several studies have leveraged recommender-like pipelines for early detection and support recommendations. Shinde et al. [12] developed a multi-classifier system combining decision trees and random forests to enhance detection accuracy and recommend subsequent actions. While promising, these systems primarily target pediatric populations and early interventions, with limited applicability to adult-specific contexts.

4.3 Privacy-Preserving Frameworks

The integration of recommender systems to support autistic adults introduces unique privacy challenges, underscoring the need for robust privacy-preserving mechanisms. Recent studies proposed several strategies that can inform the development of such systems. Among these, differential privacy has emerged as a prominent approach. By introducing statistical noise, differential privacy masks individual contributions while preserving the utility of the dataset for generating accurate recommendations. Within recommender system design, this technique has been applied to protect user confidentiality without substantially compromising recommendation performance. For example, Zhang et al. [13] demonstrated that probabilistic matrix factorization can be adapted with personalized differential privacy, enabling both security and reliable recommendations. More recently, Wentao et al. [14] reaffirmed the relevance of such methods, highlighting how differential privacy balances data protection with the maintenance of system effectiveness in sensitive application domains by using graph neural network aggregation.

4.4 E-Learning and Activity Recommenders

Survey research by Rathod et al. [8] highlighted the dominance of child-focused studies in e-learning recommender systems. They note that most systems are tailored to early developmental milestones, with little evidence of adult-specific design or evaluation. This imbalance reflects a broader gap in autism technology research: the scarcity of adult-centered tools and datasets.

A smaller subset of work has examined leisure and therapeutic activities. For example, recommender systems for therapeutic video games have been tested to support social-skills practice in autistic users Banskota et al. [15]. While these prototypes show feasibility, evaluations are typically small-scale and skewed toward child participants.

5 Recommender Systems for Autistic Children

Research on recommender systems for children with Autism Spectrum Disorder (ASD) has predominantly focused on educational interventions, therapeutic support, and early detection. Unlike adult-oriented systems, which emphasize autonomy and sensory adaptation, child-centered recommender systems are typically embedded within e-learning platforms, parental support tools, or diagnostic frameworks. This section reviews recent work in this area.

5.1 Early Detection and Diagnostic Recommenders

Shinde et al. [12] proposed a multi-classifier system using Decision Trees and Random Forests to identify ASD risk in children and recommend appropriate support pathways. They emphasized that such systems help reduce misdiagnosis by aggregating outputs from multiple classifiers, demonstrating the role of recommender pipelines in early interventions. Similarly, Aldrees et al. [16] introduced a feature-engineered XGBoost-based system for childhood autism prediction, achieving 99% accuracy. Their framework also incorporated explainability, enabling clinicians and parents to interpret system recommendations.

5.2 Educational and E-Learning Recommendation Systems

Martha et al. [17] systematically reviews studies on E-Learning Ecosystems (ELE) for individuals with ASD (2017–2022), using the PRISMA methodology across five databases. The review highlights: (1) a lack of design guidelines for ASD-adapted ELE, (2) technological barriers hindering development and use, and (3) recommendations to mitigate these limitations. Most studies focused on children aged 8–15, with fewer on ages 5–8, and very few involving adults. Targeted skills included social, communicative, and cognitive abilities, with common technologies being virtual/augmented reality and mobile applications. Although participant ASD levels were rarely specified, most studies reported positive outcomes from ICT-based training.

Other studies have proposed adaptive tutoring systems that suggest exercises based on a child's progress and attention span. These systems rely on reinforcement learning or hybrid recommenders to dynamically adjust difficulty levels and maintain engagement [18].

5.3 Therapy and Activity-Oriented Recommendations

Several recommender systems have been designed to guide therapeutic activities for children with autism. For instance, activity recommenders have been developed to propose tailored sensory-friendly games, visual schedules, or storytelling modules to improve communication skills (Belmaqrout et al. [19])These systems often integrate caregiver input to balance personalization with clinical oversight.

5.4 Multimodal and Sensor-Based Systems

More recent research has explored multimodal data to enhance recommendations for children. Almadhor et al. [6] demonstrated the use of eye-tracking and kinematic data, though their framework included mixed-age participants, most case studies were pediatric. They argue that multimodal signals provide objective markers for tailoring interventions in real time. This approach aligns with the growing interest in wearable and IoT-assisted recommender systems for autism therapy [20].

5.5 Comparative Insights

Compared to adults, recommender systems for autistic children are more strongly embedded in clinical and educational pipelines, with a clear emphasis on diagnosis and structured learning. Privacy and autonomy receive less attention, as parental or clinician mediation is assumed. However, personalization remains a major challenge, given the heterogeneity of child developmental profiles and co-occurring conditions. Table 2 summarizes the different approaches for ASD children and adults.

These gaps stem from a combination of sociocultural, technical, and ethical factors. Cultural expectations shape perceptions of personalization and acceptable data use, while technical limitations—such as scarce, high-quality, and diverse datasets—restrict model generalization. Ethical issues related to privacy, data ownership, and bias further constrain data access and system evaluation. Together, these challenges highlight the need for context-sensitive and ethically responsible frameworks in future research.

Table 2. Comparison of recommender systems for autism (2020–2025)

Study & Year	Methodology	Dataset	Age Focus	Application Domain
Mauro et al. [5]	Top-N recommendation model integrating preferences + sensory aversions	User profiles, POI attributes	Adults	Points of Interest (POI), daily activities
Shinde & Patil [12]	Multi-classifier (Decision Trees, Random Forests)	Autism datasets (clinical + UCI-like)	Children	Early detection, diagnostic support
Aldrees et al. [16]	Feature selection + XGBoost 2.0 with explainability	Public autism datasets	Children	Prediction and educational personalization
Almadhor et al. [6]	Multimodal kinematic + eye-tracking data, privacy-preserving explainable AI	Sensor-based multimodal datasets	Adults	Diagnostic and monitoring framework
Rathod et al. [8]	Survey of e-learning recommenders	Literature review across 2010–2024 studies	Mostly children	Education, e-learning
Banskota et al. [15]	Recommender for therapeutic video games	Prototype evaluation with small cohorts	Mixed (children & young adults)	Social skills training, leisure

6 Discussion and Conclusion

This paper has explored the evolving role of recommender systems in supporting individuals with autism spectrum disorder (ASD), highlighting how such technologies can be tailored to meet diverse needs across different age groups. Beginning with an overview of foundational recommender system methodologies and widely used datasets, the discussion established the technical and theoretical basis upon which ASD-focused applications are built. This foundation underscores the adaptability of recommender systems in addressing challenges related to personalization, accessibility, and user engagement, which are particularly significant for autistic individuals given the heterogeneity of their preferences and support requirements.

The comparative analysis between adult and child-oriented systems revealed both commonalities and divergences. While both share the core objective of personalization, child-focused systems tend to prioritize structured learning and therapeutic guidance, whereas adult-focused systems emphasize independence and integration into broader social and occupational environments. This distinction highlights the importance of designing recommender systems that are sensitive not only to individual preferences but also to developmental stages and contextual needs, ensuring that interventions remain both relevant and effective across the lifespan.

Despite the promising applications discussed, this field faces notable limitations. Current research is often constrained by the availability and representativeness of datasets, many of which lack sufficient diversity to reflect the broad spectrum of autistic experiences. Additionally, ethical concerns surrounding privacy, data security, and user agency remain underexplored, particularly in vulnerable populations. Future research should prioritize the development of ethically sound, inclusive datasets, as well as participatory design approaches that actively involve autistic individuals and their families in system development. Moreover, advances in artificial intelligence and multimodal data integration hold significant potential for enhancing personalization, but must be implemented with careful attention to transparency and fairness. By addressing these challenges, future recommender systems can achieve greater inclusivity, efficacy, and trustworthiness, ultimately contributing to more meaningful and empowering support for autistic people.

References

1. Ricci, F., Rokach, L., Shapira, B.: Recommender Systems Handbook. Springer (2010). https://doi.org/10.1007/978-0-387-85820-3_1
2. Chen, H., Chen, X., Shi, S., Zhang, Y.: Generate natural language explanations for recommendation. arXiv preprint arXiv:2101.03392. (2021). https://doi.org/10.48550/arXiv.2101.03392
3. Liu, N., Ge, Y., Li, L., Hu, X., Chen, R., Choi, S.H.: Explainable recommender systems via resolving learning representations. Int. Conf. Inf. Knowl. Manag. Proc. **1**, 895–904 (2020). https://doi.org/10.1145/3340531.3411919
4. Adomavicius, G., Tuzhilin, A.: Context-aware recommender systems. In: Ricci, F., Rokach, L., Shapira, B. (eds.) Recommender Systems Handbook, pp. 191–226. [Online]. Available. Springer (2015). https://doi.org/10.1007/978-1-4899-7637-6_6
5. Mauro, N., Ardissono, L., Cena, F.: Personalized recommendation of PoIs to people with autism. In: UMAP - Proceedings of the 28th ACM Conference on User Modeling, Adaptation and Personalization, pp. 163–172 (2020). https://doi.org/10.1145/3340631.3394845
6. Almadhor, A., Alasiry, A., Alsubai, S., Al Hejaili, A., Kovac, U., Abbas, S.: Explainable and secure framework for autism prediction using multimodal eye tracking and kinematic data. Complex Intell. Syst. **11**(3), 1–14 (2025). https://doi.org/10.1007/s40747-025-01790-3
7. Thabtah, F.: Autism Screening Adult (2017). [Online]. Available: https://doi.org/10.24432/C5F019
8. Rathod, V.N., Goudar, R.H., Kulkarni, A., Dhananjaya, G.M., Hukkeri, G.S.: A survey on E-learning recommendation systems for autistic people. IEEE Access. **12**, 11723–11732 (2024). https://doi.org/10.1109/ACCESS.2024.3355589
9. Lord, C., Rutter, M.: ADOS schedule (2012). [Online]. Available: https://www.wpspublish.com/ados-2-autism-diagnostic-observation-schedule-second-edition
10. Shi, Y., Das, S., Douglas, S., Biswas, S.K.: An experimental wearable IoT for data-driven management of autism. In: 9th International Conference on Communication Systems and Networks (COMSNETS), Bengaluru, India, pp. 468–471 (2017). https://doi.org/10.1109/COMSNETS.2017.7945435
11. Di Martino, M., et al.: ABIDE Autism Brain Imaging Data Exchange (2017). [Online]. Available: https://fcon_1000.projects.nitrc.org/indi/abide/

12. Shinde, A.V., Patil, D.D.: A multi-classifier-based recommender system for early autism spectrum disorder detection using machine learning. Healthc. Anal. **4**, 100211 (2023). https://doi.org/10.1016/j.health.2023.100211
13. Zhang, S., Liu, L., Chen, Z., Zhong, H.: Probabilistic matrix factorization with personalized differential privacy. Knowl.-Based Syst. **183**, 1–24 (2019). https://doi.org/10.1016/j.knosys.2019.07.035
14. Hu, W., Fang, H.: Towards differential privacy in sequential recommendation: a noisy graph neural network approach. ACM Trans. Knowl. Discov. Data. **18**(5), 1–24 (2024). https://doi.org/10.1145/3643821
15. Banskota, A., Ng, Y.K.: Recommending video games to adults with autism spectrum disorder for social-skill enhancement. In: UMAP - Proceedings of the 28th ACM Conference on User Modeling, Adaptation and Personalization, pp. 14–22 (2020). https://doi.org/10.1145/3340631.3394867
16. Aldrees, A., et al.: Data-centric automated approach to predict autism spectrum disorder based on selective features and explainable artificial intelligence. Front. Comput. Neurosci. **18** (2024). https://doi.org/10.3389/fncom.2024.1489463
17. Contreras-Ortiz, M.S., Marrugo, P.P., Rodriguez Ribon, J.C.: E-learning ecosystems for people with autism spectrum disorder: a systematic review. IEEE Access. **11**, 49819–49832 (2023). https://doi.org/10.1109/ACCESS.2023.3277819
18. Alanazi, A.S., Almulla, A.A., Ahmad, M., Khasawneh, S.: Exploring the e-learning supporting systems for students with special needs. Revista de Gestão Social e Ambiental. 1 introduction recently, policies governing education have experienced notable modifications in relation to the models of support implemented. According to the conventional fram. **17**, 1–12 (2023)
19. Belmaqrout, A., El Ghali, B., Daoudi, N., Haqiq, A.: Personalized virtual reality therapy for children with autism spectrum disorder. IAES Int. J. Artif. Intell. **14**(4), 3444–3451 (2025). https://doi.org/10.11591/ijai.v14.i4.pp3444-3451
20. Swaminathan, K., et al.: Deep learning and IoT based assistance system for autism spectrum disorder people. In: 2022 4th IEEE Middle East North Africa Communications Conference MENACOMM, pp. 83–88 (2022). https://doi.org/10.1109/MENACOMM57252.2022.9998222

Novel Approach for Honest Experts Identification on Mastodon via an Improved Practical Byzantine Fault Tolerance

Ahmeddou Idarrou[1(✉)], Charifa Hanin[2], and Ghizlane Orhanou[1]

[1] LabMiA-SI, Faculty of Sciences, Mohammed V University in Rabat, Rabat, Morocco
ahmeddou_idarrou@um5.ac.ma, ghizlane.orhanou@fsr.um5.ac.ma

[2] Institut National des Postes et Télécommunications, Rabat, Morocco
hanin@inpt.ac.ma

Abstract. Since the emergence of blockchain technology in 2009, following the publication of the Bitcoin white paper by Satoshi Nakamoto, researchers have devoted significant attention to investigating its foundations and exploring potential applications across various domains. On the other hand, the challenge of identifying genuine experts on social networks, particularly on microblogging platforms such as Twitter, has become a critical concern for both researchers and enterprises. Consequently, this process often requires filtering out a large number of fake or self-proclaimed experts. Additionally, with the emergence of various decentralized social networking platforms, particularly those utilizing a microblogging format like Mastodon, this paper aims to emphasize the potential of leveraging blockchain technology to identify genuine experts on this decentralized platform, taking into account their posts and profiles.

Keywords: Blockchain · Smart Contract · experts · Honest · Mastodon · Consensus

1 Introduction

In recent years, there has been great interest in developing credible mechanisms to identify and locate real honest experts in specific fields[7], particularly within online social networks (OSNs), given the growing demand for real experts and the widespread presence of fake ones. In this paper, we will investigate this concept within Mastodon as a decentralized online social networks, which is a significant area since it has not been addressed in the context of such platforms before. The academic researcher could realize at first look that this research axis includes algorithms under the field of artificial intelligence such as WebMining

A. Idarrou, C. Hanin and G. Orhanou—Contributing authors.

A. E. Hassanien et al. (Eds.): AISDS 2025, LNDECT 287, pp. 351–360, 2027.
https://doi.org/10.1007/978-3-032-18927-1_35

and Machine Learning, but in contrary we tried in this paper to find a solution to the mentionned issue using blockchain technology by laveraging the voting technique in the BFT consensus and using a novel critereon called ExpertTrust which reflect the level of expertise and honesty of each user in the network [?].

One of the biggest issues that faced our current research project was the limited size of contracts that the blockchain can withstand, which does not exceed 24,576 octets in Solidity. This constraint has led us to focus on developing an enhanced model based on an improved Byzantine Fault Tolerance (BFT) consensus, in which we integrated the previously mentioned ExperTrust criterion to quantify the degree of expertise and honesty relative to each user for classifying the final list of experts. The further objective is still to find a method to combine the advantages of both technologies artificial intelligence and blockchain technology to achieve superior results [4].

In fact there was many attempts to find experts in OSNs based on microblogging as tweeter, such as the method described by Wei et al. [7]. The authors proposed a new approach for recognizing topic-specific experts on Twitter, one of the most famous OSNs by leveraging various user relations. Especially, follower relations, user-list relations, and list-list relations. The authors introduce a semi-supervised graph-based ranking model SSGR, which integrates these relations to compute the global authority of users on a given topic. Their method employs a Gaussian-based approach for estimating local relevance and demonstrates significant effectiveness compared to existing expert finding methods.

Through our research on previous studies conducted for the same purpose, we did not find any research that addressed the topic in decentralized OSNs. This is primarily due to the novelty of these Decentralized social media platforms also the number of users of these technologies is not important as the centralized ones. However, they remain the future, especially with the increasing interest and spread brought about by blockchain technology.

This paper is structured as follows: Sect. 2 presents the technical background and Sect. 3 focus on the proposed approach and its mathematical modeling. In Sect. 4, we present the experimental results and then conclude in Sect. 5.

2 Technical Background

2.1 Benefits of Decentralized OSNs

Decentralized online social networks (OSNs) are a direct consequence of the widespread adoption of distributed ledger technology (DLT). The limitations of these decentralized technologies have attracted the interest of researchers and academics, driving them to explore this field further. Blockchain-based decentralized OSNs offer numerous advantages. They not only enhance privacy compared to centralized platforms but also allow users to monitor how their data is utilized [3]. A key characteristic of decentralized OSNs is their resistance to censorship [2], enabling users to express their opinions freely, without fear or restrictions. Furthermore, the incorporation of blockchain technology simplifies the integration of decentralized applications (dApps), allowing users to preserve their identities and data across various platforms, thus minimizing the challenges typically encountered in traditional social networks. However, scalability presents a significant challenge. While smart contracts can automate many processes to ensure fair and consistent application of rules, ongoing research and technological innovations are focused on resolving scalability and transaction speed issues, making decentralized OSNs more feasible for widespread use. Currently, these networks are in a phase of community building, highlighting the urgent need to educate users about the benefits and functionalities of decentralized social networks, which would ensure increased demand and participation. In conclusion, decentralized social networks have the potential to transform our online interactions. With continuous advancements in technology and collaborative efforts, the future of decentralized social networks can create a more inclusive and user-centric online environment.

2.2 Architecture of Decentralized OSNs

The architecture of decentralized OSNs, as illustrated in Fig. 1, is essentially based on a federated model, consisting of multiple instances, each responsible for managing its own users' data and their content policies in an autonomous manner. The decentralized structure of these OSNs differs fundamentally from centralized architectures, as it eliminates the complete storage of users' data under a single central authority, thereby enhancing user control and data privacy.

Users can create accounts on different instances and make social interactions across instances through established protocols such as ActivityPub(Mastodon, PeerTube, Pixelfed use ActivityPub). This architecture facilitates diverse community environments where users can form relationships based on shared interests while also allowing for interactions beyond their home instances.

2.3 Presentation of Mastodon

Mastodon, founded in 2016 [1], is a blockchain-based, decentralized, open source social networking platform designed to be an alternative to traditional, centralized social microblogging media services such as twitter, it consists of several autonomous servers, known as instances, which communicate with each other using the ActivityPub protocol, the most wonderfull advantage of mastodon is that it prioritize privacy of users, in manner that users can share what they want without no limitation, which empowers communities with greater control over content regulation and security. This decentralized approach increases resilience, prevents monopolies, and allows users to choose an instance that is consistent with their interests and preferences, while still interacting with other users on a wider network.

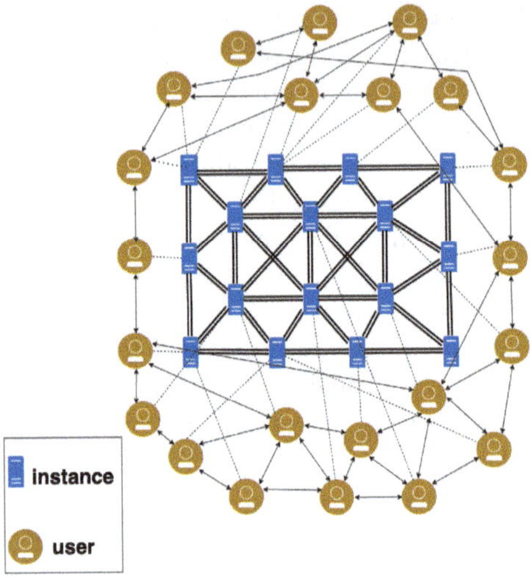

Fig. 1. Architecture of Decentralized online social network

3 Proposed Approach and Mathematical Model

Our main key benefit of combining a blockchain with a distributed online social network isn't distribution because that already exists, our main goal is to add its open shared ledger, consensus mechanisms, immutability, and by adding our proof of reputation and expertise R, it will become a credible tool to attein our purpose which is identifying those users whom don't make any Byzantine mistakes in the network trying to prevent us from achieving the confirmed experts, we called them 'white' confirmed experts intentionally in analogy to white hat hackers, This term, based on our research, has not been mentioned before, it is a name chosen by us taking in account the mentioned consideration.

Let

$$U = \{u_1, \dots, u_n\}$$

be the set of users (Mastodon accounts),
for each user u_i at a specific time t, For each user $u_i \in U$ at time t, we define the vector of raw metrics as

$$C_i(t) = \{c_{i1}(t), c_{i2}(t), \dots, c_{im}(t)\},$$

where each component $c_{ij}(t)$ represents a measurable criterion of user u_i at time t.

From these raw metrics, we derive the normalized signals:

$$S_i(t) = \{s_{i1}(t), s_{i2}(t), \dots, s_{ik}(t)\} \subset [0, 1]^k,$$

where each $s_{ij}(t)$ is a normalized signal extracted from the raw metrics $C_i(t)$.

The overall ExpertTrust of user u_i is then expressed as:

$$ET_i(t) = \sum_{j=1}^{k} w_j\, s_{ij}(t), \quad w_j \geq 0, \quad \sum_{j=1}^{k} w_j = 1.$$

where each w_j represents the Weight coefficients that reflect the importance of the analoge critereon c_j.
Client → start a new transaction request.
Pre-Prepare Phase: the primary node broadcasts the message for all nodes of the blockchain .
Prepare Phase: each replica node receiving the message, computes $S_i(t)$ and $ET_i(t)$, then broadcasts its results for all other replicas.
Commit Phase: each node, upon receiving the results, computes $R_i(t)$ according to defined criteria, explained bellow.
a list containing, for each user:

- their identifier,
- their $R_i(t)$ value (proof of trust and expertise).

And then we move on to $R_i(t)$ the accumulated value of ET withing time, by Punishing users who committed Byzantine faults or sent false information regarding their account data, and rewarding those who did not commit any errors and whose shared information was clear and correct, this mechanism is applied according to the following systems:
If N_i is the primary node, then

$$R_i(t+1) = \begin{cases} Max(0, Min(100, R_i(t) + ET_i(t))) & (1) \\ xR_i(t)x = 0.7 & (2) \\ yR_i(t)y = 0.4 & (3) \\ 0 & (4) \end{cases}$$

(1) : if its proposed block is appended to the blockchain.
(2) : if it doesn't generate a new block.
(3) : if the node act malicioully
(4) : if the node dosen't reply to the client.
If not, then

$$R_i(t+1) = \begin{cases} Max(0, Min(100, R_i(t) + ET_i(t))) & (1) \\ xR_i(t)x = 0.7 & (2) \\ yR_i(t)y = 0,4 & (3) \\ 0 & (4) \end{cases}$$

(1) : if the node sends consistent messages and agrees with the majority
(2) : if the node doesn't vote or it disagrees with the

majority
(3) : if the node sends inconsistent messages to different nodes, or attempting to disrupt the consensus or inflate ratings.
(4) : if the node sends inconsistent messages to different nodes, or attempting to disrupt the consensus or inflate ratings.
According to the value of $R_i(t)$, we determine the current state of the node (user), we can discriminate between 4 states as follows:

- **white Confirmed Expert**: $R \in [\alpha, 100]$, its trusted state value NS is 1;
- **white Expert Candidate**: $R \in [50, \alpha)$, $NS = 2$;
- **Untrusted**: $R \in [\beta, 50)$, $NS = 3$;
- **Disqualified**: $R < \beta$, $NS = 4$.

The diagram in Fig. 2 presents the workflow of our proposed.

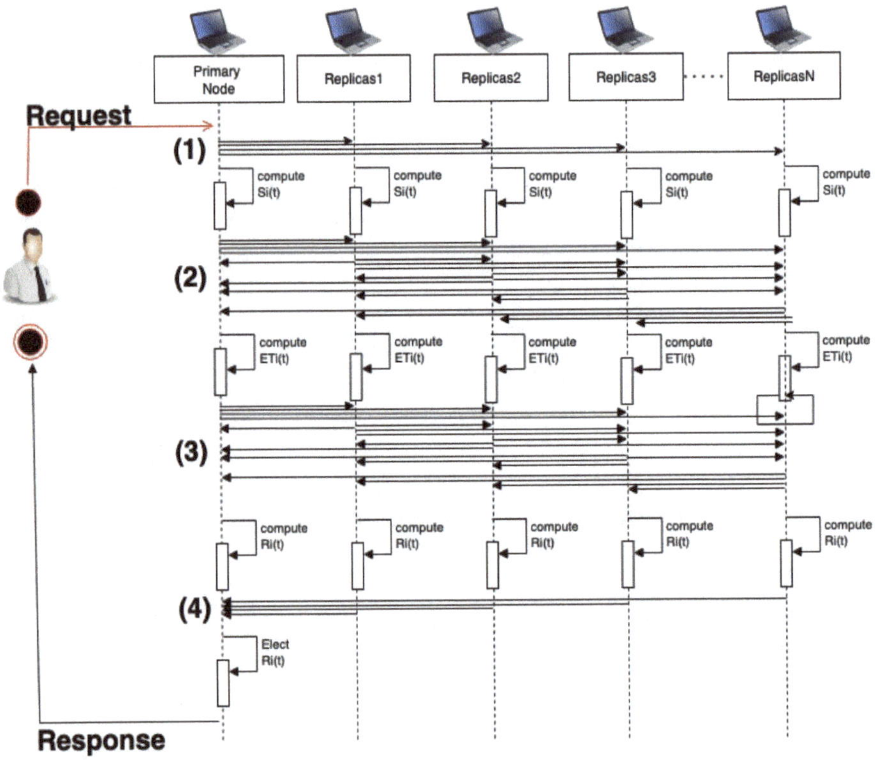

Fig. 2. Expert identification Algorithm Sequence Diagram

(1): The primary Node Broadcast the request message to all nodes of network.
(2): Each node in the network, compute locally the value of $S_i(t)$ and brodcast it to all other nodes of network.

(3): Each node in the network, compute locally the value of $ET_i(t)$ and brodcast it to all other nodes of network.
(4): The primary node collects a list of calculated $R_i(t)$ values rom all replicas, and to this list its own computed $R_i(t)$.

It should be noted, as we discuss our approach for finding a solution to the issue, that the problem of scalability is a general issue that blockchain technology suffers from untill the writing of this article. There have been and continue to be several attempts to find an effective solution, but the proposed solutions have remained lacking in efficiency, either in security or decentralization. In order to ensure the best possible scalability, we have chosen Ethereum as blockchain, as it allows for the use of smart contracts, in which we can calculate the various necessary parameters, and it also provides a reasonable level of scalability.

4 Ethical Foundations

The experimentation of our system is based on the gathering the necessaries informations directly from Mastodon, this act of collecting data carries significant responsibilities, especially concerning privacy and algorithmic bias, these two fundamental pillars for the success of any similar system are guaranteed firstly in our case by the use of blockchain technology, because this techniqe offers a very strong level of privacy and very high level of transparency, especially with our orientation to store the data locally in the local memory of each user's participating of the network, furthermore our system with the integration of the parameter called ExpertTrust with BFT consensus is totally unbiased for anyone.

5 Experimental Results

In this section, we will present the practical implementation results of our proposed methodology. Initially, we will utilize a Python-based program designed to establish a direct connection with the Mastodon API, allowing for interaction with a user-selected instance. The program will identify the most valuable profiles in a specific field, based on six predefined parameters previously discussed in the earlier sections of this paper. These parameters include credentials, the number of tweets, retweeted tweets, interaction levels, follower count, and the presence of the user profile containing the keyword specified by the user. The collected data will subsequently be compiled into a CSV file. To evaluate the effectiveness of our proposed approach, we conducted a series of experiments, we will present the experimental setup, datasets, evaluation metrics, and results. our aim is to demonstrate the performance, validate the correctness of the method, and highlight its credible results. To facilitate the use of our system we created a simple Back-End API application that interacts with mastodon via Token Access to collect the data relative to users, then we applicated our algorithm between

Fig. 3. Dataset Generator

all these users (by a simulation) to elect the experts one according to six parameters choosen as instance, in our experimental tests we choosed: credentials, tweets number, followers counter, profile, tweets retweeted and interactions.

Using our dataset generator illustred in Fig. 3 we obtained necessaries data in a csv file composed 6 columns, with a differents values arranged in decreasing order, as we can see that these values reflect the huge traffic in mastodon as these big number of tweets is generated just in this last months, in the Fig. 4 we captured a subset of nine users of our datset:

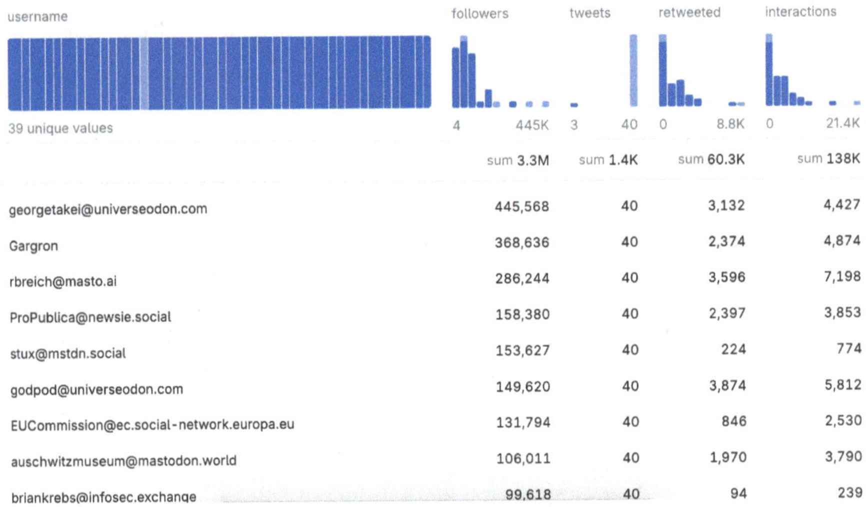

username	followers	tweets	retweeted	interactions
39 unique values	4 – 445K	3 – 40	0 – 8.8K	0 – 21.4K
	sum 3.3M	sum 1.4K	sum 60.3K	sum 138K
georgetakei@universeodon.com	445,568	40	3,132	4,427
Gargron	368,636	40	2,374	4,874
rbreich@masto.ai	286,244	40	3,596	7,198
ProPublica@newsie.social	158,380	40	2,397	3,853
stux@mstdn.social	153,627	40	224	774
godpod@universeodon.com	149,620	40	3,874	5,812
EUCommission@ec.social-network.europa.eu	131,794	40	846	2,530
auschwitzmuseum@mastodon.world	106,011	40	1,970	3,790
briankrebs@infosec.exchange	99,618	40	94	239

Fig. 4. A subset of our dataset generated using python

Once the necessary data is extracted directly from the Masodon API, we start the second stage, what we call exploitation phase. We create a blockchain where every user that is acquired is a node. We then add two more nodes, one representing the client and sending the initial request to start the operation, and one being the primary node. We will then apply our consensus on this blockchain. Finally we obtained as result a table (Treeview) with the number of experts found, classified according to their Score R.

It should be noted that the six parameters we selected are not definitive, they can be adapted to meet the client's specific requirement that should be present in the profile of the experts that we look for.

6 Conclusion

In this work we presented our genius method for identifynig white and real experts in distributed online social network, by leveraging the public API of Mastodon, we dynamically collect user data and perform a quantitative analysis based on behavioral indicators such as credentials, the number of followers, recent tweets, total interactions for all tweets and the frequency of posts being boosted by others (tweets retweeted). The results from the proposed simulation and implementation highlight both strong contributions, the approach demonstrates the feasibility of combining our consensus with expertise evaluation metrics in a Mastodon, by using a composite set of criteria—such as credentials, profile relevance, tweets retweeted, recent activity, and content quality, the system can find the most valuable profiles that shows a hight indicators, a real and honest expert in specific field selected by the client or user.

As future work, we aim to expand this system into a fully operational and scalable framework that can be integrated into both decentralized and centralized social platforms like Twitter or Reddit, this enhanced framework would support a wide range of applications, including expert recommendation, the system can serve as a credible tool for real-time trust and expertise assessment.

References

1. Bendig, D., Charlet, M.: Opportunities and challenges of blockchain for multi-sided platforms. Electron. Markets **35**(1), 25 (2025)
2. Shilina, S.: The future of social networking: Decentralization for user empowerment, privacy, and freedom from censorship [Internet], Medium Site (2023). https://medium.com/@sshshln/the-future-of-social-networking-decentralization-for-user-empowerment-privacy-and-freedom-from-a0a8f74790cb
3. Chen, N., D.S., Cho, -Y.: A blockchain based autonomous decentralized online social network. In: 2021 IEEE International Conference on Consumer Electronics and Computer Engineering (ICCECE), pp. 186–190 (2021)
4. Tanwar, S., Bhatia, Q., Patel, P., Kumari, A., Singh, P.K., Hong, W.-C.: Machine learning adoption in blockchain-based smart applications: the challenges, and a way forward. IEEE Access **8**, 474–488 (2019)
5. Horne, B.D., Nevo, D., Adalı, S.: Recognizing experts on social media: A heuristics-based approach. In: ACM SIGMIS Database: The DATABASE for Advances in Information Systems, vol. 50, pp. 66–84 (2019)

6. Lei, K., Zhang, Q., Xu, L., Qi, Z.: Reputation-based byzantine fault-tolerance for consortium blockchain. In: 2018 IEEE 24th International Conference on Parallel and Distributed Systems (ICPADS), pp. 604–611 (2018)
7. Wei, W., Cong, G., Miao, C., Zhu, F., Li, G.: Learning to find topic experts in Twitter via different relations. IEEE Trans. Knowl. Data Eng. **28**, 1764–1778 (2016)
8. Nakamoto, S.: Bitcoin: A peer-to-peer electronic cash system (2008). https://bitcoin.org/bitcoin.pdf. Accessed 29 Sept 2025

A Design Framework for Diagnosis and Treatment of Dyscalculia Using Machine Learning and Recommender Systems

Fatima Zohra Lhafra[1(✉)] and Otman Abdoun[2]

[1] Department of Computer Science and Interdisciplinary Education, Faculty of Legal, Economic, and Social Sciences, Abdelmalek Essaadi University, Tetouan, Morocco
lhafra.fatimazohra@gmail.com

[2] Computer Science Department, Faculty of sciences, Abdelmalek Essaadi University, Tetouan, Morocco

Abstract. Dyscalculia is a neurodevelopmental disorder (NDD) that affects learners' ability to acquire mathematical skills. It hinders the development of numerical processing, analytical, and problem-solving abilities. In this context, early and accurate diagnosis is essential to provide a treatment process tailored to each learner. The objective of this paper is to present an interdisciplinary approach combining artificial intelligence (AI), neuroscience, and educational sciences for the diagnosis and an adaptive treatment of dyscalculia. The first phase of the proposed approach aims to establish an early and precise diagnostic strategy using supervised machine learning algorithms, such as Random Forest, Naive Bayes, and Support Vector Machine (SVM), while taking into account the cognitive processes involved in dyscalculia. The second phase focuses on an adaptive treatment process based on the obtained diagnosis, leveraging recommender systems, particularly the collaborative filtering technique. The diagnostic test measures the risk of dyscalculia based on learners' test responses in order to offer them an adaptive treatment process. A real-world experiment will be conducted to measure the effectiveness of the proposed approach. This will be done by monitoring the progress of learners' achievements during the processing process.

Keywords: neurodevelopmental disorder · dyscalculia · recommender systems · machine learning

1 Introduction

Neurodevelopmental disorders are a group of disorders that appear in childhood and make each child different in the way they learn and interact with their environment. They generally affect cognitive, social, emotional, and motor skills. These disorders include attention deficit hyperactivity disorder (ADHD), autism spectrum disorder (ASD), dyslexia, dyscalculia, and others. Despite normal intelligence, dyscalculia leads to difficulties in basic mathematical learning, such as understanding and manipulating numbers. Learning mathematics affects several aspects of the learner's development.

A. E. Hassanien et al. (Eds.): AISDS 2025, LNDECT 287, pp. 361–370, 2027.
https://doi.org/10.1007/978-3-032-18927-1_36

Cognitively, it enhances logical and analytical thinking, improves problem-solving skills, and strengthens memory. Functionally, it has a major impact on the learner's daily and professional life. It is estimated to affect approximately 5% to 7% of the population[1].

Although several studies have explored various approaches to diagnosing and treating mathematical learning disorders, the existing literature does not provide sufficient evidence for a coherent model that integrates both phases. This fragmentation between diagnosis and intervention limits the effectiveness of current practices and highlights the need for a comprehensive framework that connects assessment outcomes to targeted treatment strategies. For this reason, we offer an approach based on early diagnosis and a learning process tailored to the difficulties identified during diagnosis. The implementation of the proposed solution is based on an interdisciplinary approach combining artificial intelligence, educational sciences, and neuroscience in order to offer an intelligent learning process tailored to the needs of learners with dyscalculia and help them to overcome their difficulties.

This paper aims to present an adaptive learning approach for learners with or at risk of dyscalculia. The approach is based on two main stages: the first aims to establish an early diagnosis and consists of six exercises covering the cognitive process involved in dyscalculia. The results of this test will be the subject of classification using: Naive Bayes, Support Vector Machine (SVM), and Random Forest algorithms in order to predict whether or not the learner has dyscalculia, as well as to specify the class or classes of difficulty. A comparative study will be conducted between these three algorithms in order to identify the most appropriate solution. The second step involves the use of recommendation systems, particularly collaborative filtering technique, to propose a series of math learning activities aimed at remedying the identified disorders and helping the learner to progress in their learning process. The effectiveness of the proposed approach is evaluated experimentally by systematically measuring the degree of progress in learners' mathematical skills.

The structure of this article is as follows: Sect. 2 is devoted to a literature review to provide an overview of neurodevelopmental disorders and the studies conducted in this field. Section 3 focuses on dyscalculia and the integration of AI as a solution. Section 3 describes the proposed approach, and we end with a conclusion and some perspectives for this work.

2 Literature Review

Neurodevelopmental disorders are defined in the DSM-5-TR [2] as a group of disorders that appear during the developmental period, often before the child even enters primary school. They are characterized by developmental deficits that affect personal, social, academic, or professional performance. They include disorders such as: intellectual developmental disorder, autism spectrum disorder (ASD), attention deficit disorder with or without hyperactivity (ADHD), and learning disorders (dyslexia, dysorthographia, dysphasia, dyscalculia, etc.).

With the emergence of AI technologies, several efforts have been made to offer intelligent solutions and help children with different characteristics to learn, communicate, interact, and integrate into society more effectively. For example, for ASD numerous studies have explored the use of AI as a methodological framework for advancing diagnostic practices. For instance, Nasser et al. have introduced a model employing neural networks to identify ASD [3]. The evaluation of test data indicated that the ANN model achieved a diagnostic accuracy of 100%. Several studies have focused on proposing adaptive learning processes for this category of learners. This includes the study by Boussakssou et al. [4], which explored various possibilities for implementing adaptive learning processes through the use of the Q-learning algorithm.

Dyslexia has also been the subject of several studies. For example, Shalileh et al. [5] presented an approach for identifying dyslexia in students using AI, based on eye-tracking and demographic data. Another study highlighted the role of AI in dyslexia research through a review article focusing on the main applications and challenges encountered [6]. The study discussed the application in terms of diagnosis, personalised learning, language processing and neuroimaging analysis.

Dysgraphia has been the subject of numerous research investigations, especially regarding its integration with AI technologies aimed at developing AI-driven screening approaches. One example is the work of Rangasrinivasan et al., who presented a tool for identifying behavioural indicators of dyslexia and dysgraphia in children's handwriting. The proposed solution also offers a methodology for collecting offline and online writing samples from children [7].

One difficulty that influences the entire learning process by affecting academic performance, self-confidence and school integration is dyscalculia. This disorder affects the learner's ability to make progress in mathematics, which can extend to other subjects that require numerical skills. For this reason, certain research studies have been developed in order to establish effective solutions dedicated to this category of learners. For example, Bhushan et al. highlighted the methods and applications of dyscalculia screening using artificial intelligence, while focusing on the importance of these methods and future directions [7]. Another study describes the process of developing, validating, and standardizing a dyscalculia test designed to identify mathematical difficulties in students [8]. Adriano et al. conducted a study based on AI to analyze the interaction between learning techniques, learning difficulties, and academic performance. The study identifies the fundamental mechanisms of academic performance and proposes adapted interventions based on neural networks and fuzzy logic [9]. Another study conducted by Kaser et al. focused on the implementation of an adaptive learning system to improve mathematics acquisition among learners with dyscalculia or difficulties in mathematics. Based on the student's knowledge, the system selects the most appropriate tasks for their needs in order to implement a personalized learning process based on Bayesian networks [10]. The study conducted by Nupur et al. [11] developed an automated approach for detecting dyscalculia, based on machine learning techniques, in particular Random Forest and Decision Tree algorithms. Another study conducted by Vaughn aimed to evaluate the effectiveness of a virtual math teacher based on AI, named "Mathias," with students with disorders such as dyslexia, dysgraphia, and dyscalculia. The learning process is based on the GPT model developed by OpenAI, allowing learners to interact with "Mathias" via a chatbot and progress through predefined topics aimed at developing specific math skills [12].

Based on the analysis of research conducted in the context of integrating AI to propose solutions for neurodevelopmental disorders, we have found that these disorders have attracted considerable scientific attention. Despite this attention, studies on dyscalculia are still very limited, compared to other disorders. Furthermore, analysis of this research revealed a significant gap between the diagnostic phase and the treatment phase. This discontinuity makes the intervention process more complex, as it is essential to take into account the specific difficulties identified during diagnosis in order to offer treatment that is effectively tailored to learners' needs and thus avoid the risk of cognitive overload. In addition, one of the major limitations identified in these studies is the poor integration of the pedagogical framework and principles derived from neuroscience in the design of intelligent remediation solutions. To address these limitations, we have designed a coherent approach that provides a direct transition from the dyscalculia diagnosis phase to the adaptive treatment process. This approach is based on a balanced combination of artificial intelligence, neuroscience, and pedagogical foundations of the learning process, with the aim of offering a more personalized, effective, and scientifically grounded solution.

3 Interdisciplinary Approach to Dyscalculia

Dyscalculia is a disorder that has benefited from emerging AI technologies, particularly in early detection and personalized interventions. Using AI, intelligent diagnostic systems can be implemented that are designed to identify and analyze learner interactions and responses to facilitate the early and accurate identification of dyscalculia. In addition, within the framework of adaptive learning systems, AI makes it possible to offer an adaptive and personalized learning process. Nevertheless, in order to ensure the effectiveness of the proposed solution, it is essential that an interdisciplinary approach be taken, combining AI technologies, neurosciences, and educational sciences. Neurosciences are involved in identifying the cognitive processes underlying dyscalculia, as well as the proposed treatments and intervention strategies. Educational sciences provide the pedagogical framework on which the teaching-learning process is based, AI provides the technological and intelligent aspect to ensure an adaptive solution. In this sense, dyscalculia should not be considered an isolated learning difficulty, but is rather a neurodevelopmental disorder that requires a multidisciplinary approach.

3.1 Understanding Dyscalculia

Several definitions have been attributed to dyscalculia. For example, Poornappriya et al. mentioned that dyscalculia has always been defined as a difficulty in retaining mathematical information, leading to frequent use of simple and undeveloped calculation strategies [13]. Dyscalculia has also been defined in the Diagnostic and Statistical Manual of Mental Disorders (DSM-5) as a specific learning disability in mathematics characterized by difficulties in processing numerical data, learning formulas, and performing calculations with accuracy and fluency [2].

The main symptoms of dyscalculia include:

- Impaired sense of numbers and memory for arithmetic facts;

- Inaccurate calculations and mathematical reasoning;
- Difficulty understanding mathematical terms, operations, or concepts;
- Difficulty recognizing or reading numerical symbols or arithmetic signs;
- Difficulty learning, observing, and memorizing operational signs;
- Long resolution times;
- Inability to follow and memorize the sequence of steps used in various mathematical operations;
- Inability to concentrate on mentally demanding tasks;

3.2 Cognitive Mechanisms Underlying Dyscalculia

Dyscalculia is a disorder that does not result from a lack of general intelligence or motivation of the child, but from a dysfunction of certain mental mechanisms related to numerical processing and mathematical reasoning. For this reason, the cognitive process plays an important role in dyscalculia through several fundamental cognitive functions that are involved. The analysis of this process is essential for implementing an early and accurate diagnostic mechanism, as well as for developing a personalized intervention strategy. Table 1 outlines the cognitive mechanisms of dyscalculia.

Table 1. Cognitive mechanisms underlying dyscalculia

Cognitive process	Characteristics
Number sense	1. Approximate Number System (ANS) – Alterations in intuitive understanding of quantities – Difficulty in comparing numerical values without resorting to counting 2. Symbolic representation of numbers – Difficulty in associating numerical symbols with their corresponding quantities
Working memory	– Difficulty in memorizing the intermediate steps of calculations – Follow the various steps necessary to solve a problem. – Remember arithmetic facts.
Visuospatial Skills	– Inability to imagine and manipulate objects mentally – Problems related to the layout and spatial organization of numbers.
Executive Functions	– Difficulties in planning multi-step problems – The effective transition between different mathematical strategies.
Attention	– Difficulty in concentrating on math exercises, which leads to errors and unfinished work
Language processing	– Impairing the ability to understand and deal with mathematical language, including written problems, instructions, and specific terminology

3.3 Treatment Approaches for Dyscalculia Across Age Groups

The treatment approach adopted for dyscalculia varies according to age group. Bhushan et al. classified the treatment process into three categories: 3–6 years, 7–12 years, and 13–18 years [14].This classification is based on learners' cognitive needs ,and their ability to acquire knowledge and progress through the learning process. For learners in the first category, the objective is to reinforce the basics of number sense and intuitive understanding of quantities. To this end, interventions are mainly based on integrating a playful aspect through educational games, concrete manipulations, and visual aids. For the second category, which corresponds to adolescents, the objective is to consolidate basic arithmetic skills and enhance mathematical reasoning. The developmental process at this stage focuses on fostering problem-planning abilities and promoting the use of diagrams and visual representations, supported by adaptive applications and interactive exercises to facilitate learning and sustain motivation.

The third category focuses on targeted remediation and educational adaptation to help learners to overcome their difficulties and guide them towards better integration. The following Table 2 presents the classes of the treatment.

Table 2. Treatment approaches

3 to 6 years old	7 to 12 years old	Teenagers aged 13 to 18
– Game-based learning (digital game + manipulable materials) – Interactive stories – Parental involvement in daily activities	– Structured mathematics learning – Use of: • educational software • Aide memory • Intelligent tutoring system	– Executive function training – Assistive technologies (graphing calculators) – Peer tutoring – Functional math instruction based on daily activities – Cognitive behavioral therapy

4 Proposed Approach

4.1 Early and Accurate Diagnosis

The implementation of the proposed approach relies primarily on the diagnostic phase, which is considered a key step in ensuring effective treatment of dyscalculia. The diagnosis is usually made by neuropsychologists using manual tests, which are often tedious and can lead to errors and inaccurate assessments. These tests cover the entire cognitive process, making the diagnostic task particularly complex and laborious. In addition, patients often have to wait a long time to receive their assessment reports, as the data is stored in physical format in hospital records. The treatment process is also slowed down by delayed diagnosis, as in most cases families are unaware that dyscalculia exists. Children's difficulties with mathematics are therefore often interpreted as a lack of intelligence or simply as learning difficulties.

Early diagnosis in primary and preschool schools is essential in order to facilitate the treatment process and help learners to build self-confidence despite the difficulties identified. To this end, we aimed to implement an intelligent approach based primarily on an accurate diagnosis according to the age classifications mentioned in Sect. 3. The diagnosis addresses the entire cognitive process of dyscalculia: number sense, working memory, visuospatial skills, executive functions, attention, and language processing (see Fig. 1). It is based on a test consisting of six exercises, each targeting a specific function of the cognitive process. Based on parameters such as response time, number of errors, number of attempts, and type of errors, a classification is performed using supervised learning algorithms: Random Forest, SVM, and Naive Bayes to determine whether the learner has or is likely to have dyscalculia. This classification also makes it possible to accurately identify the cognitive functions requiring appropriate intervention. The adoption of these three algorithms aims to conduct a comparative study to determine which one is most suitable based on the results obtained during the treatment process.

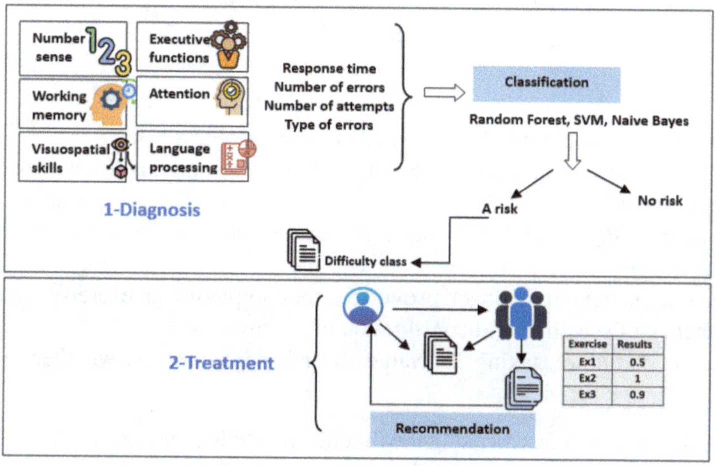

Fig. 1. Proposed approach

4.2 Treatment Process

Once the diagnostic phase is complete, it is essential to implement tailored, personalized treatment to help learners to overcome the difficulties identified. This treatment is part of an adaptive learning approach based on artificial intelligence technologies.

We have chosen to use a recommendation system, specifically collaborative filtering, which is based on the principle of neighborhood. This approach allows us to benefit from the experience of other learners who have faced the same difficulties and undergone the same treatment. To implement the collaborative filtering technique, we adopted the following steps:

- **Step 1**: Construction of the matrix (learner × exercise) in order to assign a specific weight to each exercise based on the learner's profile.
- **Step 2**: Calculate the similarity measure between different learners.
- **Step 3**: Select the K learners with the highest degree of similarity.

- **Step 4**: Generate a prediction based on a weighted combination of the assessments provided by the selected neighbors.

To assign a weight to each treatment exercise, we considered several properties that characterize them. The scoring function *S* (Eqs. 1 and 2) is used to calculate this weight. To calculate the score for each exercise, we took into account the parameters characteristic of these exercises, reflecting both the degree of success and the level of preference of the learners, such as: duration, frequency of selection, learner's feedback, difficulty level, and obtained result. The aim is to obtain a more detailed and representative measure of the pedagogical effectiveness of each activity.

$$\mathrm{S} = \frac{1}{2}(App + I) \tag{1}$$

$$\begin{cases} I = S + LD + R + 2B + 2F * App \\ B = e^{-D} \end{cases} \tag{2}$$

Where:

- **S**: exercise associated with the identified cognitive process.
- **LD**: difficulty level, taking the values 1, 2, or 3 corresponding to an easy, medium, or difficult activity.
- **D**: duration during which the learner performs the selected activity, used to calculate function *B*.
- **App**: learner's explicit feedback provided after completing an exercise.
- **F**: frequency of selection and consultation of the exercise.
- **R**: success or failure, taking the values 0 or 1 depending on whether the learner succeeded in the activity.

The evaluation of exercises using the function *S* enabled the construction of a matrix M (learner × exercise), consisting of n rows (learners) and n columns (exercises), as illustrated in Table 3.

Table 3. Exercise-Learner Combination Matrix

	Exercise 1	Exercise 2	Exercise 3	Exercise 4
Learner 1	–	5	–	–
Learner 2	7	8	3	–
Learner 3	–	9	–	–
Learner 4	1	–	7	4

The resulting matrix reflects the degree of appreciation and effectiveness of the exercises, represented by the weight calculated using the function *S*. However, some learners do not provide feedback for certain exercises, such as learner 1 for activities 1, 3, and 4. In this case, collaborative filtering, based on the system's user community, allows leveraging the experience of other learners to recommend these activities to those who have not evaluated them. This leads to the next step: calculating the similarity between learner

u and the active learner a, using Pearson correlation (Eq. 3). The choice of Pearson's correlation coefficient is justified by its proven effectiveness in calculating the similarity between variables. Indeed, this measure is particularly well suited to identifying similar learner profiles in an intelligent learning context [15].

$$\text{Corr } a, u = \frac{\sum_{i \in I}((r_{ai} - \overline{r}_a)(r_{ui} - \overline{r}_u))}{\sqrt{\left(r_{ai - \overline{r}_a}\right)^2 (r_{ui} - \overline{r}_u)^2}} \tag{3}$$

Based on the results of the similarity measurement, we selected K learners with the closest values in order to compute a weighted average of their evaluations, thereby providing a relevant prediction. The formula for this prediction is given in Eq. 4.

$$Pred \text{ a, u} = \overline{r}_a + \frac{\sum_{u \in K}(r_{ui} - \overline{r_u}) * corr_{a,u}}{\sum_{u \in k}\left|corr_{a,u}\right|} \tag{4}$$

The prediction process provides learners with a personalized list of activities for treating dyscalculia, arranged in order of priority. This prioritization is based on the calculated similarity between learner profiles, so that the most relevant activities are recommended. The approach aims to optimize the learning experience by tailoring exercises to each learner's specific difficulties and needs.

4.3 Evaluation and Experimentation Process

In order to verify the effectiveness of the proposed approach, an evaluation phase will first be conducted on the diagnostic component, using performance metrics (F1-score and accuracy). These indicators will measure the approach's accuracy in identifying learners presenting signs of dyscalculia. Subsequently, a real experiment will be conducted with a sample of first year primary school students in a Moroccan public school. This experiment will initially focus on six targeted exercises designed to assess the fundamental cognitive processes related to dyscalculia, while taking into account the ethical aspects of adopting AI technologies. This population was chosen because of the importance of early diagnosis in children, in order to prevent the negative consequences that dyscalculia can have on academic and cognitive development. The treatment phase will focus exclusively on unmastered skills, with the aim of ensuring a personalized adaptive process and avoiding cognitive overload in learners.

5 Conclusion

Dyscalculia is among the least studied neurodevelopmental disorders in research. Given the importance of learning mathematics for developing learners' analytical and problem-solving skills, as well as its connection with other disciplines, we have developed an interdisciplinary approach for the diagnosis and treatment of dyscalculia. This approach is based on artificial intelligence, neuroscience, and educational sciences, aiming to provide a strategy for early and accurate diagnosis along with an adapted treatment process. The experimental phase is a crucial step in the proposed approach, as it allows for empirical validation of the results obtained and evaluation of the model's effectiveness in a real learning context. It also forms part of a broader effort to develop a coherent and integrated system for the diagnosis and treatment of other neurodevelopmental disorders.

References

1. Butterworth, B., Varma, S., Laurillard, D.: Dyscalculia: from brain to education. Science. **332**, 1049–1053 (2011)
2. American Psychiatric Association, Crocq, M.-A., Boehrer, A.E., Guelfi, D.G.: DSM-5-TR Manuel Diagnostique et Statistique Des Troubles Mentaux, Texte Révisé. American Psychiatric Association (2023)
3. Thomas, M., Chandran, A.: Artificial neural network for diagnosing autism spectrum disorder. In: 2nd International Conference on Trends in Electronics and Informatics (ICOEI), pp. 930–933. IEEE (2018). https://doi.org/10.1109/ICOEI.2018.8553781
4. Boussakssou, M., Hssina, B., Erittali, M.: Towards an adaptive e-learning system based on q-learning algorithm. Procedia Comput. Sci. **170**, 1198–1203 (2020)
5. Shalileh, S., Ignatov, D., Lopukhina, A., Dragoy, O.: Identifying dyslexia in school pupils from eye movement and demographic data using artificial intelligence. PLoS One. **18**, e0292047 (2023)
6. Yap, J.R., Aruthanan, T., Chin, M.: Artificial intelligence in dyslexia research and education: a scoping review. IEEE Access. **13**, 7123–7134 (2025)
7. Rangasrinivasan, S., et al.: AI-enhanced child handwriting analysis: a framework for the early screening of dyslexia and dysgraphia. SN Comput. Sci. **6**, 399 (2025)
8. Bhushan, S., et al.: AI-enhanced dyscalculia screening: a survey of methods and applications for children. Diagnostics. **14**, 1441 (2024)
9. Eteng-Uket, S.: The development, validation, and standardization of a new tool: the dyscalculia test. Numeracy. **16** (2023). https://doi.org/10.5038/1936-4660.16.2.1417
10. Bressane, A., et al.: Understanding the role of study strategies and learning disabilities on student academic performance to enhance educational approaches: a proposal using artificial intelligence. Comput. Educ.: Artif. Intell. **6**, 100196 (2024)
11. Käser, T., et al.: Modelling and optimizing mathematics learning in children. Int. J. Artif. Intell. Educ. **23**, 115–135 (2013)
12. Giri, N., et al.: Detection of dyscalculia using machine learning. In: 2020 5th International Conference on Communication and Electronics Systems (ICCES), pp. 1–6. IEEE (2020). https://doi.org/10.1109/ICCES48766.2020.9137871
13. Holmes, V.: Designing an AI math tutor for children with dyslexia, dysgraphia, and dyscalculia. Res. Arch. Ris. Sch.. Preprint at. (2024). https://doi.org/10.58445/rars.2035
14. Poornappriya, T.S., Gopinath, R.: Application of machine learning techniques for improving learning disabilities. Int. J. Electr. Eng. Technol. (IJEET). **11**, 403 (2020)
15. Lhafra, F.Z., Abdoun, O.: Hybrid approach to recommending adaptive remediation activities based on assessment results in an e-learning system using machine learning. In: Advanced Intelligent Systems for Sustainable Development, pp. 679–696 (2022)

Author Index

A. E. Hassanien et al. (Eds.): AISDS 2025, LNDECT 287, pp. 371–372, 2027.
https://doi.org/10.1007/978-3-032-18927-1

Kolibri 360 GmbH
Ferdinand-Jühlke-Straße 7
99095 Erfurt, Deutschland
produktsicherheit@kolibri360.de